AF698482

Selected Titles in This Series

175 **S. K. Godunov,** Modern aspects of linear algebra, 1998

174 **Ya-Zhe Chen and Lan-Cheng Wu,** Second order elliptic equations and elliptic systems, 1998

173 **Yu. A. Davydov, M. A. Lifshits, and N. V. Smorodina,** Local properties of distributions of stochastic functionals, 1998

172 **Ya. G. Berkovich and E. M. Zhmud′,** Characters of finite groups. Part 1, 1998

171 **E. M. Landis,** Second order equations of elliptic and parabolic type, 1998

170 **Viktor Prasolov and Yuri Solovyev,** Elliptic functions and elliptic integrals, 1997

169 **S. K. Godunov,** Ordinary differential equations with constant coefficient, 1997

168 **Junjiro Noguchi,** Introduction to complex analysis, 1998

167 **Masaya Yamaguti, Masayoshi Hata, and Jun Kigami,** Mathematics of fractals, 1997

166 **Kenji Ueno,** An introduction to algebraic geometry, 1997

165 **V. V. Ishkhanov, B. B. Lur′e, and D. K. Faddeev,** The embedding problem in Galois theory, 1997

164 **E. I. Gordon,** Nonstandard methods in commutative harmonic analysis, 1997

163 **A. Ya. Dorogovtsev, D. S. Silvestrov, A. V. Skorokhod, and M. I. Yadrenko,** Probability theory: Collection of problems, 1997

162 **M. V. Boldin, G. I. Simonova, and Yu. N. Tyurin,** Sign-based methods in linear statistical models, 1997

161 **Michael Blank,** Discreteness and continuity in problems of chaotic dynamics, 1997

160 **V. G. Osmolovskiĭ,** Linear and nonlinear perturbations of the operator div, 1997

159 **S. Ya. Khavinson,** Best approximation by linear superpositions (approximate nomography), 1997

158 **Hideki Omori,** Infinite-dimensional Lie groups, 1997

157 **V. B. Kolmanovskiĭ and L. E. Shaĭkhet,** Control of systems with aftereffect, 1996

156 **V. N. Shevchenko,** Qualitative topics in integer linear programming, 1997

155 **Yu. Safarov and D. Vassiliev,** The asymptotic distribution of eigenvalues of partial differential operators, 1997

154 **V. V. Prasolov and A. B. Sossinsky,** Knots, links, braids and 3-manifolds. An introduction to the new invariants in low-dimensional topology, 1997

153 **S. Kh. Aranson, G. R. Belitsky, and E. V. Zhuzhoma,** Introduction to the qualitative theory of dynamical systems on surfaces, 1996

152 **R. S. Ismagilov,** Representations of infinite-dimensional groups, 1996

151 **S. Yu. Slavyanov,** Asymptotic solutions of the one-dimensional Schrödinger equation, 1996

150 **B. Ya. Levin,** Lectures on entire functions, 1996

149 **Takashi Sakai,** Riemannian geometry, 1996

148 **Vladimir I. Piterbarg,** Asymptotic methods in the theory of Gaussian processes and fields, 1996

147 **S. G. Gindikin and L. R. Volevich,** Mixed problem for partial differential equations with quasihomogeneous principal part, 1996

146 **L. Ya. Adrianova,** Introduction to linear systems of differential equations, 1995

145 **A. N. Andrianov and V. G. Zhuravlev,** Modular forms and Hecke operators, 1995

144 **O. V. Troshkin,** Nontraditional methods in mathematical hydrodynamics, 1995

143 **V. A. Malyshev and R. A. Minlos,** Linear infinite-particle operators, 1995

142 **N. V. Krylov,** Introduction to the theory of diffusion processes, 1995

141 **A. A. Davydov,** Qualitative theory of control systems, 1994

(Continued in the back of this publication)

Modern Aspects of Linear Algebra

Translations of

MATHEMATICAL MONOGRAPHS

Volume 175

Nova Southeastern
University Libraries
3301 College Avenue
Ft. Lauderdale, FL 33314

Modern Aspects of Linear Algebra

S. K. Godunov

American Mathematical Society
Providence, Rhode Island

EDITORIAL COMMITTEE

AMS Subcommittee
Robert D. MacPherson
Grigorii A. Margulis
James D. Stasheff (Chair)
ASL Subcommittee Steffen Lempp (Chair)
IMS Subcommittee Mark I. Freidlin (Chair)

С. К. ГОДУНОВ

СОВРЕМЕННЫЕ АСПЕКТЫ ЛИНЕЙНОЙ АЛГЕБРЫ

НАУЧНАЯ КНИГА, НОВОСИБИРСК, 1997

Translated from the Russian by Tamara Rozhkovskaya
with the participation of Scientific Books (RIMIBE NSU),
Novosibirsk, Russia

1991 *Mathematics Subject Classification.* Primary 15–01, 65–01;
Secondary 47Axx, 34A40, 35A40.

ABSTRACT. Systematic exposition of the spectral theory of nonselfadjoint operators in finite-dimensional linear spaces is presented. Formulation of the computational problems related to the analysis of the spectrum and brief description of certain algorithms, including some non-standard ones, is given. Examples of operators approximating ordinary and partial differential operators are studied.

The book can be used by researchers and graduate students working in linear algebra, differential equations, applied mathematics, and computational physics.

Library of Congress Cataloging-in-Publication Data

Godunov, S. K. (Sergeĭ Konstantinovich)
[Sovremennye aspekty lineĭnoĭ algebry. English]
Modern aspects of linear algebra / S. K. Godunov ; [translator, Tamara Rozhkovskaya with the participation of Scientific Books (RIMIBE NSU), Novosibirsk, Russia].
p. cm. — (Translations of mathematical monographs, ISSN 0065-9282 ; v. 175)
Includes bibliographical references and index.
ISBN 0-8218-0888-5 (alk. paper)
1. Algebras, Linear. 2. Nonselfadjoint operators. 3. Spectral theory (Mathematics) I. Title. II. Series.
QA184.G5813 1998
512′.5—dc21 98-13024
CIP

Copying and reprinting. Individual readers of this publication, and nonprofit libraries acting for them, are permitted to make fair use of the material, such as to copy a chapter for use in teaching or research. Permission is granted to quote brief passages from this publication in reviews, provided the customary acknowledgment of the source is given.

Republication, systematic copying, or multiple reproduction of any material in this publication (including abstracts) is permitted only under license from the American Mathematical Society. Requests for such permission should be addressed to the Assistant to the Publisher, American Mathematical Society, P. O. Box 6248, Providence, Rhode Island 02940-6248. Requests can also be made by e-mail to reprint-permission@ams.org.

© 1998 by the American Mathematical Society. All rights reserved.
The American Mathematical Society retains all rights
except those granted to the United States Government.
Printed in the United States of America.

∞ The paper used in this book is acid-free and falls within the guidelines
established to ensure permanence and durability.
Visit the AMS home page at URL: http://www.ams.org/

10 9 8 7 6 5 4 3 2 1 03 02 01 00 99 98

Contents

Preface

This book, intended for specialists,[1] discusses fundamental ideas of linear algebra from the point of view the author had formed during more that 30 years of working with computational algorithms. Actually, only one chapter (Chapter 14) is devoted to algorithms per se (there are several additional examples in Chapters 2 and 3 where the theory of orthogonal and unitary transformations is discussed). However, even in Chapter 14 we consider not specific algorithms but rather basic principles of constructing such algorithms.

My goal in this book was not to give a detailed survey but to concentrate on some methods used by our research group. The main concepts presented in the book was elaborated in the analysis of algorithms described in Chapter 14.

There is a significant difference between the qualitative behavior of spectral problems for self-adjoint and for nonselfadjoint matrix operators. In particular, the spectra of two close operators are close if the operators are Hermitian, whereas this is far from being true if the operators are not Hermitian. This leads to some well-known paradoxes. For example, the computation of eigenvalues of an integer-valued matrix and of the transposed matrix, performed on the same computer using the same standard software can lead to different results (see §1.3). To understand the nature of such paradoxes, it is necessary to study the so-called ε-spectrum, i.e., the set of all eigenvalues of all matrices that are ε-close to the matrix under consideration.

It is clear that the ε-spectrum consists not of discrete points but of spots containing the exact eigenvalues. It turns out that the diameters of these spots are not necessarily small even for a very small ε. This lead us (see [**13, 16, 28**]) and simultaneously Trefethen [**32**] to the introduction of spectral portraits, i.e., the graphic representation of ε-spectra (pseudospectra in the terminology used by Trefethen). Different domains of the complex plane are colored by different colors depending on the value ε for which points of these domains belong to the ε-spectrum of the matrix A. The boundaries of domains are level lines for the norm of the resolvent $\|(\lambda I - A)^{-1}\|$. Thus, studying the spectral portrait of a matrix A, we pay attention not only to the eigenvalues (poles of the resolvent) themselves, but to the behavior of the resolvent in a neighborhood of the poles.

The study of the resolvent is a usual tool in the study of operators in infinite-dimensional function spaces, e.g., in Hilbert spaces. However, many results obtained in the infinite-dimensional case are not represented in textbooks on linear algebra. In particular, we mention the estimate for the resolvent in terms of singular values of operators. As early as 1950s, Keldysh [**20, 21**] suggested to use some results in the theory of entire functions in the study of the location of spectral spots for

[1] We recommend that the reader who is not familiar with the theory of nonselfadjoint operators start reading consecutively starting with Chapter 1.

matrix operators (other terminology was used at that time). This idea formed a basis for the study (see Chapters 12 and 13) of finite-dimensional models of elliptic differential operators. Such models are obtained by a widely used method known as the finite element method. In Chapters 12 and 13 we prove the completeness of eigenvectors and adjoined vectors. The proof is a simplified version of the proof suggested by Lidskii [**25**]. We note that Lidskii's work was significantly influenced by the discussions with Keldysh [**21**].

If we restrict ourselves (as in Chapter 13) to the study of spots corresponding only to the extremal (with the smallest module) points of the spectrum, then the conditions on the operator can be weakened and the proof can be simplified.

In the proof we use an estimate for the resolvent given in §8.6. The justification of this estimate follows the monograph by Gohberg and Krein [**19**], where the versions of the estimate obtained by Lidskii and Matsaev were improved.

Spectral portraits are particularly useful in theoretical studies, whereas the real construction of spectral portraits requires huge computational resources even for matrices of moderate size. It seems that a much more practical approach consists in using criteria for the absence of eigenvalues in certain parts of the complex plane. The study of such criteria occupies the central place in this book. In Chapters 8–10, 14 and §13.6 we study efficient division of the spectrum into disjoint parts. The division is realized by certain curves in such a way that the norm of the resolvent in each curve is not large. In fact, the search of an appropriate formulation of this problem was the main goal of our research. We tried to find a formulation that would be convenient in applications and simultaneously would admit a solution with guaranteed accuracy of the result. These investigations, started by the author as well as Sarybekov [**31**] and Bulgakov [**4**], were used in applications to differential equations. An analogy with the known ideas of Hermite and Lyapunov were useful in our research. Eigenvalues of a matrix A are often used in the study of the asymptotic behavior of solutions to differential equations of the form $\dot{x} = Ax$ as $t \to \infty$. In particular, this is a subject of the stability theory which develops the Lyapunov approach.

As is known, for stability it is necessary that the spectrum lies in the left half-plane of the complex plane. However, this condition is not sufficient if we want to have a practical estimate for the rate of decrease of solutions. Furthermore, we must guarantee that the stability is preserved when coefficients vary in the limits determined by the given accuracy of computations or measurements. It is also necessary to require that the norm of the resolvent $\|(\lambda I - A)^{-1}\|$ would be not too large on the imaginary axis. This condition is equivalent to the existence of the quadratic Lyapunov function (Hx, x). The Lyapunov function is admissible if it is constant on the ellipsoid for which the ratio of the largest and the smallest semiaxes is small.

The matrix H corresponding to the quadratic form in the definition of the Lyapunov function can be represented as an integral along the imaginary axis $\lambda = it$ of the product $(-itI - A^*)^{-1}(itI - A)^{-1}$. It is also represented as a solution to the classical matrix Lyapunov equation $HA + A^*H + I = 0$. The solution admits the integral representation

$$\int_0^\infty e^{tA^*} e^{tA}\, dt,$$

where the integrand is the product of the matrix exponential e^{tA} and the conjugate matrix e^{tA^*}. It is very important that for $t > 0$ the norm of the matrix exponential satisfies the estimate

$$\|e^{tA}\| \le \sqrt{\kappa(A)}e^{-t\|A\|/\kappa(A)},$$

where $\kappa(A) = \|H\|\cdot 2\|A\|$. With this estimate in mind, Bulgakov [**4**] suggested using $\kappa(A)$ as a dimensionless parameter (the dichotomy quality) that characterizes how "deeply" the spectrum of the matrix A is located in the left half-plane.

Similar constructions can be used to study and describe the asymptotic behavior of the matrix powers A^n as $n \to \infty$. Some generalizations were also studied, e.g., to the cases where the spectrum is divided into parts lying in the right and left half-planes, or inside and outside a disk, or even inside and outside a domain bounded by a certain curve dividing the plane into two parts. Such problems are known as spectrum dichotomy problems. The use of quadratic forms in solving similar questions was suggested by Jacobi and Hermite back in the middle of the 19th century. This idea was described and developed in the small (45 page) monograph by M. G. Krein and M. A. Naimark, "The method of symmetric and Hermitian forms for separating roots of algebraic equations". This book was published in Khar′kov in 1936 and is practically inaccessible at present.

We note that the quadratic Lyapunov function (Hx, x) can be regarded as a version of the Jacobi–Hermite form that is specifically adapted for formulation of the fact that the spectrum of a matrix A lies in one (left) half-plane.

For the numerical analysis of the spectrum dichotomy by the imaginary axis, it was suggested in [**12**] to use quadratic forms with matrices admitting the integral representation

$$H = \int_{-\infty}^{+\infty} G^*(t)G(t)\,dt,$$

where $G(t)$ is the Green function for the boundary-value problem $\dot{x} - Ax = f(x)$ on the real line.

It should be pointed out that to study the dichotomy of the spectrum by the imaginary axis, Daletskii and Krein suggested ([**6**], §7, Chapter 1, §7) using an indefinite quadratic form with the matrix W admitting the integral representation

$$W = -\int_{-\infty}^{0} G^*(t)G(t)\,dt + \int_{0}^{\infty} G^*(t)G(t)\,dt,$$

which is similar to the integral form of H. I believe that the choice of a positive definite matrix H is more convenient in this case. However, indefinite forms play an important role in the study of the orthogonal-power algorithm (see Chapter 14).

Bulgakov [**5**] showed that if, together with a matrix H generalizing the Lyapunov matrix, we consider the (commutating with A) projections P and $I - P$ onto the invariant subspaces of A corresponding the points of the spectra lying in disjoint domains of the complex plane, then H and P satisfy a system of matrix equations that generalizes the classical Lyapunov equation and has similar properties. This allows us to develop algorithms for solving dichotomy problems with subsequent verification of accuracy.

It is interesting to note that the arguments of Bulgakov in the study of the matrices H and P were close to the analysis given in [**6**].

Actually, the suggestion developed in our group consists in using the dichotomy criteria rather than eigenvalues themselves in questions involving numerical solutions of spectral problems. This allows us to treat spectral problems for nonselfadjoint operators as problems admitting a solution with guaranteed accuracy. The dichotomy criteria and the corresponding matrix equations are considered in Chapters 9 and 10.

Introductory Chapters 1–4 present standard facts required in further considerations. These facts are illustrated by many important and not too trivial examples (see e.g., §1.3). Special orthogonal and unitary transformations are described. These transformations are used in the known computational algorithms (see §§2.1–2.3 and §§3.1–3.3). In the subsequent chapters we refer to the technique described in these introductory chapters. The detailed description of algorithms of orthogonal transformations and the use of these algorithms for solving symmetric spectral problem are given in [**15**]. In this book we concentrate mainly on nonselfadjoint problems. Self-adjoint problems (for example, variational principles in Chapter 7) are mentioned only when we want to make the exposition self-contained.

It should be noted that the introductory chapters of the monograph of Gohberg and Krein [**19**] significantly influenced the choice of the material. These chapters in [**19**] were often used in the special courses presented by the author. Once, a part in such a course was taught by A. N. Malyshev. In particular, he gave a lecture with a proof of the Horn theorem which is the converse of the Weyl theorem about the location of eigenvalues and singular values. His notes were used to plan topics of Chapters 5 and 6.[2]

The basic idea of the spectral theory is to use unitarily invariant characteristics of operators. As is known, such characteristics can be expressed in terms of singular values. Thus, in [**25**], where the infinite-dimensional case is considered, the author used Keldysh's assumption that the sum of certain powers of singular values converges. To formulate and prove the theorem about the annular spectrum dichotomy (see §13.5), the notion of the spectral condition numbers is introduced (see §7.8). These numbers give the quantitative criteria for an operator to belong to a certain Keldysh class. Chapter 12 deals with estimates for the spectral conditionality of operators used in the applications of finite element methods to the approximations of second order elliptic differential operators.

It is important to emphasize that we do not use the normal canonical Jordan form anywhere. We can avoid this because the whole space is decomposed into invariant subspaces corresponding to eigenvalues lying in disjoint spectral spots; moreover, in our approach the study of the detailed structure of the operator on the subspace corresponding to one of these spots can be avoided.

In Chapter 14, computational algorithms are discussed. The chapter begins with the quickly convergent Davison–Man method of solving the Lyapunov equation. Three sections are devoted to the orthogonal exhaustion method due to Malyshev [**26**]–[**28**] and to applications to spectrum dichotomy problems. Using

[2]The representation of the geometry of Hausdorff sets in Chapter 11 was prepared by the author together with Gordienko and was based on the article S. K. Godunov and V. M. Gordienko, *The Hausdorff sets of matrices and estimates for the angle between invariant subspaces*, Sibirsk. Mat. Zh., **26** (1995), no. 3, 531–533.

Malyshev's algorithm, we can solve some matrix equations generalizing the Lyapunov equation. Such equations are considered in Chapter 10. Malyshev's algorithm can be regarded as a modification of the matrix sign-function algorithm that was introduced by Roberts **[30]** (see also Abramov **[1]**) and has been widely used in recent years. However, the Malyshev modification has the advantage that it does not require the computation of the inverse matrix at each step of the iteration process.

The last three sections of Chapter 14 present the orthogonal-power method and its modification, the QR-algorithm. We apply this method to find a sequence of invariant subspaces with spectra separated by circles $|\lambda| = \text{const}$ on which the norm of the resolvent is not too large. Such circles realize the circle dichotomy of the spectrum. The exposition is based on the article **[18]**, where the stability of the process with respect to computation errors is proved using the ideas of the orthogonal sweep method of solving boundary-value problems. The appearance of a boundary-value problem in theory of iteration procedures is quite unexpected.

The orthogonal sweep method was developed in 1957 (and first published in 1961). Originally it was used in computing the critical parameters of nuclear reactors **[2, 10]**.

We have already mentioned the book **[15]**, which is devoted to the question of guaranteed accuracy for linear algebra problems. In **[15]** we focused on the algorithms for solving linear systems that are not too large and on the spectral problems for symmetric matrices. The study of these questions does not require significant changes in the point of view and modifications in the formulations of the problems. In fact, the only new aspect of **[15]** is the detailed exposition of the algorithm of two-sided Sturm sequences. This algorithm, suggested in **[17]**, is used in the precision computation of eigenvectors of symmetric three-diagonal matrices.

Both this book and **[15]** complete the detailed exposition of ideas that appeared as a result of investigations of the seminar organized by the author in Novosibirsk. The main results of these works were included in the talk **[11]** at the 1986 International Congress of Mathematicians in Berkeley.

At present it is not clear how to formulate computational problems for large sparse matrices. However, I hope that Chapters 11–13 will be useful to specialists working in this area. It seems necessary to develop closer contacts between specialists in applications and specialists in functional analysis. I began to study aspects of linear algebra in connection with the study of sparse matrices. So far, I have no definite opinion about statements of problems in which such matrices appear. My reasoning is presented in Chapters 11–13. However, they cannot be taken as a final understanding of all ideas required in the analysis of the algorithms.

Work on this book started three yeas ago when T. N. Rozhkovskaya insisted that I allow her to edit the notes of my lectures. Later this material was repeatedly rewritten and supplemented. It formed a basis of the first ten chapters and a part of Chapter 14. Chapters 11–13 appeared later in the process of rewriting. In particular, I understood the role of the works of Keldysh and Lidskii in overcoming the difficulties connected with the necessity to use spots of the ε-spectrum but not a discrete sequence of eigenvalues. In many practical problems, such spots do not glue together, but fall into smaller spots containing spectral clusters of small multiplicity. I think that this fact will be useful in constructing algorithms of spectral analysis for nonselfadjoint operators appearing in a number of important applications.

For three years T. N. Rozhkovskaya arranged and edited numerous versions of my notes and then prepared the book for publication. I am deeply grateful to her. However, I should mention that it was not always easy to be the subject of her pressing. Of course, without this pressure the book would have never been written.

I am grateful to A. A. Abramov, K. B. Brushlinskii, G. V. Demidenko, V. D. Mazurov, A. N. Konovalov, and V. I. Kostin for a number of remarks on the preliminary version of the book or of some chapters. Some of their useful remarks were taken into account in the preparation of the final version.

I am thankful to all my colleagues, the participants of the seminars at which the main structure of the book was developed. The book could not appear without their investigations and discussions.

I would like to thank the Russian Fund for Fundamental Research for the financial support of the publication of this book (grant N 096-01-140006) and my scientific work (grant N 93-011-1515).

S. K. Godunov
Akademgorodok, Novosibirsk
August 9, 1996

Part 1

Introduction

CHAPTER 1

Euclidean Linear Spaces

Beginning the study of linear algebra in Euclidean spaces, we must recall some known facts. We do not discuss axioms of a linear space in detail, starting with the notion of an inner product.

1.1. The simplest properties

Axioms of inner product. The complex case. Bases. Coordinates of a vector in a fixed basis. The length (norm) of a vector. The Cauchy–Schwarz–Bunyakovskii inequality. The triangle inequality.

A linear space is said to be N*-dimensional* if it contains N linearly independent vectors and any system of more than N vectors is linearly dependent. We recall that vectors $x^{(1)}$, $x^{(2)}, \dots, x^{(k)}$ are said to be *linearly independent* if

$$\alpha_1 x^{(1)} + \alpha_2 x^{(2)} + \cdots + \alpha_k x^{(k)} \neq 0$$

for any real scalars $\alpha_1, \alpha_2, \dots, \alpha_k$, not all of them being zero.

A Euclidean N-dimensional space E_N is a linear space equipped with the inner product $(\cdot,\cdot)$ satisfying the following axioms:

$$\begin{aligned} &\text{for any } x, y, z \in E_N \\ &(x, y) = (y, x) \text{ is a real number}, \\ &(x, x) > 0 \text{ for } x \neq 0, \\ &(\alpha x, y) = \alpha(x, y) \text{ for any real scalar } \alpha, \\ &(x + y, z) = (x, z) + (y, z). \end{aligned} \tag{1.1.1}$$

Sometimes, we will consider a complex Euclidean space H_N which differs from E_N in that complex scalars can be also used and the axioms (1.1.1) are replaced by the following axioms:

$$\begin{aligned} &\text{for any } x, y, z \in H_N \\ &(x, y) = \overline{(y, x)} \text{ is a complex number}, \\ &(x, x) \text{ is a real number, } (x, x) > 0 \text{ for } x \neq 0, \\ &(\alpha x, y) = \alpha(x, y) \text{ for any complex scalar } \alpha, \\ &(x + y, z) = (x, z) + (y, z). \end{aligned} \tag{1.1.2}$$

Here the bar means complex conjugation.

Let $l^{(1)}, l^{(2)}, \dots, l^{(N)}$ be linearly independent vectors in E_N (or H_N). For any vector $x \in E_N$ ($x \in H_N$) there exist real (complex) scalars $\alpha_1, \alpha_2, \dots, \alpha_N$, and $\alpha \neq 0$ such that

$$\alpha_1 l^{(1)} + \alpha_2 l^{(2)} + \cdots + \alpha_N l^{(N)} + \alpha x = 0.$$

In other words, an arbitrary vector x is represented as a linear combination of $l^{(1)}, l^{(2)}, \dots, l^{(N)}$, i.e.,

$$x = \xi_1 l^{(1)} + \xi_2 l^{(2)} + \cdots + \xi_N l^{(N)}, \tag{1.1.3}$$

where $\xi_1 = -\alpha_1/\alpha, \xi_2 = -\alpha_2/\alpha, \dots, \xi_N = -\alpha_N/\alpha$.

The vectors $l^{(1)}, l^{(2)}, \dots, l^{(N)}$ form a *basis* for the space E_N (H_N).

The coefficients $\xi_1, \xi_2, \dots, \xi_N$ are called the components of the vector x in the basis $l^{(1)}, l^{(2)}, \dots, l^{(N)}$. It is easy to check that for a fixed basis $l^{(1)}, l^{(2)}, \dots, l^{(N)}$ the coefficients $\xi_1, \xi_2, \dots, \xi_N$ in (1.1.3) are uniquely determined by the vector x.

A basis $e^{(1)}, e^{(2)}, \dots, e^{(N)}$ is said to be *orthonormal* if

$$(e^{(i)}, e^{(j)}) = 0 \text{ for } i \neq j, \qquad (e^{(j)}, e^{(j)}) = 1.$$

Let $l^{(1)}, l^{(2)}, \dots, l^{(N)}$ be a basis. We set

$$\begin{aligned} e^{(1)} &= \rho_{11} l^{(1)}, \\ e^{(2)} &= \rho_{21} l^{(1)} + \rho_{22} l^{(2)}, \\ &\cdots\cdots\cdots\cdots\cdots\cdots \\ e^{(N)} &= \rho_{N1} l^{(1)} + \rho_{N2} l^{(2)} + \cdots + \rho_{NN} l^{(N)}. \end{aligned}$$

Choosing ρ_{ij}, $i,j = 1, 2, \dots, N$, so that $\rho_{11} \neq 0, \rho_{22} \neq 0, \dots, \rho_{NN} \neq 0$, we easily verify that the vectors $e^{(1)}, e^{(2)}, \dots, e^{(N)}$ also form a basis. Since the choice of ρ_{ij} is arbitrary, it is possible to choose an orthonormal basis $e^{(1)}, e^{(2)}, \dots, e^{(N)}$.

If $\xi_1, \xi_2, \dots, \xi_N$ and $\eta_1, \eta_2, \dots, \eta_N$ are the components of vectors $x \in E_N$ and $y \in E_N$ in the orthonormal basis $e^{(1)}, e^{(2)}, \dots, e^{(N)}$, i.e.,

$$x = \sum_{j=1}^{N} \xi_j e^{(j)}, \qquad y = \sum_{j=1}^{N} \eta_j e^{(j)},$$

then the inner product (x, y) can be found by the formula

$$(x, y) = \sum_{j=1}^{N} \xi_j \eta_j, \quad x, y \in E_N. \tag{1.1.4}$$

In the complex case, for $x \in H_N$ and $y \in H_N$, formula (1.1.4) takes the form

$$(x, y) = \sum_{j=1}^{N} \xi_j \overline{\eta}_j, \quad x, y \in H_N.$$

The components $\xi_1, \xi_2, \dots, \xi_N$ of a vector x in an orthonormal basis can be found in terms of the inner products, namely, $\xi_i = (x, e^{(i)})$.

Usually, we will consider spaces with fixed bases. Thereby, to each vector x we associate the column vector with components $x_1, x_2, \dots, x_N$ symbolically written $x = (x_1, x_2, \dots, x_N)^T$.

By the *length* (*norm*) of a vector x we mean the number

$$\|x\| = \sqrt{(x, x)} = \left(\sum_{i=1}^{N} |x_i|^2 \right)^{1/2}.$$

We present some important inequalities. They can be proved with the help of the axioms of inner product.

The inequality

$$|(x, y)| \leq \|x\| \, \|y\|$$

is called the *Cauchy–Schwarz–Bunyakovskii inequality.* The equality takes place only if x and y are collinear (linearly dependent) vectors.

The inequality

$$|\, \|x\| - \|y\| \,| \leq \|x + y\| \leq \|x\| + \|y\|$$

is called the *triangle inequality.*

By introducing a metric in a Euclidean space, we can regard the theory of Euclidean spaces as a part of mathematical analysis, where the notions of continuous dependence, convergence, and so on play an important role.

1.2. Linear mappings and matrices. Determinants

Linear mappings and matrices. Identification of linear mappings and matrices. Adjoint mappings. Self-adjoint (symmetric) mapping and matrices. The norm of a linear mapping. Determinant as a function of a matrix. Properties of the determinant. The Cramer formulas. Homogeneous equations. Eigenvalues and eigenvectors.

The study of linear mappings is a central object of linear algebra.

A mapping (transformation) A: $E_N \to E_M$ ($H_N \to H_M$) taking vectors $x \in E_N$ ($x \in H_N$) to vectors $y \in E_M$ ($y \in H_M$) is said to be linear if the equalities

$$y^{(1)} = Ax^{(1)}, \quad y^{(2)} = Ax^{(2)}$$

imply the equality

$$A(\alpha_1 x^{(1)} + \alpha_2 x^{(2)}) = \alpha_1 y^{(1)} + \alpha_2 y^{(2)},$$

where α_1, α_2 are arbitrary real (complex) scalars.

If vectors x and y have the components $x_1, x_2, \ldots, x_N$ and $y_1, y_2, \ldots, y_M$ in some fixed bases for the spaces E_N and E_M (H_N and H_M), i.e.,

$$x = (x_1, x_2, \ldots, x_N)^T, \quad y = (y_1, y_2, \ldots, y_M)^T,$$

then the mapping $y = Ax$ can be given by the formula

$$\begin{pmatrix} y_1 \\ y_2 \\ \vdots \\ y_M \end{pmatrix} = \begin{bmatrix} a_{11} & a_{12} & \cdots & a_{1N} \\ a_{21} & a_{22} & \cdots & a_{2N} \\ \vdots & \vdots & \ddots & \vdots \\ a_{M1} & a_{M2} & \cdots & a_{MN} \end{bmatrix} \begin{pmatrix} x_1 \\ x_2 \\ \vdots \\ x_N \end{pmatrix} \tag{1.2.1}$$

with some matrix, denoted by the same letter A:

$$A = \begin{bmatrix} a_{11} & a_{12} & \cdots & a_{1N} \\ a_{21} & a_{22} & \cdots & a_{2N} \\ \vdots & \vdots & \ddots & \vdots \\ a_{M1} & a_{M2} & \cdots & a_{MN} \end{bmatrix}.$$

Formula (1.2.1) is the standard notation for a system of equalities

$$y_i = \sum_{j=1}^{N} a_{ij} x_j, \quad 1 \leq i \leq M.$$

We emphasize that such an identification of a mapping and the corresponding matrix leads to some inconvenience because a mapping can be represented by different matrices in different bases.

Transformations $A : H_N \xrightarrow{A} H_M$ and $B : H_M \xrightarrow{B} H_N$ are said to be *adjoint* if for any vectors $x \in H_N$ and $y \in H_M$ the following equality holds:

$$(Ax, y) = (x, By). \tag{1.2.2}$$

On the left- and right-hand side of (1.2.2) there are the inner products (Ax, y) and (x, By) which are usually computed by different formulas. Indeed, if $N \neq M$, then the vectors Ax and y are M-dimensional, whereas the vectors x and By have different dimension N. For a pedantic approach, we should denote the inner products by different symbols, e.g., $(\cdot,\cdot)_{H_M}$ and $(\cdot,\cdot)_{H_N}$, and write $(Ax, y)_{H_M} = (x, By)_{H_N}$ instead of (1.2.2). However, we will not follow strict rules in the hope that some liberty in notation will not cause serious problems for the reader.

It is not hard to verify that the inner products (x, u) for $x, u \in H_N$ and (v, y) for $v, y \in H_M$ can be computed by the formulas

$$(x, u) = \sum_{i=1}^{N} x_i \overline{u}_i, \quad (v, y) = \sum_{i=1}^{M} v_i \overline{y}_i;$$

i.e., in the case of orthonormal bases, the equality $(Ax, y) = (x, By)$ is valid for all x, y if and only if the elements a_{ij} and b_{kl} of the matrices corresponding to the transformations A and B are connected by the relations

$$a_{ij} = \overline{b_{ji}} \quad (b_{ij} = \overline{a_{ji}}),$$

so that the matrices A and B can be obtained, one from another, by transposition and replacement of each element with the conjugate complex ($A = \overline{B^T}$, $B = \overline{A^T}$).

If A and B are real matrices corresponding to the transformations $E_N \to E_M$ and $E_M \to E_N$ in the orthonormal bases, then the mapping B is adjoint to the mapping A if $a_{ij} = b_{ji}$, i.e., if B is obtained from A by transposition: $B = A^T$ ($A = B^T$).

The transformation B adjoint to A is denoted by A^*. It is obvious that $A = A^{**}$ and $(AB)^* = B^* A^*$.

A square $N \times N$-matrix A is said to be *self-adjoint* or *Hermitian* (in the real case, *symmetric*) if $A = A^*$.

To the linear mappings $E_N \to E_M$ we associate matrices with real elements, whereas matrices associated to the mappings $H_N \to H_M$ may have complex elements.

By the *norm* $\|A\|$ of a linear mapping A we mean the number

$$\|A\| = \sup_{x \neq 0} \frac{\|Ax\|}{\|x\|}.$$

Let matrices A and B have the same size. For the sum $A + B$ we have the following inequalities:

$$|\,\|A\| - \|B\|\,| \leq \|A + B\| \leq \|A\| + \|B\|.$$

In the theory of linear transformations and the corresponding matrices, the notion of a determinant plays an important role. Any textbook on linear algebra describes it in detail. We assume that the reader is familiar with this notion and restrict ourselves to the axiomatic definition and the most important properties which immediately follow from the definition.

It is convenient to regard the columns of a square $N \times N$-matrix

$$A = \begin{bmatrix} a_{11} & a_{12} & \dots & a_{1N} \\ a_{21} & a_{22} & \dots & a_{2N} \\ \vdots & \vdots & \dots & \vdots \\ a_{N1} & a_{N2} & \dots & a_{NN} \end{bmatrix}$$

as the column vectors

$$a^{(1)} = \begin{pmatrix} a_{11} \\ a_{21} \\ \vdots \\ a_{N1} \end{pmatrix}, \quad a^{(2)} = \begin{pmatrix} a_{12} \\ a_{22} \\ \vdots \\ a_{N2} \end{pmatrix}, \quad a^{(N)} = \begin{pmatrix} a_{1N} \\ a_{2N} \\ \vdots \\ a_{NN} \end{pmatrix}$$

and use the notation

$$A = (a^{(1)} \vdots a^{(2)} \vdots \dots \vdots a^{(N)}).$$

By the *determinant*

$$\det A = \begin{vmatrix} a_{11} & a_{12} & \dots & a_{1N} \\ a_{21} & a_{22} & \dots & a_{2N} \\ \vdots & \vdots & \dots & \vdots \\ a_{N1} & a_{N2} & \dots & a_{NN} \end{vmatrix}$$

of a matrix A we mean the function of a_{ij} satisfying the following three axioms:

⟨1⟩ The value of $\det A$ does not change if a column $a^{(i)}$ of the matrix A is replaced by the sum of columns $a^{(i)} + a^{(k)}$ $(k \neq i)$.

⟨2⟩ $\det A$ becomes $\alpha \det A$ if a column $a^{(i)}$ of the matrix A is replaced by a collinear column $\alpha a^{(i)}$, where α is a complex number.

⟨3⟩ The determinant of the identity matrix is 1, i.e.,

$$\det I = \det \begin{bmatrix} 1 & & & 0 \\ & 1 & & \\ & & \ddots & \\ 0 & & & 1 \end{bmatrix} = 1.$$

One can prove that the function det satisfying the above axioms exists and is unique.

We indicate some useful properties of the determinant.

⟨4⟩ If the columns $a^{(1)}, a^{(2)}, \dots, a^{(N)}$ of a matrix A are linearly dependent, then $\det A = 0$.

⟨5⟩ If we exchange the columns $a^{(i)}$ and $a^{(k)}$ $(i \neq k)$ of a matrix A leaving the remaining columns unchanged, then the determinant is multiplied by -1.

⟨6⟩ $\det A^T = \det A$ and $\det A^* = \overline{\det A}$. In other words, the value of the determinant of a matrix A does not change under the replacement of A by the transposed matrix A^T $(A \to A^T)$. Under the replacement of a matrix A by the adjoint matrix A^* $(A \to A^*)$ the determinant is replaced by the conjugate complex number.

⟨7⟩ If A and B are square $N \times N$-matrices, then the determinant of the product of A and B is equal to the product of their determinants, i.e., $\det(AB) = \det A \det B$.

⟨8⟩ The determinant of an upper triangular matrix (lower triangular matrix) is equal to the product of its diagonal elements. In general, for block triangular square matrices of the form

$$A = \begin{bmatrix} A_{11} & A_{12} \\ 0 & A_{22} \end{bmatrix}, \quad B = \begin{bmatrix} B_{11} & 0 \\ B_{21} & B_{22} \end{bmatrix}$$

with square diagonal blocks A_{11}, A_{22} and B_{11}, B_{22}, the following equalities hold:

$$\det \begin{bmatrix} A_{11} & A_{12} \\ 0 & A_{22} \end{bmatrix} = \det A_{11} \det A_{22},$$

$$\det \begin{bmatrix} B_{11} & 0 \\ B_{21} & B_{22} \end{bmatrix} = \det B_{11} \det B_{22}.$$

⟨9⟩ For any square $N \times N$-matrix A and number λ the following formula holds:

$$\det(\lambda I_N - A) = \lambda^N - P_1\lambda^{N-1} + P_2\lambda^{N-2} + \cdots + (-1)^{N-1}P_{N-1}\lambda + (-1)^N P_N,$$

where I_N denotes the identity $N \times N$-matrix,

$$P_N = \det A, \quad P_1 = a_{11} + a_{22} + \cdots + a_{NN},$$

$$P_l = \sum_{\substack{j_{i_1}, j_{i_2}, \ldots, j_{i_l} \\ (j_{i_1} < j_{i_k} < j_{i_l})}} \det \begin{bmatrix} a_{j_1 j_1} & a_{j_1 j_2} & \cdots & a_{j_1 j_l} \\ a_{j_2 j_1} & a_{j_2 j_2} & \cdots & a_{j_2 j_l} \\ \cdots & \cdots & \cdots & \cdots \\ a_{j_l j_1} & a_{j_l j_2} & \cdots & a_{j_l j_l} \end{bmatrix}.$$

The polynomial $\det(\lambda I_N - A)$ is called the *characteristic polynomial of the matrix* A. The coefficient P_1 is called the *trace of the matrix* A and is denoted by $\operatorname{tr} A$, i.e., $\operatorname{tr} A = P_1 = \sum_{i=1}^{N} a_{ii}$.

⟨10⟩ The determinant of a matrix A vanishes ($\det A = 0$) if and only if the columns (rows) of the matrix A are linearly dependent.

With the help of determinants, we can write explicit formulas for the components $x_1, x_2, \ldots, x_N$ of the vector x satisfying the equation $Ax = b$. These formulas, called the *Cramer formulas*, have the form

$$x_i = \frac{\Delta_i}{\Delta}, \quad i = 1, 2, \ldots, N,$$

where

$$\Delta = \det A = \det(a^{(1)} \vdots a^{(2)} \vdots \ldots \vdots a^{(N)}),$$

$$\Delta_1 = \det(b \vdots a^{(2)} \vdots \ldots \vdots a^{(N)}),$$

$$\Delta_2 = \det(a^{(1)} \vdots b \vdots a^{(3)} \vdots \ldots \vdots a^{(N)}),$$

$$\cdots\cdots\cdots\cdots\cdots\cdots\cdots\cdots\cdots\cdots\cdots$$

$$\Delta_N = \det(a^{(1)} \vdots a^{(2)} \vdots \ldots \vdots a^{(N-1)} \vdots b).$$

To apply the Cramer formulas, the condition $\Delta \neq 0$ is necessary. This condition ensures the existence and uniqueness of a solution to the equation $Ax = b$ for any right-hand side b. In particular, if $\det A \neq 0$, then the homogeneous system $Ax = 0$ has only the zero solution $x = 0$.

We emphasize that the Cramer formulas cannot be always applied for the numerical solution of a vector equation $Ax = b$ (i.e., a system of linear equations with respect to x_j), especially in the case of matrices of large order N.

The homogeneous equation $Ax = 0$ has a nonzero solution if (and only if) $\det A = 0$. If $Ax = 0$, then for any α the vector $y = \alpha x$ satisfies the equation $Ay = 0$. Therefore, a solution always is not unique. A solution to a nonhomogeneous vector equation $Ax = b$ with $\det A = 0$ might exist (it is not unique a priori) or not exist. We indicate a vector equation that has no solutions and a vector equation having more than one solution:

$$\begin{bmatrix} 1 & 1 \\ 3 & 3 \end{bmatrix} \begin{pmatrix} x_1 \\ x_2 \end{pmatrix} = \begin{pmatrix} 1 \\ 2 \end{pmatrix}, \quad \begin{bmatrix} 1 & 1 \\ 3 & 3 \end{bmatrix} \begin{pmatrix} x_1 \\ x_2 \end{pmatrix} = \begin{pmatrix} 3 \\ 9 \end{pmatrix}.$$

A complex number λ is called an *eigenvalue* of a linear mapping $A{:}H_N \to H_N$ if there exists a nonzero vector $x \in H_N$, called an *eigenvector*, such that $Ax = \lambda x$. Eigenvalues of a mapping A are roots of the characteristic equation $\det(\lambda I - A) = 0$. Even for a real matrix A, the roots can be complex numbers. Therefore, to an eigenvalue of a mapping $E_N \to E_N$ we cannot always associate a real eigenvector $x \in E_N$. However, embedding the space E_N into the space H_N in a natural way, to a complex eigenvalue λ we can always associate an eigenvector in $H_N \supset E_N$.

If $\det T \neq 0$, then there exists the matrix $B = TAT^{-1}$ corresponding to the same linear transformation as the matrix A but written in another basis. In addition, $\det(\lambda I - A) = \det(\lambda I - B)$, i.e., the characteristic polynomials of the matrices A and B coincide.

The eigenvalues $\lambda_1, \lambda_2, \dots, \lambda_N$ of a self-adjoint matrix A ($A = A^*$) are real. Indeed, if x is an eigenvector corresponding to an eigenvalue λ, then

$$\begin{aligned} (Ax, x) &= (x, A^*x) = (x, Ax), \\ (Ax, x) &= (\lambda x, x) = \lambda(x, x), \\ (x, Ax) &= (x, \lambda x) = \overline{\lambda}(x, x), \\ \lambda(x, x) &= \overline{\lambda}(x, x). \end{aligned}$$

Since $x \neq 0$, we have $(x, x) = \|x\|^2 > 0$. Hence the equality $\lambda(x, x) = \overline{\lambda}(x, x)$ yields the equality $\lambda = \overline{\lambda}$, which means that λ is real.

1.3. The accuracy problems in computations

Computation errors and the accuracy parameter. An unexpected example. ε-Spectrum of a matrix. Spectral portraits of matrices. Illustrations.

In practical problems, one usually meets matrices obtained in some measurements or preliminary computations. Measurements and computations are always preformed with errors. If, instead of an exact matrix A, we are given an approximate matrix $A + \Delta$, then it is natural to characterize the accuracy of measurement or computation by the dimensionless accuracy parameter ε such that the following estimate holds:

$$\frac{\|(A+\Delta)-A\|}{\|A\|} = \frac{\|\Delta\|}{\|A\|} < \varepsilon.$$

The following question arises. How to choose the accuracy parameter ε in order to provide a suitable accuracy of a final solution if in the computation procedure the approximate matrix $A+\Delta$ instead of the initial matrix A was used? Answers to different problems can differ.

To illustrate the situation, we give one unexpected example.

Eigenvalues of the matrix

$$C = \begin{bmatrix} 289 & 2064 & 336 & 128 & 80 & 32 & 16 \\ 1152 & 30 & 1312 & 512 & 288 & 128 & 32 \\ -29 & -2000 & 756 & 384 & 1008 & 224 & 48 \\ 512 & 128 & 640 & 0 & 640 & 512 & 128 \\ 1053 & 2256 & -504 & -384 & -756 & 800 & 208 \\ -287 & -16 & 1712 & -128 & 1968 & -30 & 2032 \\ -2176 & -287 & -1565 & -512 & -541 & -1152 & -289 \end{bmatrix}$$

were computed by a standard program. The following result was obtained (we indicate only first three digits):

Approximate spectrum of the matrix C:

$5.59 \pm 2.21i, \quad 1.33 \pm 5.02i, \quad -6.18, \quad -3.83 \pm 3.96i.$

With the help of the matrix

$$P = \begin{bmatrix} 0 & 0 & 0 & 1 & 0 & 0 & 0 \\ 0 & 0 & 0 & 0 & 1 & 0 & 0 \\ 0 & 0 & 0 & 0 & 0 & 1 & 0 \\ 0 & 0 & 0 & 0 & 0 & 0 & 1 \\ 1 & 0 & 0 & 0 & 0 & 0 & 0 \\ 0 & 1 & 0 & 0 & 0 & 0 & 0 \\ 0 & 0 & 1 & 0 & 0 & 0 & 0 \end{bmatrix} \qquad (P^{-1} = P^T),$$

which permutates the basis vectors, the following matrices were obtained:

$$D = P^{-1}CP, \quad F = P^{-1}DP = P^{-1}(P^{-1}CP)P.$$

These matrices realize the same transformation as the matrix C, but in the bases obtained from the initial basis by permutations of the basis vectors. Using the same standard program, eigenvalues of the matrices D and F were computed. Eigenvalues of the matrix D, as well as eigenvalues of the matrix F, must coincide with eigenvalues of the matrix C. However, the following result was obtained:

Approximate spectrum of the matrix D:

$7.36 \pm 3.22i, \quad 1.76 \pm 7.21i, \quad -5.07 \pm 5.70i, \quad -8.10.$

Approximate spectrum of the matrix F:

$5.34, \quad 3.21 \pm 3.09i, \quad -1.15 \pm 3.89i, \quad -4.73 \pm 1.63i.$

In the computation of the matrices D and F no arithmetic errors were made because the computation reduces to permutations of integer-valued columns and rows of the matrix C.

However faultless a standard program might be, arithmetic operations performed are necessarily accompanied with round-off errors. The magnitude of these errors depends on the grid used for the representation of numbers in a computer. In the computation of eigenvalues of the matrices C, D, and F, it is impossible to avoid arithmetic errors even though these matrices are known precisely. Errors can be simulated by a perturbation of the initial matrix whose spectrum is computed. Thus, the difference between the computed spectra of the matrices C, D, and F can be explained by saying that the computed λ_i's are exact eigenvalues of one of the matrices

$$C + \Delta, \quad D + \widetilde{\Delta}, \quad F + \widetilde{\widetilde{\Delta}}$$

under the conditions

$$\frac{\|\Delta\|}{\|C\|} < \varepsilon, \quad \frac{\|\widetilde{\Delta}\|}{\|D\|} < \varepsilon, \quad \frac{\|\widetilde{\widetilde{\Delta}}\|}{\|F\|} < \varepsilon, \tag{1.3.1}$$

where the number ε characterizes the accuracy and depends on the algorithm and the digit grid in the computer. The above example was computed using a computer which provides the representation of numbers with relative error $\asymp 10^{-16}$. It is clear that the accuracy parameter ε in the inequalities (1.3.1) must be not less than 10^{-16}. The detailed verification confirms this assumption.

We note that the eigenvalues of the matrices C, D, and F are integers. This follows from the representation

$$C = L^{-1}RL$$

of the matrix C in terms of the lower triangular matrix

$$L = \begin{bmatrix} 1 & 0 & 0 & 0 & 0 & 0 & 0 \\ 0 & 1 & 0 & 0 & 0 & 0 & 0 \\ 1 & 0 & 1 & 0 & 0 & 0 & 0 \\ 0 & 0 & 0 & 1 & 0 & 0 & 0 \\ 0 & 0 & 1 & 0 & 1 & 0 & 0 \\ 1 & 0 & 0 & 0 & 0 & 1 & 0 \\ 0 & 1 & 1 & 0 & 1 & 0 & 1 \end{bmatrix}$$

and the upper triangular matrix

$$R = \begin{bmatrix} 1 & 2048 & 256 & 128 & 64 & 32 & 16 \\ 0 & -2 & 1024 & 512 & 256 & 128 & 32 \\ 0 & 0 & 4 & 512 & 1024 & 256 & 64 \\ 0 & 0 & 0 & 0 & 512 & 512 & 128 \\ 0 & 0 & 0 & 0 & -4 & 1024 & 256 \\ 0 & 0 & 0 & 0 & 0 & 2 & 2048 \\ 0 & 0 & 0 & 0 & 0 & 0 & -1 \end{bmatrix}.$$

The spectrum of the matrix C coincides with the spectrum of the matrix R and consists of the numbers located on the main diagonal.

Exact spectrum of the matrix C:

$0, \quad \pm 1, \quad \pm 2, \quad \pm 4.$

To explain similar situations, it is instructive to use the notion of the ε-spectrum of a matrix and to study the graphic representations of the ε-spectrum.

We say that a complex number λ *belongs to the* ε*-spectrum* $\Lambda_\varepsilon(A)$ *of a matrix* A if there exists a perturbation Δ such that $\|\Delta\| \le \varepsilon\|A\|$ and $\det(\lambda I - A - \Delta) = 0$.

It is possible to suggest an algorithm that allows us, starting with a given matrix A and an eigenvalue λ, to find ε such that $\lambda \in \Lambda_\varepsilon(A)$, i.e., λ belongs to the ε-spectrum of the matrix A. We will not describe the algorithm here.

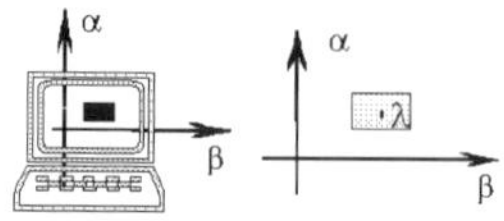

By using a modern computer, it is possible to obtain visual representations of ε-spectra on the screen of the monitor at least for matrices of not too large order. The screen of the monitor consists of a large but finite number of points (pixels). Considering some rectangle of the complex plane in the screen, to each pixel we associate the center λ of the rectangle. For example, if the resolution of the screen is 640×350 points, we associate to pixels 224.000 points of the complex plane.

Using the above-mentioned algorithm, for every point we can find the smallest ε for which $\lambda \in \Lambda_\varepsilon(A)$ and choose a color of the corresponding pixel depending on ε. Having colored all pixels, we obtain a picture on the screen. This picture is referred to as the spectral portrait of the matrix A.

To illustrate the above arguments, we give examples of spectral portraits of two matrices.

The level lines of the spectral portrait of the matrix

$$B = \begin{bmatrix} -2 & 25 & 0 & 0 & 0 & 0 & 0 \\ 0 & -3 & 10 & 3 & 3 & 3 & 0 \\ 0 & 0 & 2 & 15 & 3 & 3 & 0 \\ 0 & 0 & 0 & 0 & 15 & 3 & 0 \\ 0 & 0 & 0 & 0 & 3 & 10 & 0 \\ 0 & 0 & 0 & 0 & 0 & -2 & 25 \\ 0 & 0 & 0 & 0 & 0 & 0 & -3 \end{bmatrix}$$

are shown in Figure 1.3.1.

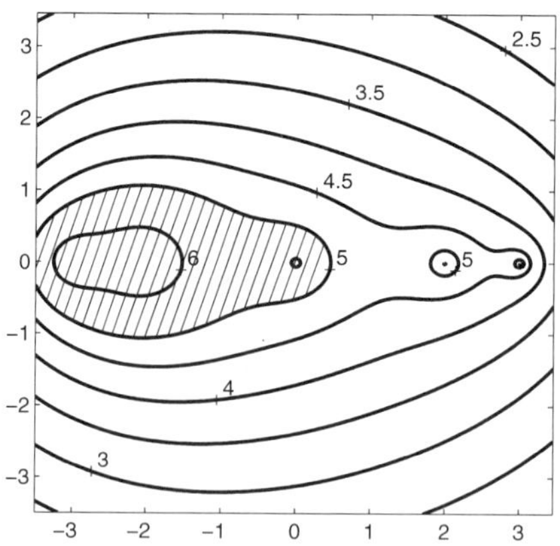

FIGURE 1.3.1

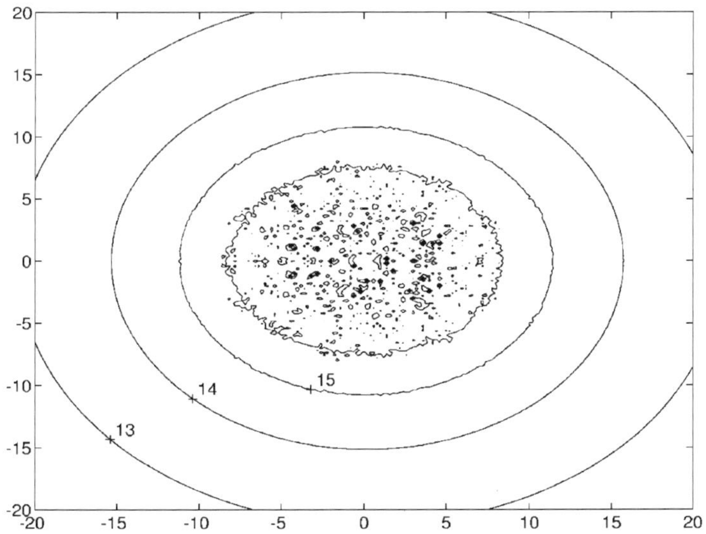

FIGURE 1.3.2. The spectral portrait of the matrix C

It shows very significant perturbations of eigenvalues of a matrix B although the matrix B was perturbed by a small matrix, with the norm less than $10^{-6}\|B\|$.

The spectral portrait of the matrix C is shown in Figure 1.3.2.

The above examples put us on guard. They suggest that large sensitivity to small perturbations depreciates conclusions of the spectral analysis of matrices.

In the following chapters, we will distinguish classes of matrices that are stable under perturbations of certain spectral characteristics and outline methods for computing such characteristic parameters.

CHAPTER 2

Orthogonal and Unitary Linear Transformations

2.1. Orthogonal transformations

Equivalent definitions of orthogonal transformations. Unitary transformations. The complex case.

By an *orthogonal linear transformation* we mean a transformation under which the norms of vectors do not change. In Chapter 7, we will introduce several variants of the notion of "conditionality". This notion is connected with the metric of a space. In order to give an exact definition and to derive estimates for conditionality, it is natural to reduce matrices to a canonical form by an orthogonal transformation. A linear transformation $Q : E_N \to E_N$ is said to be *orthogonal* if $\|y\| = \|Qx\| = \|x\|$ for any $x \in E_N$. It is obvious that the definition can be expressed by the equality $\|Qx\|^2 = \|x\|^2$ since $\|Qx\|^2 = (Qx, Qx) = (x, x) = \|x\|^2$.

Let $x = a+b$. In view of properties of an inner product, the following equalities hold:

$$\begin{aligned}(Q(a+b), Q(a+b)) &= (Qa, Qa) + (Qa, Qb) + (Qb, Qa) + (Qb, Qb),\\ (a+b, a+b) &= (a,a) + (a,b) + (b,a) + (b,b),\\ (Qa, Qb) &= (Qb, Qa), \quad (a,b) = (b,a).\end{aligned}$$

By the definition of an orthogonal transformation, we have

$$\begin{aligned}(Q(a+b), Q(a+b)) &= (a+b, a+b),\\ (Qa, Qa) &= (a,a), \quad (Qb, Qb) = (b,b).\end{aligned}$$

Hence $(Qa, Qb) = (a, b)$ for an orthogonal transformation Q and $a, b \in E_N$.

The above arguments justify the second (equivalent) definition of an orthogonal transformation. A linear transformation $Q{:}E_N \to E_N$ is said to be *orthogonal* if $(Qa, Qb) = (a, b)$ for any $a \in E_N$, $b \in E_N$. We recall that $(Qa, Qb) = (Q^TQa, b)$.

Let an orthogonal transformation Q be represented in an orthonormal basis by the matrix

$$Q = \begin{bmatrix} q_{11} & q_{12} & \cdots & q_{1N}\\ q_{21} & q_{22} & \cdots & q_{2N}\\ \vdots & \vdots & \ddots & \vdots\\ q_{N1} & q_{N2} & \cdots & q_{NN}\end{bmatrix}.$$

One can show that the orthogonality of the matrix Q is equivalent to the equalities

$$\sum_{j=1}^{N} q_{ji}q_{jk} = \begin{cases} 0 & \text{if } i \neq k,\\ 1 & \text{if } i = k,\end{cases}$$

which can be written in the matrix form

$$Q^T Q = QQ^T = I,$$

where I denotes the unit matrix in E_N, i.e.,

$$I \equiv I_N = \begin{bmatrix} 1 & & & 0 \\ & 1 & & \\ & & \ddots & \\ 0 & & & 1 \end{bmatrix}.$$

In other words, if the matrix Q is orthogonal, then $Q^{-1} = Q^T$.

If a transformation Q acts in the complex Euclidean space H_N and does not change the norms of vectors ($\|x\| = \|Qx\|$), then, arguing as before, we can show that $Q\overline{Q^T} = I$, where the matrix $\overline{Q^T}$ is obtained from the matrix Q by taking the transpose matrix and replacing each element with the conjugate complex. The notation $\overline{Q^T} = Q^*$ is traditionally used. We recall that Q^* denotes the matrix adjoint to Q (see §1.2). If a matrix Q is real, then $Q^* = Q^T$.

A transformation Q of the complex Euclidean space H_N is said to be *unitary* if it preserves the norms of vectors, i.e., $\|Qx\| = \|x\|$. We note that the equality $Q^*Q = QQ^* = I$ can be regarded as a criterion for unitarity.

In the following sections of this chapter we will describe some classes of the simplest orthogonal (or unitary) transformations which will be used as the main tool in the study of matrices. We begin with those orthogonal transformations of the space E_N that are reflections in some hyperplanes. After that we will consider the so-called chains of two-dimensional rotations of a Euclidean space.

2.2. Orthogonal reflections

Reflection of a vector with respect to a hyperplane. The complex case: unitary reflections. The main property of reflections.

Let the space E_N be equipped with an orthonormal basis so that vectors of the space E_N are represented by column vectors in this basis, e.g.,

$$p = \begin{pmatrix} p_1 \\ p_2 \\ \vdots \\ p_N \end{pmatrix}, \quad x = \begin{pmatrix} x_1 \\ x_2 \\ \vdots \\ x_N \end{pmatrix}.$$

Vectors x such that $(x, p) = 0$ form a hyperplane which is orthogonal to p. For any x the module of $\dfrac{(x,p)}{\|p\|}$ is equal to the distance from the point x to the hyperplane $(x,p) = 0$ and the vector $\dfrac{(x,p)}{\|p\|}\dfrac{1}{\|p\|}p$ is orthogonal to this hyperplane. In addition,

$$\left(p, x - \frac{(x,p)}{\|p\|^2}p\right) = 0,$$

i.e., the vector $z = x - \dfrac{(x,p)}{\|p\|^2}p$ lies in the hyperplane $(z,p) = 0$. The point $x - \dfrac{(x,p)}{\|p\|^2}p$ is the base of the perpendicular dropped from the endpoint of the vector x to the hyperplane $(x,p) = 0$. The vector $y = x - 2\frac{(x,p)}{\|p\|^2}p$ is called the *reflection of the vector x in the hyperplane $(z,p) = 0$.*

The transformation of the space E_N that takes the vector $x \in E_N$ to the reflection y of x with respect to the hyperplane $(z,p) = 0$ is denoted by $\mathcal{P}$, i.e.,

$$y = \mathcal{P}x = x - 2\frac{(x,p)}{\|p\|^2}p.$$

It is easy to see that $\mathcal{P}$ is linear. Furthermore,

$$\mathcal{P}p = p - 2\frac{(p,p)}{\|p\|^2}p = -p,$$

$$\mathcal{P}^2x = \mathcal{P}\left[x - 2\frac{(x,p)}{\|p\|^2}p\right] = \mathcal{P}x - 2\frac{(x,p)}{\|p\|^2}\mathcal{P}p = \left[x - 2\frac{(x,p)}{\|p\|^2}p\right] - 2\frac{(x,p)}{\|p\|^2}(-p) = x,$$

i.e., $\mathcal{P}^2 = I$. To the transformation $\mathcal{P}$ we associate the matrix, denoted by the same letter $\mathcal{P}$, with elements

$$\mathcal{P}_{ij} = \delta_{ij} - 2\frac{p_i p_j}{\|p\|^2}, \quad i,j = 1,\dots,N,$$

where $\delta_{ij} = 0$ for $i \neq j$ and $\delta_{ij} = 1$ for $i = j$. It is obvious that $\mathcal{P}_{ij} = \mathcal{P}_{ji}$, i.e., $\mathcal{P} = \mathcal{P}^T$ (or $\mathcal{P} = \mathcal{P}^*$, which is the same in the real space E_N), and the matrix $\mathcal{P}$ is orthogonal, i.e., $\mathcal{P}^*\mathcal{P} = \mathcal{P}\mathcal{P}^* = \mathcal{P}^2 = I$.

Let q and r be linearly independent (noncollinear) vectors of the same length ($\|q\| = \|r\|$). Then there exists a reflection $\mathcal{P}$ such that $\mathcal{P}q = r$. Indeed, let $p = q - r$ ($p \neq 0$). Then

$$\begin{aligned}
\mathcal{P}q &= q - 2\frac{(q,p)}{\|p\|^2}p = q - 2\frac{(q,q-r)}{(q-r,q-r)}(q-r) \\
&= q - \frac{(q-r,q-r)}{(q-r,q-r)}(q-r) - \frac{(q+r,q-r)}{(q-r,q-r)}(q-r) \\
&= q - \frac{(q-r,q-r)}{(q-r,q-r)}(q-r) = r.
\end{aligned}$$

We have used the fact that the vectors $q-r$ and $q+r$ are orthogonal and the norms of q and r are equal ($\|q\| = \|r\|$). Indeed,

$$(q+r,q-r) = (q,q) + (r,q) - (q,r) - (r,r) = \|q\|^2 - \|r\|^2 = 0.$$

In the complex Euclidean space H_N, by a *unitary reflection* we mean a linear transformation $\mathcal{P}\colon H_N \to H_N$ defined by the same formula as in the real case, i.e.,

$$y = \mathcal{P}x = x - 2\frac{(x,p)}{\|p\|^2}p.$$

Generally speaking, the complex case (p,x) is equal to $\overline{(x,p)}$, i.e., (p,x) does not coincide with (x,p), and the elements $\mathcal{P}_{ij}$ of the matrix $\mathcal{P}$ have the form

$$\mathcal{P}_{ij} = \delta_{ij} - 2\frac{\overline{p}_i p_j}{\|p\|^2}, \quad i,j = 1,\dots,N.$$

The last formula shows that $\mathcal{P}$ is a Hermitian matrix, i.e., $\mathcal{P} = \mathcal{P}^*$. Arguing as in the real case, we can verify that $\mathcal{P}^2 = I$. Therefore, $\mathcal{P}^*\mathcal{P} = \mathcal{P}\mathcal{P}^* = \mathcal{P}^2 = I$. Consequently, a unitary reflection is given by a unitary matrix.

In spite of the indicated analogy, there are some significant differences from the real case. For example, in the complex space, there exist a pair of vectors $q \in H_N$ and $r \in H_N$ of the same norm ($\|q\| = \|r\|$) such that there is no unitary reflection $\mathcal{P}$ mapping q to r. For the existence of such a reflection it is necessary that the inner product (q, r) be real. If $\operatorname{Im}(q, r) = 0$, i.e., $(q, r) = \overline{(r, q)} = (r, q)$, then we put $p = q - r$ and acting as in the real case prove that $(q + r, q - r) = \|q\|^2 - \|r\|^2 = 0$ and $\mathcal{P}q = r$.

Under the multiplication of the vector p defining the reflection $\mathcal{P}$ by a factor of the form $e^{i\alpha}$, reflections do not change since the products $\overline{p}_i p_j$, $i, j = 1, 2, \ldots, N$, and the elements $\mathcal{P}_{ij}$ of the matrix $\mathcal{P}$ computed from these products remain unchanged.

Now we are able to establish the main property of reflections in a Euclidean (real or complex) space.

The main property of reflections is that for any linearly independent vectors a and b there is a reflection $\mathcal{P}$ such that the vector $\mathcal{P}a$ is collinear to the vector b, i.e., $\mathcal{P}a = \rho b$.

Indeed, for $(a, b) = me^{i\mu}$ with m and μ real, we set

$$q = \frac{1}{\|a\|} a, \qquad r = \frac{e^{i\mu}}{\|b\|} b.$$

Then $\|q\| = \|r\| = 1$, the inner product $(q, r) = \dfrac{m}{\|a\|\,\|b\|}$ is a real number, and for the reflection $\mathcal{P}$ constructed by the vector $p = q - r$ the equality $\mathcal{P}q = r$ holds. This implies the formulas

$$\mathcal{P}a = e^{i\mu}\frac{\|a\|}{\|b\|} b = \rho b, \qquad \rho = e^{i\mu}\frac{\|a\|}{\|b\|}.$$

If the vectors a and b are collinear, then the reflection generated by the vector $p = \dfrac{a}{\|a\|}$ transforms each of the vectors a and b to a vector collinear to a. Thus, for any vectors $a \neq 0$ and $b \neq 0$ there exists a reflection $\mathcal{P}$ that transforms a to a vector collinear to b.

For a vector u we intend to choose a reflection $\mathcal{P}$ such that the vector $v = \mathcal{P}u$ satisfies the conditions described below. We choose one of N components $u_1, u_2, \ldots, u_N$ of the vector u, say u_{k_0}, and divide the remaining components into two groups. Let the first group contain the components $u_{k_1}, u_{k_2} \cdots u_{k_l}$ and the second group contain the remaining components $u_{k_{l+1}}, u_{k_{l+2}}, \ldots, u_{k_{N-1}}$. We want the components of the vector u from the second group to remain unchanged under the reflection, i.e.,

$$v_{k_{l+1}} = u_{k_{l+1}}, \quad v_{k_{l+2}} = u_{k_{l+2}}, \ldots, v_{k_{N-1}} = u_{k_{N-1}},$$

and the components from the first group to vanish, i.e., $v_{k_1} = 0, v_{k_2} = 0, \ldots, v_{k_l} = 0$. To this end, the reflection vector p must be constructed in a special way.

We restrict ourselves to the case where the first group consists of the components $u_1, u_2, \ldots, u_l$, whereas the second group consists of the components $u_{l+2}, u_{l+3}, \ldots, u_N$. It is clear that a general case is reduced to this special case by renumbering components (if each group contains at least one component). If the first or second group is empty ($l + 1 = 1$ or $l + 1 = N$), the obvious necessary modifications are left to the reader.

By the equality $\|u\| = \|\mathcal{P}u\| = \|v\|$, we have

$$\begin{aligned}\sum_{k=1}^{l} |u_k|^2 + |u_{l+1}|^2 + \sum_{k=l+2}^{N} |u_k|^2 &= |v_{l+1}|^2 + \sum_{k=1}^{l} |v_k|^2 + \sum_{k=l+2}^{N} |v_k|^2 \\ &\equiv |v_{l+1}|^2 + \sum_{k=l+2}^{N} |u_k|^2,\end{aligned}$$

$$\sum_{k=1}^{l} |u_k|^2 + |u_{l+1}|^2 = |v_{l+1}|^2, \quad |v_{l+1}| = (|u_{l+1}|^2 + \sum_{k=1}^{l} |u_k|^2)^{1/2}.$$

Thus, the vector v must have the following components:

$$\begin{aligned}&v_1 = v_2 = \cdots = v_l = 0, \\ &v_{l+1} = e^{i\beta}\left(|u_{l+1}|^2 + \sum_{k=1}^{l} |u_k|^2\right)^{1/2}, \quad \beta \in \mathbb{R}, \\ &v_{l+2} = u_{l+2}, v_{l+3} = u_{l+3}, \ldots, v_N = u_N.\end{aligned}$$

$$v_{l+1} = \begin{cases} \mp\dfrac{u_{l+1}}{|u_{l+1}|}\left(|u_{l+1}|^2 + \sum\limits_{k=1}^{l} |u_k|^2\right)^{1/2} & \text{if } u_{l+1} \neq 0, \\ \mp\left(\sum\limits_{k=1}^{l} |u_k|^2\right)^{1/2} & \text{if } u_{l+1} = 0. \end{cases}$$

By the choice of v, the vector $p = u - v$ has the components

$$\begin{aligned}&p_1 = u_1, p_2 = u_2, \ldots, p_l = u_l, \\ &p_{l+1} = \begin{cases} u_{l+1} \pm \dfrac{u_{l+1}}{|u_{l+1}|}\left(u_{l+1}^2 + \sum\limits_{k=1}^{l} |u_k|^2\right)^{1/2} & \text{if } u_{l+1} \neq 0, \\ \pm\left(\sum\limits_{k=1}^{l} |u_k|^2\right)^{1/2} & \text{if } u_{l+1} = 0, \end{cases} \\ &p_{l+2} = p_{l+3} = \cdots = p_N = 0.\end{aligned}$$

The choice of the sign ($+$ or $-$) is arbitrary. It is convenient to choose the plus sign since the computed value of p_{l+1} (for $u_{l+1} \neq 0$) is less sensitive to computation errors.

Let $\overline{x}$ and $\overline{y}$ be N-dimensional column vectors with components $\overline{x}_i$ and $\overline{y}_i$, respectively. If Q is a unitary reflection ($Q^* = Q$, $Q^2 = I$) such that $Q\overline{x} = \rho\overline{y}$, then

$$x^T Q = \overline{x}^* Q^* \equiv [x_1, x_2, \ldots, x_N]Q^* = \overline{\rho}[y_1, y_2, \ldots, y_N] = \overline{\rho} y^T,$$

where x^T and y^T are row vectors (transposed columns x and y). Assume that we can choose a reflection Q that transforms $\overline{x}$ into a column vector collinear to $\overline{y}$. Thereby, we can find the same reflection using the condition that the row vector x^T multiplied from the right by Q goes to a row vector that is proportional to a given row vector y^T.

2.3. Chain of two-dimensional rotations

The definition of a chain of two-dimensional rotations. Inverse chain of rotations. The problem about the polarization of a vector.

We fix an orthogonal basis for a Euclidean space E_N.

By a *chain of two-dimensional rotations with parameters*

$$c_N, s_N, c_{N-1}, s_{N-1}, \ldots, c_2, s_2$$

we mean the composition

$$C = C_N C_{N-1} \ldots C_3 C_2$$

of successive transformations $C_2, C_3, \ldots, C_N$ represented by the matrices

$$C_2 = \begin{bmatrix} c_2 & -s_2 & & & & \\ s_2 & c_2 & & & \mathbf{0} & \\ & & 1 & & & \\ & & & 1 & & \\ & \mathbf{0} & & & \ddots & \\ & & & & & 1 \end{bmatrix},$$

$$C_j = \begin{bmatrix} 1 & & & & & & & & \\ & \ddots & & & & & \mathbf{0} & & \\ & & 1 & & & & & & \\ & & & c_j & -s_j & & & & \\ \cdots & \cdots & \cdots & s_j & c_j & \cdots & \cdots & \cdots & \cdots \\ & & & & & 1 & & & \\ & & & & & & 1 & & \\ & \mathbf{0} & & & & & & \ddots & \\ & & & & & & & & 1 \end{bmatrix} \quad j\text{th row},$$

$$j = 3, \ldots, N-1,$$

$$C_N = \begin{bmatrix} 1 & & & & \\ & \ddots & & \mathbf{0} & \\ & & 1 & & \\ & & & c_N & -s_N \\ \mathbf{0} & & & s_N & c_N \end{bmatrix}.$$

The parameters c_j and s_j are the cosine and the sine of the angle of rotations C_j, $j = 2, \ldots, N$, under which only $(j-1)$th and jth components change.

Every chain of two-dimensional rotations is completely determined by its parameters $c_2, s_2, c_3, s_3, \ldots, c_N, s_N$ satisfying the conditions $c_i^2 + s_j^2 = 1$, $j = 2, \ldots, N$.

The sequence of parameters $c_2, -s_2, c_3, -s_3, \ldots, c_N, -s_N$ corresponds to the inverse transformation $C^{-1} = C^* = C_2^* C_3^* \ldots C_{N-1}^* C_N^*$. The inverse transformation also can be regarded as a chain of two-dimensional rotations. Under the action of this chain, the components of a vector are transformed in the order opposite to that in the case of the chain C.

We again emphasize that the transformation corresponding to the matrix C_j is the rotation at some angle α_j in the two-dimensional plane spanned by the basis vectors $e^{(j-1)}$ and $e^{(j)}$. In addition, $c_j = \cos\alpha_j$ and $s_j = \sin\alpha_j$. We usually assume that

$$0 \le \alpha_j < \pi, \quad j = 2, \dots, N,$$

or, equivalently, $0 \le s_j \le 1, -1 < c_j \le 1$.

Using chains of two-dimensional rotations, we can solve the following problem.

PROBLEM (about the polarization of a vector x). Let x be a real vector of non-zero length ($\|x\| \ne 0$). Construct a chain of two-dimensional rotations $C = C_N C_{N-1} \dots C_2$ which transforms x to a vector that is proportional to the basis vector $e^{(N)} = (0, 0 \dots, 0, 1)^T$, i.e., $\rho e^{(N)} = Cx$, where ρ is a number.

The chain C in this problem is called the *chain polarizing* x.

LEMMA 2.3.1. *Let a chain of two-dimensional rotations $C = C_N \dots C_2$ polarize a vector x ($\|x\| \ne 0$). Then the components $x_1, x_2, \dots, x_N$ of the vector x and the parameters $c_2, c_3, \dots, c_N$ of the chain C are connected by the following relations:*

$$\begin{aligned} x_1 &= \rho s_2 s_3 \dots s_{N-1} s_N, \\ x_2 &= \rho c_2 s_3 s_4 \dots s_{N-1} s_N, \\ x_3 &= \rho c_3 s_4 \dots s_{N-1} s_N, \\ &\dots\dots\dots\dots\dots\dots \\ x_{N-1} &= \rho c_{N-1} s_N, \\ x_N &= \rho c_N. \end{aligned} \tag{2.3.1}$$

PROOF. We solve the polarization problem for x and simultaneously prove Lemma 2.3.1. We assume that the first r components of the vector x equal zero. Consider two cases. First we give a rule for computing parameters c_j and s_j. Then we check if they define the required transformation C.

Case 1. $r > 0$, i.e., $x_1 = x_2 = \dots = x_r = 0, x_{r+1} \ne 0$. We set $s_1 = s_2 = \dots = s_r = s_{r+1} = 0$ and $c_1 = c_2 = \dots = c_r = c_{r+1} = 1$.

Case 2. $r = 0$, i.e., $x_1 \ne 0$. We set $s_1 = s_{r+1} = 0$, $c_1 = c_{r+1} = 1$.

In both cases, for $j = r+2, r+3, \dots, N$ the values s_j and c_j are found by the following recurrence formulas:

$$z_{r+1} = x_{r+1},$$

$$z_j = \varepsilon_x \sqrt{z_{j-1}^2 + x_j^2}, \quad j = r+2, r+3, \dots, N,$$

where

$$\varepsilon_x = \begin{cases} +1 & \text{if } x_{r+1} > 0, \\ -1 & \text{if } x_{r+1} < 0, \end{cases}$$

$$s_j = \frac{z_{j-1}}{z_j}, \quad c_j = \frac{x_j}{z_j}, \ j = r+2, r+3, \dots, N.$$

It is obvious that all z_{j-1} and z_j, $j = r+2, r+3, \dots, N$, differ from zero and have the same sign. Therefore, $s_j > 0$ for $j = r+2, r+3, \dots, N$. Since $z_j^2 = z_{j-1}^2 + x_j^2$, for $j = r+2, r+3, \dots, N$, we have

$$|x_j| < |z_j|, \quad -1 < c_j \le 1, \quad j = r+2, r+3, \dots, N.$$

It is also obvious that

$$z_j^2 = x_1^2 + x_2^2 + \cdots + x_j^2, \; j = r+2, r+3, \ldots, N-1,$$
$$z_N^2 = x_1^2 + x_2^2 + \cdots + x_N^2,$$
$$x_N = c_N z_N.$$

Since $z_{j-1} = s_j z_j = s_j s_{j+1} \ldots s_N z_N$, $j = r+2,\, r+3, \ldots, N$, we have

$$\begin{aligned} z_{r+1} &= s_{r+2} s_{r+3} \ldots s_N z_N, \\ z_{r+2} &= s_{r+3} s_{r+4} \ldots s_N z_N, \\ \ldots \quad &\ldots \quad \ldots \\ z_{N-1} &= s_N z_N, \\ z_N &= z_N. \end{aligned}$$

Moreover, $x_j = c_j z_j$, $j = r+2, r+3, \ldots, N$. Consequently,

$$\begin{aligned} x_{r+1} &= c_{r+1} s_{r+2} s_{r+3} \ldots s_N z_N, \\ x_{r+2} &= c_{r+2} s_{r+3} s_{r+4} \ldots s_N z_N, \\ \ldots \quad &\ldots \quad \ldots \quad \ldots \\ x_{N-1} &= c_{N-1} s_N z_N, \\ x_N &= c_N z_N. \end{aligned} \tag{2.3.2}$$

In Case 1, i.e., for $r > 0$, we have

$$x_1 = x_2 = \cdots = x_r = 0,$$
$$s_1 = s_2 = \cdots = s_r = s_{r+1} = 0.$$

It is obvious that

$$\begin{aligned} x_1 &= s_2 s_3 \ldots s_{r+1} s_{r+2} \ldots s_N z_N, \\ x_2 &= c_2 s_3 \ldots s_{r+1} s_{r+2} \ldots s_N z_N, \\ \ldots \quad &\ldots \quad \ldots \quad \ldots \\ x_r &= c_r s_{r+1} s_{r+2} \ldots s_N z_N. \end{aligned} \tag{2.3.3}$$

In Case 2, i.e., for $r = 0$, we have

$$x_1 = s_2 z_2 = s_2 s_3 \ldots s_N z_N. \tag{2.3.4}$$

It remains to set

$$\rho = z_N = \varepsilon_x \sqrt{x_1^2 + x_2^2 + \cdots + x_N^2},$$

where $\varepsilon_x = +1$ for $x_1 > 0$ and $\varepsilon_x = -1$ for $x_1 < 0$. Formulas (2.3.3) and (2.3.2) in the first case ($r > 0$) and formulas (2.3.4) and (2.3.2) in the second case ($r = 0$) provide the representations (2.3.1) of the components of the vector x ($\|x\| \neq 0$) in terms of parameters ρ, c_j, s_j, $j = 2, 3, \ldots, N$, satisfying the conditions $\rho \neq 0$, $c_j^2 + s_j^2 = 1$, $0 \leq s_j \leq 1$, $-1 < c_j \leq 1$.

To solve the polarization problem, it remains to show that having constructed the chain $C = C_N C_{N-1} \ldots C_2$ from the parameters ρ, c_j, s_j, $j = 2, 3, \ldots, N$, we arrive at the equality

$$Cx = \rho e^{(N)}. \tag{2.3.5}$$

We show that $x = C^{-1}\rho e^{(N)} = C_2^* C_3^* \dots C_N^* \rho e^{(N)}$, which is equivalent to (2.3.5). Indeed,

$$C_N^* \begin{pmatrix} 0 \\ 0 \\ 0 \\ 0 \\ \vdots \\ 0 \\ \rho \end{pmatrix} = \begin{pmatrix} 0 \\ 0 \\ 0 \\ \vdots \\ 0 \\ \begin{bmatrix} c_N & s_N \\ -s_N & c_N \end{bmatrix} \begin{matrix} 0 \\ \rho \end{matrix} \end{pmatrix} = \begin{pmatrix} 0 \\ 0 \\ 0 \\ \vdots \\ 0 \\ \rho s_N \\ \rho c_N \end{pmatrix},$$

$$C_{N-1}^* \begin{pmatrix} 0 \\ 0 \\ 0 \\ \vdots \\ 0 \\ \rho s_N \\ \rho c_N \end{pmatrix} = \begin{pmatrix} 0 \\ 0 \\ \vdots \\ 0 \\ s_{N-1}\rho s_N \\ c_{N-1}\rho s_N \\ \rho c_N \end{pmatrix},$$

$$C_{N-2}^* \begin{pmatrix} 0 \\ 0 \\ \vdots \\ 0 \\ \rho s_{N-1} s_N \\ \rho c_{N-1} s_N \\ \rho c_N \end{pmatrix} = \begin{pmatrix} 0 \\ \vdots \\ 0 \\ s_{N-2}\rho s_{N-1} s_N \\ c_{N-1}\rho s_{N-1} s_N \\ \rho c_{N-1} s_N \\ \rho c_N \end{pmatrix},$$

$$\dots \quad \dots \quad \dots \quad \dots \quad \dots \quad \dots \quad \dots \quad \dots \quad \dots \; .$$

Continuing this chain of equalities, we find that the components x_j of the vector x are expressed in terms of ρ, c_j, and s_j by formulas (2.3.1). Thereby, the equality (2.3.5) is established. The proof of Lemma 2.3.1 is complete. □

PROBLEM 2.3.1. Construct another orthogonal transformation realizing the polarization of a nonzero vector; namely, polarize it with the help of orthogonal reflections. Construct the polarization of a vector with complex coordinates by unitary reflections.

CHAPTER 3

Orthogonal and Unitary Transformations. Singular Values

3.1. Representation of a rectangular matrix

Annihilation of columns of a rectangular matrix. Annihilation of rows of a rectangular matrix. Reduction of a rectangular matrix to a block matrix with a single nonzero block. The block form of a matrix of rank r. The QR decomposition of a rectangular matrix. The complex case.

Consider a rectangular $M \times N$-matrix

$$A = \begin{bmatrix} a_{11} & a_{12} & \dots & a_{1N} \\ a_{21} & a_{22} & \dots & a_{2N} \\ \vdots & \vdots & \ddots & \vdots \\ a_{M1} & a_{M2} & \dots & a_{MN} \end{bmatrix}.$$

Suppose that the columns (vectors of the M-dimensional space) of A are linearly dependent, i.e., there exists a nonzero N-dimensional vector $x = (x_1, x_2, \dots, x_N)^T$ ($\|x\| \neq 0$) such that $Ax = 0$. Let a chain of rotations C of the space E_N polarize (see §2.3) the vector x, i.e.,

$$C^* e^{(N)} = \frac{1}{\rho} C_2^* C_3^* \dots C_N^* e^{(N)} = \frac{1}{\rho} x.$$

It is obvious that

$$0 = A\left(\frac{1}{\rho} x\right) = AC^* e^{(N)} = AC^* \begin{pmatrix} 0 \\ 0 \\ \vdots \\ 1 \end{pmatrix}.$$

Thus, the last column of the matrix AC^* is zero. Therefore, the $M \times N$-matrix AC^* can be represented as an $M \times N$-matrix with the zero last column, i.e.,

$$AC^* = \left[\begin{array}{c:c} & 0 \\ A^{(1)} & 0 \\ & \vdots \\ & 0 \end{array}\right],$$

where

$$A^{(1)} = \begin{bmatrix} a_{11}^{(1)} & a_{12}^{(1)} & \dots & a_{1\ N-1}^{(1)} \\ a_{21}^{(1)} & a_{22}^{(1)} & \dots & a_{2\ N-1}^{(1)} \\ \vdots & \vdots & \ddots & \vdots \\ a_{M1}^{(1)} & a_{M2}^{(1)} & \dots & a_{M\ N-1}^{(1)} \end{bmatrix}.$$

If the columns of the matrix $A^{(1)}$ are also linearly dependent, then we choose the chain of rotations $C^{(1)}$ so that

$$AC^*C^{(1)*} = \begin{bmatrix} & \vdots & 0 \\ A^{(1)} & \vdots & 0 \\ & \vdots & \vdots \\ & \vdots & 0 \end{bmatrix} C^{(1)*} = \begin{bmatrix} & \vdots & 0 & 0 \\ A^{(2)} & \vdots & 0 & 0 \\ & \vdots & \vdots & \vdots \\ & \vdots & 0 & 0 \end{bmatrix},$$

where the matrix $A^{(2)}$ has only $N-2$ columns. Continuing the procedure, we obtain the orthogonal matrix Q_1 corresponding to the product of the chain of rotations so that

$$AQ_1^* = \Bigg[\underbrace{\quad B_0 \quad}_{r \text{ columns}} \vdots \underbrace{\quad 0 \quad}_{N-r \text{ columns}} \Bigg] \Big\} \ M \text{ rows},$$

where B_0 is a rectangular $M \times r$-matrix with linearly independent columns.

Consider the $r \times M$-matrix B_0^*. It has r linearly independent rows and M ($M \geq r$) columns. For B_0^* there is an orthogonal transformation Q composed similarly to Q_1 of chains of two-dimensional rotations so that the product $B_0^*Q^*$ has the form

$$B_0^*Q^* = \Bigg[\underbrace{\quad B_1^* \quad}_{r_1 \text{ columns}} \vdots \underbrace{\quad 0 \quad}_{M-r_1 \text{ columns}} \Bigg] \Big\} \ r \text{ rows}.$$

Moreover, all columns of B_1^* are linearly independent. This is possible only if $r_1 \le r$. Introducing the orthogonal block diagonal matrix

$$Q_2 = \begin{bmatrix} Q & \vdots & 0 \\ \dots & \vdots & \dots \\ 0 & \vdots & I_{N-M} \end{bmatrix},$$

we can represent $Q_2 A Q_1^*$ in the form

$$Q_2 A Q_1^* = \begin{bmatrix} B_1 & \vdots & 0 \\ \dots & \vdots & \dots \\ 0 & \vdots & 0 \end{bmatrix} \begin{matrix} \} \, r_1 \text{ rows} \\ \\ \} \, M - r_1 \text{ rows.} \end{matrix}$$

$$\underbrace{\qquad}_{r \text{ columns}} \quad \underbrace{\qquad}_{N-r \text{ columns}}$$

We show that all columns of the matrix B_1 are linearly independent. Indeed, otherwise there exists an r-dimensional vector $z \neq 0$ such that $B_1 z = 0$. Taking the vector $y \neq 0$ such that

$$y = \begin{pmatrix} z \\ 0 \\ \vdots \\ 0 \end{pmatrix} \begin{matrix} \} \, r \text{ components} \\ \\ \} \, N - r \text{ components} \end{matrix}$$

we arrive at the equalities

$$Q_2^* Q_2 A Q_1^* y = A Q_1^* y = 0,$$
$$A Q_1^* y = \begin{bmatrix} B_0 & \vdots & 0 \end{bmatrix} \begin{pmatrix} z \\ 0 \end{pmatrix} = B_0 z,$$
$$B_0 z = 0.$$

The equality $B_0 z = 0$ is equivalent to the assertion that the columns of the matrix B_0 are linearly dependent. We recall that all columns of the matrix B_0 are linearly independent by construction. Since the matrix B_1 has r_1 rows, and r columns of B_1 are linearly independent, we conclude that $r \le r_1$. However, as we have shown, $r_1 \le r$. Thus, B_1 is a square $r \times r$-matrix.

Therefore, the following theorem holds.

THEOREM 3.1.1. *For any $M \times N$-matrix A there is an orthogonal $N \times N$-matrix Q_1 and $M \times M$-matrix Q_2 such that*

$$Q_2 A Q_1^* = \begin{bmatrix} B_1 & \vdots & 0 \\ \dots & \vdots & \dots \\ 0 & \vdots & 0 \end{bmatrix} \begin{matrix} \} \, r \text{ rows} \\ \\ \} \, M - r \text{ rows,} \end{matrix}$$

$$\underbrace{\qquad}_{r \text{ columns}} \quad \underbrace{\qquad}_{N-r \text{ columns}}$$

where B_1 is a square nonsingular matrix ($\det B_1 \neq 0$).

Theorem 3.1.1 means that the number r of linearly independent rows coincides with the number of linearly independent columns of a matrix A. The number r is called the *rank* of the matrix A.

A matrix A of rank r can be represented in the form

$$A = Q_2^* \begin{bmatrix} B_1 & \vdots & 0 \\ \dots & \vdots & \dots \\ 0 & \vdots & 0 \end{bmatrix} Q_1. \tag{3.1.1}$$

Two $N \times M$-matrices A and B are said to be *orthogonally* (*unitarily*) *congruent* if there exists an orthogonal (unitary) $N \times N$-matrix Q_2 ($Q_2^* Q_2 = I_N$) and an $M \times M$-matrix Q_1 ($Q_1^* Q_1 = I_M$) such that $A = Q_2^* B Q_1$.

It is natural to refer to Theorem 3.1.1 as the theorem about the orthogonal congruent reduction of a rectangular matrix of rank r to a block matrix that contains a nonsingular $r \times r$-block, whereas the remaining blocks are zero.

We consider one more variant of the congruent transformation of a rectangular $M \times N$-matrix. We will use the procedure which is known as the QR decomposition. It is often used in practical applications.

We regard the first column of the matrix A as the vector with components $a_{11}, a_{21}, \dots a_{N1}$. There exists an orthogonal reflection $\mathcal{P}^{(1)}$ that takes this vector to the vector $a^{(1)}$ with all components except $a_{11}^{(1)}$ equal to zero. The reflection $\mathcal{P}^{(1)}$ transforms the first column of the matrix A to a vector that is collinear to the basis vector $e^{(1)}$ (see §§2.2 and 2.3).

Consider the $M \times N$-matrix $A^{(1)} = \mathcal{P}^{(1)} A$. The columns of $A^{(1)}$ are obtained from the corresponding columns of the matrix A by applying the reflection $\mathcal{P}^{(1)}$, i.e.,

$$\begin{pmatrix} a_{1j}^{(1)} \\ a_{2j}^{(1)} \\ \vdots \\ a_{Mj}^{(1)} \end{pmatrix} = \mathcal{P}^{(1)} \begin{pmatrix} a_{1j} \\ a_{2j} \\ \vdots \\ a_{Mj} \end{pmatrix}, \qquad j = 1, 2, \dots, N.$$

The matrix $A^{(1)}$ has the following form:

$$A^{(1)} = \begin{bmatrix} \times & \times & \times & \cdots & \times \\ 0 & \times & \times & \cdots & \times \\ 0 & \times & \times & \cdots & \times \\ \vdots & \vdots & \vdots & & \vdots \\ 0 & \times & \times & \cdots & \times \end{bmatrix}.$$

(Hereinafter, the cross denotes an element that might, in general, be nonzero. The value of such an element is not essential for our purposes.)

Further, we choose reflections $\mathcal{P}^{(2)}, \mathcal{P}^{(3)}, \dots$ so that the number of zero elements in the structure of matrices

$$A^{(2)} = \mathcal{P}^{(2)} A^{(1)},\ A^{(3)} = \mathcal{P}^{(3)} A^{(2)}, \dots, A^{(k)} = \mathcal{P}^{(k)} A^{(k-1)}, \dots$$

increases, i.e.,

$$
A^{(k)} = \underbrace{\begin{bmatrix}
\times & \times & \times & \cdots & \times & \times & \times & \cdots & \times \\
0 & \times & \times & \cdots & \times & \times & \times & \cdots & \times \\
0 & 0 & \times & \cdots & \times & \times & \times & \cdots & \times \\
 & & \ddots & \vdots & \vdots & \vdots & \vdots & & \vdots \\
 & & & & \times & \times & \times & \cdots & \times \\
 & & & & 0 & \times & \times & \cdots & \times \\
 & & \mathbf{0} & & 0 & \times & \times & \cdots & \times \\
 & & & & \vdots & \vdots & \vdots & & \vdots \\
 & & & & 0 & \times & \times & \cdots & \times
\end{bmatrix}}_{N \text{ columns}}
\begin{matrix} k \text{ rows} \\ \\ M-k \text{ rows.} \end{matrix}
$$

To construct the reflection $\mathcal{P}^{(i)}$, the reader can use the procedure described in §2.2. The process stops when we obtain the matrix $A^{(l)}$ with elements below the diagonal starting from the upper left corner being zero. In other words, the process is complete when we compute the matrix $A^{(l)}$ with $l = \min\{N, M\}$. Denoting the matrix $A^{(l)}$ by R, we can write

$$R = \mathcal{P}^{(l)}\mathcal{P}^{l-1}\dots\mathcal{P}^{(1)}A, \quad A = \mathcal{P}^{(1)}\mathcal{P}^{(2)}\dots\mathcal{P}^{(l)}R.$$

(We note that R is an upper triangular matrix for $N = M$.) We used the fact that a reflection coincides with its inverse. As usual, Q denotes a linear orthogonal transformation or the corresponding matrix of the form

$$Q = \mathcal{P}^{(1)}\mathcal{P}^{(2)}\dots\mathcal{P}^{(l)} \quad (Q^*Q = I_N).$$

As a result, we obtain the so-called QR representation of the matrix A:

$$A = QR. \tag{3.1.2}$$

If A is a square matrix, then (3.1.2) represents the matrix A as the product of a square triangular matrix R and an orthogonal matrix Q.

If the number of rows M is less than the number of columns M, then Q is a square matrix, whereas R is a trapezoidal matrix. Nonzero elements of the matrix R form a trapezoid such that the upper base is greater than the lower base, for example,

$$\begin{bmatrix} a_{11} & a_{12} & a_{13} & a_{14} \\ a_{21} & a_{22} & a_{23} & a_{24} \end{bmatrix} = \begin{bmatrix} q_{11} & q_{12} \\ q_{21} & q_{22} \end{bmatrix} \begin{bmatrix} r_{11} & r_{12} & r_{13} & r_{14} \\ 0 & r_{22} & r_{23} & r_{24} \end{bmatrix}.$$

If the number of columns N is less than the number of rows M, then several lower rows of the matrix R consist of zero elements, i.e.,

$$\begin{bmatrix} a_{11} & a_{12} & a_{13} \\ a_{21} & a_{22} & a_{23} \\ a_{31} & a_{32} & a_{33} \\ a_{41} & a_{42} & a_{43} \\ a_{51} & a_{52} & a_{53} \end{bmatrix} = \begin{bmatrix} q_{11} & q_{12} & q_{13} & q_{14} & q_{15} \\ q_{21} & q_{22} & q_{23} & q_{24} & q_{25} \\ q_{31} & q_{32} & q_{33} & q_{34} & q_{35} \\ q_{41} & q_{42} & q_{43} & q_{44} & q_{45} \\ q_{51} & q_{52} & q_{53} & q_{54} & q_{55} \end{bmatrix} \begin{bmatrix} r_{11} & r_{12} & r_{13} \\ 0 & r_{22} & r_{23} \\ 0 & 0 & r_{33} \\ 0 & 0 & 0 \\ 0 & 0 & 0 \end{bmatrix}.$$

In this case, we can eliminate the last zero rows of the matrix R and the same number of the last columns of the matrix Q. We arrive at the representation

$$\begin{bmatrix} a_{11} & a_{12} & a_{13} \\ a_{21} & a_{22} & a_{23} \\ a_{31} & a_{32} & a_{33} \\ a_{41} & a_{42} & a_{43} \\ a_{51} & a_{52} & a_{53} \end{bmatrix} = \begin{bmatrix} q_{11} & q_{12} & q_{13} \\ q_{21} & q_{22} & q_{23} \\ q_{31} & q_{32} & q_{33} \\ q_{41} & q_{42} & q_{43} \\ q_{51} & q_{52} & q_{53} \end{bmatrix} \begin{bmatrix} r_{11} & r_{12} & r_{13} \\ 0 & r_{22} & r_{23} \\ 0 & 0 & r_{33} \end{bmatrix},$$

where R is a square upper triangular matrix and the columns of the matrix Q are orthonormal. Thus, in the case $N < M$, by the *QR decomposition of an $M \times N$-matrix* A, we mean the representation $A = QR$, where the matrix Q ($Q^*Q = I_N$) has the same number of the columns N as the matrix A, whereas R is a square upper triangular $N \times N$-matrix.

Considering the representations of a rectangular matrix in §3.1, we used real matrices and orthogonal (real) transformations composed of chains of two-dimensional rotations or reflections. For a complex matrix A corresponding to a transformation $H_N \xrightarrow{A} H_M$, we have a similar representation

$$A = U_2^* \begin{bmatrix} B_1 & \vdots & 0 \\ \dots & \vdots & \dots \\ 0 & \vdots & 0 \end{bmatrix} U_1,$$

where U_1 and U_2 are square unitary matrices of order N and M, respectively. The only difference with the real case is that to polarize a vector x (possibly, with complex components) we use unitary reflections instead of a chain of two-dimensional rotations. Indeed, we have shown that for any nonzero vector x there exists a unitary reflection $\mathcal{P}$ that transforms x to a vector collinear to a given vector (say, $e^{(N)}$), i.e., for any nonzero vector x there exists a unitary reflection that polarizes the vector x.

In the real case, in the proof of Theorem 3.1.1, we could also use polarizations by orthogonal reflections (see Problem 2.3.1) instead of chains of rotations. The QR decomposition can be generalized naturally to the complex case.

3.2. Simplification of matrices. Hessenberg matrices

Reduction of a square matrix to a two-diagonal matrix. Reduction of a rectangular matrix to a two-diagonal matrix. Reduction of a symmetric square matrix to a three-diagonal matrix. Application to the numerical solution of a system of linear equations. Upper Hessenberg matrices and their properties. A remark about the QR algorithm.

Given an arbitrary $r \times r$-matrix

$$B = \begin{bmatrix} b_{11} & b_{12} & \dots & b_{1r} \\ b_{21} & b_{22} & \dots & b_{2r} \\ \vdots & \vdots & & \vdots \\ b_{r1} & b_{r2} & \dots & b_{rr} \end{bmatrix},$$

we choose orthogonal (or unitary if there are complex b_{ij}) matrices (transformations) P and Q so that the matrix B takes the following two-diagonal form:

$$D = P^*BQ = \begin{bmatrix} d_1 & b_2 & & & \mathbf{0} \\ & d_2 & b_3 & & \\ & & \ddots & \ddots & \\ \mathbf{0} & & & d_{r-1} & b_r \\ & & & & d_r \end{bmatrix}.$$

The transformations P and Q are taken as compositions of (orthogonal or unitary) reflections. We use formulas from §2.2. We recommend that the reader review these formulas before the study of the algorithm described below.

Step 1. Starting with elements of the first column of the matrix B, we choose a reflection $\mathcal{P}_1$ such that

$$\mathcal{P}_1 B = B_1 = \begin{bmatrix} b'_{11} & b'_{12} & \dots & b'_{1r} \\ 0 & b'_{22} & \dots & b'_{2r} \\ 0 & b'_{32} & \dots & b'_{3r} \\ \vdots & \vdots & & \vdots \\ 0 & b'_{r2} & \dots & b'_{rr} \end{bmatrix}.$$

In other words, we choose $\mathcal{P}_1$ that annihilates the entire first column of the matrix B, except the first element. The corresponding formulas are given in §2.2.

Step 2. We choose a reflection $\mathcal{Q}_1$ such that after the multiplication from the right of B_1 by $\mathcal{Q}_1$ the first column is preserved, whereas all elements of the first row, starting with the third element, become zero, i.e.,

$$B_2 = \mathcal{P}_1 B \mathcal{Q}_1 = B_1 \mathcal{Q}_1 = \begin{bmatrix} b'_{11} & b''_{12} & 0 & \dots & 0 \\ 0 & b''_{22} & b''_{23} & \dots & b''_{2r} \\ \vdots & \vdots & \vdots & & \vdots \\ 0 & b''_{r2} & b''_{r3} & \dots & b''_{rr} \end{bmatrix}.$$

We note that b'_{11}, b''_{12} will not change under further transformations.

Step 3. We choose a reflection $\mathcal{P}_2$ such that after the multiplication of B_2 by $\mathcal{P}_2$ from the left the first element of the second column does not change, whereas the remaining elements of the second column, starting with the third element, equal zero. Constructing $\mathcal{P}_2$, we see that $\mathcal{P}_2$ preserves the first column of B_2. Indeed, the first component of the corresponding reflection vector and all elements of the first column of B_2, starting with the second element, are zero. Furthermore, $\mathcal{P}_2$ preserves the elements of the first row of B_2 because the first component of the reflection vector is zero. Consequently, after the multiplication of B_2 by $\mathcal{P}_2$ from the left, the form of the matrix B_2 does not change. By construction, the matrix

$B_3 = \mathcal{P}_2\mathcal{P}_1 B \mathcal{Q}_1 = \mathcal{P}_2 B_2$ has the following form:

$$B_3 = \left[\begin{array}{c|cccccc|} \times & \times & 0 & 0 & \dots & 0 \\ \cline{2-6} 0 & \times & \times & \times & \dots & \times \\ 0 & 0 & \times & \times & \dots & \times \\ 0 & 0 & \times & \times & \dots & \times \\ \vdots & \vdots & \vdots & \vdots & & \vdots \\ 0 & 0 & \times & \times & \dots & \times \\ \cline{2-6} \end{array}\right],$$

where we framed the submatrix that will be changed further. We recall that the cross denotes an element that might be nonzero.

PROBLEM 3.2.1. Verify the above assertions using the formulas from §2.2.

Step 4. Choose the transformation $\mathcal{Q}_2$ that acts on B_3 from the right in such a way that in the matrix $B_4 = \mathcal{P}_2\mathcal{P}_1 B \mathcal{Q}_1\mathcal{Q}_2$ the elements of the second row, starting with the fourth element, are zero, whereas the first two columns remain unchanged.

Continuing the process, we obtain matrices with an increasing number of zero elements. Further steps of the algorithm are realized as follows.

Step $2i-1$. For a matrix B_j with even number $j = 2i-2$ we construct a reflection $\mathcal{P}_i$ ($B_{2i-1} = \mathcal{P}_i B_{2i-2}$) that preserves the first $i-1$ elements and annihilates the elements of the ith column of the matrix B_{2i-2}, starting with the $(i+1)$th element. Then $\mathcal{P}_i$ preserves the first $i-1$ rows of the matrix B_{2i-2}.

Step $2i$. For a matrix B_j with odd number $j = 2i-1$ we construct a reflection $\mathcal{Q}_i$ ($B_{2i} = B_{2i-1}\mathcal{Q}_i$) that preserves the first i elements of the ith row of the matrix B_{2i-1} and annihilates the elements starting with the $(i+2)$th element, ($\mathcal{Q}_i$ preserves the first i columns of the matrix B_{2i-1}).

After constructing the matrix B_{2r-4} in the above way, we see that it has the following form:

$$B_{2r-4} = \begin{bmatrix} \times & \times & 0 & \dots & 0 & 0 & 0 & 0 \\ 0 & \times & \times & \dots & 0 & 0 & 0 & 0 \\ 0 & 0 & \times & \ddots & 0 & 0 & 0 & 0 \\ & & & \ddots & \ddots & & & \\ 0 & 0 & 0 & \dots & \times & \times & 0 & 0 \\ 0 & 0 & 0 & \dots & 0 & \times & \times & 0 \\ 0 & 0 & 0 & \dots & 0 & 0 & \times & \times \\ 0 & 0 & 0 & \dots & 0 & 0 & \times & \times \end{bmatrix}.$$

Step $2r-3$. For the matrix B_{2r-4} we choose a reflection $\mathcal{P}_{r-1}$ ($B_{2r-3} = \mathcal{P}_{r-1}B_{2r-4}$) that preserves the first $r-2$ elements and annihilates the lowest element of the next to the last column of the matrix B_{2r-4}. It is easy to check that the matrix B_{2r-3} has the form

$$B_{2r-3} = \begin{bmatrix} \times & \times & 0 & \dots & 0 & 0 & 0 & 0 \\ 0 & \times & \times & \dots & 0 & 0 & 0 & 0 \\ 0 & 0 & \times & \ddots & 0 & 0 & 0 & 0 \\ & & & \ddots & \ddots & & & \\ 0 & 0 & 0 & \dots & \times & \times & 0 & 0 \\ 0 & 0 & 0 & \dots & 0 & \times & \times & 0 \\ 0 & 0 & 0 & \dots & 0 & 0 & \times & \times \\ 0 & 0 & 0 & \dots & 0 & 0 & 0 & \times \end{bmatrix}.$$

We have completed the description of the algorithm for constructing orthogonal (unitary) transformations $P = \mathcal{P}_1\mathcal{P}_2\dots\mathcal{P}_{r-1}$ and $Q = \mathcal{Q}_1\mathcal{Q}_2\dots\mathcal{Q}_{r-2}$ such that $P^*BQ = B_{2r-1}$ ($B = PB_{2r-1}Q^*$), i.e., transformations that congruently reduce a given square matrix B to a two-diagonal matrix.

PROBLEM 3.2.2. Describe an algorithm (as above) that allows us to reduce a rectangular (not square) matrix B to a two-diagonal matrix, i.e., to a matrix of one of the following forms:

$$\begin{bmatrix} \times & \times & 0 & 0 & & & & & \\ 0 & \times & \times & 0 & & & \mathbf{0} & & \\ 0 & 0 & \times & \times & & & & & \\ & & & \ddots & \ddots & & & & \\ & \mathbf{0} & & & \times & \times & 0 & \dots & 0 \\ & & & & 0 & \times & 0 & \dots & 0 \end{bmatrix},$$

$$\begin{bmatrix} \times & \times & 0 & 0 & & & \\ 0 & \times & \times & 0 & & \mathbf{0} & \\ 0 & 0 & \times & \times & & & \\ & & & \ddots & & & \\ & & & & \times & 0 & 0 \\ & & & & \times & \times & 0 \\ & & & & 0 & \times & \times \\ & \mathbf{0} & & & 0 & 0 & \times \\ & & & & \dots & \dots & \dots \\ & & & & 0 & 0 & 0 \end{bmatrix}.$$

Show that the procedure can be chosen so that all nonzero elements of these two-diagonal matrices are nonnegative.

PROBLEM 3.2.3. For a symmetric real square $N \times N$-matrix B ($B = B^*$) describe an algorithm for constructing an orthogonal transformation P (in the form of the product of reflections) such that P^*BP is a three-diagonal symmetric matrix of the form

$$P^*BP = \begin{bmatrix} d_1 & b_2 & 0 & 0 & & & & \\ b_2 & d_2 & b_3 & 0 & & & \mathbf{0} & \\ 0 & b_3 & d_3 & b_4 & & & & \\ & & \ddots & \ddots & \ddots & & & \\ & & & b_{N-2} & d_{N-2} & b_{N-1} & 0 & \\ & \mathbf{0} & & 0 & b_{N-1} & d_{N-1} & b_N & \\ & & & 0 & 0 & b_N & d_N & \end{bmatrix}.$$

The congruent reduction of a matrix to a two-diagonal matrix is convenient for the numerical study of a system of linear equations written as a vector equation $Bx = f$. Indeed, let $PBQ = D$. Having computed $g = Pf$, we easily find the vector y from the equality

$$Dy = g$$

or, in detail,

$$Dy = \begin{bmatrix} d_1 & b_2 & & & \mathbf{0} \\ & d_2 & b_3 & & \\ & & \ddots & \ddots & \\ \mathbf{0} & & & d_{r-1} & b_r \\ & & & & d_r \end{bmatrix} \begin{pmatrix} y_1 \\ y_2 \\ \vdots \\ y_{r-1} \\ y_r \end{pmatrix} = \begin{pmatrix} g_1 \\ g_2 \\ \vdots \\ g_{r-1} \\ g_r \end{pmatrix},$$

using the formulas

$$\begin{aligned} y_r &= \frac{1}{d_r} g_r, \\ y_{r-1} &= \frac{1}{d_{r-1}}(g_{r-1} - b_r y_r), \\ y_{r-2} &= \frac{1}{d_{r-2}}(g_{r-2} - b_{r-1} y_{r-1}), \\ &\cdots\cdots\cdots\cdots\cdots\cdots \\ y_1 &= \frac{1}{d_1}(g_1 - b_2 y_2). \end{aligned}$$

These formulas can be applied provided that there are no zero elements among $d_1, d_2, \ldots, d_r$, i.e., $\det D = d_1 d_2 \ldots d_r \neq 0$. Since $\det D = \det (PBQ) = \det P \cdot \det Q \cdot \det B = \pm \det B$, this condition is equivalent to the unique solvability of the system $Bx = f$ for any right-hand side f. Having computed y, we can obtain x by the formula $x = Qy$. For a detailed discussion regarding the solution using the above scheme we refer the reader to [**15**].

A matrix of the form

$$A = \begin{bmatrix} a_{11} & a_{12} & a_{13} & \dots & a_{1\ N-2} & a_{1\ N-1} & a_{1N} \\ a_{21} & a_{22} & a_{23} & \dots & a_{2\ N-2} & a_{2\ N-1} & a_{2N} \\ 0 & a_{32} & a_{33} & \dots & a_{3\ N-2} & a_{3\ N-1} & a_{3N} \\ 0 & 0 & a_{43} & \dots & a_{4\ N-2} & a_{4\ N-1} & a_{4N} \\ 0 & 0 & 0 & \dots & a_{5\ N-2} & a_{5\ N-1} & a_{5N} \\ \vdots & \vdots & \vdots & & \vdots & \vdots & \vdots \\ 0 & 0 & 0 & \dots & a_{N-1\ N-2} & a_{N-1\ N-1} & a_{N-1\ N} \\ 0 & 0 & 0 & \dots & 0 & a_{N\ N-1} & a_{NN} \end{bmatrix},$$

where only a_{ij} for $j \geq i - 1$ can be nonzero, is called an *upper Hessenberg matrix.* Many important computational procedures are significantly simplified for Hessenberg matrices.

The following problems represent useful properties of Hessenberg matrices.

PROBLEM 3.2.4. Prove that for any square matrix B there exists a unitary matrix U ($U^*U = I$) such that $U^*BU = U^{-1}BU = A$ is an upper Hessenberg matrix.

PROBLEM 3.2.5. Show that the orthogonal transformation given by the inverse chain of two-dimensional rotations

$$C^{-1} = C^* = C_2^* C_3^* \dots C_{N-1}^* C_N^*$$

(see §2.3) can be written in terms of the upper Hessenberg matrix $A = C^{-1}$.

PROBLEM 3.2.6. Show that any orthogonal upper Hessenberg matrix Q can be represented as the product $Q = C^{-1}D$, where C is a chain of two-dimensional rotations and D is a diagonal matrix such that $D^2 = I$ ($d_{ii} = \pm 1$).

PROBLEM 3.2.7. Show that any real Hessenberg matrix A can be represented as the product $A = C^{-1}R$, where C is a chain of two-dimensional rotations and R is an upper triangular matrix. Describe an algorithm for reducing A to such a form.

PROBLEM 3.2.8. Show that the product $RC^{-1} = CAC^{-1}$ with R from Problem 3.2.7 is an upper Hessenberg matrix.

The above properties of Hessenberg matrices and chains of two-dimensional rotations form the basis of a very popular algorithm for computing eigenvalues. The algorithm is due to Kublanovskaya and Frensis. The algorithm is based on the use of a series of successive QR decompositions (see §3.1).

A square matrix A is said to be *upper block triangular* if it can be divided into blocks, i.e.,

$$A = \begin{bmatrix} A_{11} & A_{12} & A_{13} \\ 0 & A_{22} & A_{23} \\ 0 & 0 & A_{33} \end{bmatrix},$$

where the diagonal blocks A_{ii} are square matrices and the blocks A_{ij}, $j < i$, located below the main diagonal are zero. In a general case, the number of diagonal blocks is not necessarily three.

The QR algorithm. We set $A_0 = A$ and successively find $A_1, A_2, A_3, \ldots$. The jth step of the iteration begins with the computation of an orthogonal matrix Q_j and an upper triangular matrix R_j from the known matrix A_{j-1}. The matrices Q_j and R_j are matrix factors in the QR decomposition $A_{j-1} = Q_j R_j$. Multiplying these factors in the inverse order, we obtain the matrix $A_j = R_j Q_j$. It turns out that the matrices A_j become an (upper) block triangular matrix and the products of the characteristic polynomials of diagonal blocks tend to the characteristic determinant of the initial matrix A as j increases. As a rule, for every block the modules of limit eigenvalues are close. (We note that the blocks themselves can have no limits.) We will describe the idea and justify the QR algorithm in Chapter 14. Here we only note that for a general $N\times N$-matrix A_{j-1} the choice of Q_j and the computation of $R_j Q_j$ require a large number of arithmetic operations; moreover, this number increases as N^3 if N increases. However, if A_{j-1} is an upper Hessenberg matrix and Q_j is a chain of two-dimensional rotations, then the number of elementary arithmetical operations needed to compute the matrix A_j from the known matrix A_{j-1} can be significantly less; namely, the number of operations we need to perform is of order N^2, but not of order N^3 as in a general case. This explains why Hessenberg matrices and chains of two-dimensional rotations are so popular in numerical methods.

PROBLEM 3.2.9. Verify the statement about the number of arithmetic operations in the computation of the matrix A_j from the known matrix A_{j-1} by the QR algorithm.

3.3. Singular value decomposition

Reduction of a nonsingular square matrix to a diagonal matrix. Reduction of a rectangular matrix to a block matrix of special form. Singular value decomposition of a rectangular matrix.

Let B be a nonsingular $r \times r$-matrix and B^* the adjoint matrix. From B, B^*, and the zero square blocks we compose the square self-adjoint matrix

$$\mathcal{B} = \begin{bmatrix} 0 & B^* \\ B & 0 \end{bmatrix}.$$

The matrix $\mathcal{B}$ is nonsingular since $|\det \mathcal{B}| = |\det B^*|\,|\det B| = |\det B|^2 \neq 0$. Let β be an eigenvalue of the matrix $\mathcal{B}$ and let $(u^T, v^T)^T$ be the corresponding eigenvector, i.e.,

$$\mathcal{B}\begin{pmatrix} u \\ v \end{pmatrix} = \begin{bmatrix} 0 & B^* \\ B & 0 \end{bmatrix}\begin{pmatrix} u \\ v \end{pmatrix} = \beta\begin{pmatrix} u \\ v \end{pmatrix}.$$

In other words, $Bu = \beta v$ and $B^*v = \beta u$. Since the matrix B is nonsingular, we have $\beta \neq 0$. The following equalities hold:

$$(Bu, v) = (\beta v, v) = \beta(v, v),$$
$$(Bu, v) = (u, B^*v) = (u, \beta u) = \beta(u, u),$$
$$\beta(u, u) = \beta(v, v).$$

(We used the fact that the eigenvalues of a self-adjoint matrix are real.) Since $\beta \neq 0$, we have $(u, u) = (v, v)$. Without loss of generality, we can assume that the vectors u and v are normalized, i.e., $\|u\| = \|v\| = 1$. We choose two unitary reflections $\mathcal{P}_r^{(R)}$ and $\mathcal{P}_r^{(L)}$ so that $\mathcal{P}_r^{(R)}u = \varepsilon_R e^{(r)}$ and $\mathcal{P}_r^{(L)}v = \varepsilon_L e^{(r)}$, where ε_R and ε_L are complex scalars with $|\varepsilon_R| = 1$ and $|\varepsilon_L| = 1$. Since $[\mathcal{P}_r^{(R)}]^2 = I_r$ and $[\mathcal{P}_r^{(L)}]^2 = I_r$, we have $\mathcal{P}_r^{(R)}e^{(r)} = \overline{\varepsilon}_R u$ and $\mathcal{P}_r^{(L)}e^{(r)} = \overline{\varepsilon}_L v$.

Let $\widetilde{B} = \mathcal{P}_r^{(L)} B \mathcal{P}_r^{(R)}$ and $\widetilde{B}^* = \mathcal{P}_r^{(R)} B^* \mathcal{P}_r^{(L)}$. Since

$$\widetilde{B}e^{(r)} = \mathcal{P}_r^{(L)} B \mathcal{P}_r^{(R)} e^{(r)} = \mathcal{P}_r^{(L)}[\beta\overline{\varepsilon}_R]v = \beta\overline{\varepsilon}_R\varepsilon_L e^{(r)},$$
$$\widetilde{B}^*e^{(r)} = \mathcal{P}_r^{(R)} B^* \mathcal{P}_r^{(L)} e^{(r)} = \mathcal{P}_r^{(R)}[\beta\overline{\varepsilon}_L]u = \beta\varepsilon_R\overline{\varepsilon}_L e^{(r)},$$

the matrix $\widetilde{B}$ has the form

$$\widetilde{B} = \begin{bmatrix} \widetilde{b}_{11} & \widetilde{b}_{12} & \dots & \widetilde{b}_{1\ r-1} & 0 \\ \widetilde{b}_{21} & \widetilde{b}_{22} & \dots & \widetilde{b}_{2\ r-1} & 0 \\ \vdots & \vdots & \ddots & \vdots & \vdots \\ \widetilde{b}_{r-1\ 1} & \widetilde{b}_{r-1\ 2} & \dots & \widetilde{b}_{r-1\ r-1} & 0 \\ 0 & 0 & \dots & 0 & \varepsilon^{(r)}\beta \end{bmatrix},$$

where $|\varepsilon^{(r)}| = 1$ ($\varepsilon^{(r)} = \overline{\varepsilon}_R\varepsilon_L$). Similarly, by choosing successively diagonal elements we can show that there exist unitary transformations $Q^{(L)}$ and $Q^{(R)}$ such that

$$Q^{(L)}BQ^{(R)} = \begin{bmatrix} \varepsilon^{(1)}\beta_1 & & & \mathbf{0} \\ & \varepsilon^{(2)}\beta_2 & & \\ & & \ddots & \\ \mathbf{0} & & & \varepsilon^{(r)}\beta_r \end{bmatrix},$$

where $|\varepsilon^{(j)}| = 1$ and $\varepsilon^{(j)} = e^{i\varphi_j}$. These unitary transformations are constructed as the products of successive reflections $\mathcal{P}_s^{(L)}$ (or $\mathcal{P}_s^{(R)}$), where $s = r, r-1, r-2, \dots, 1$. Each reflection does not change the columns (rows), all elements of which (except the elements on the main diagonal) are annihilated by the previous reflections. The construction of such reflections was described in §2.2. We recommend that the reader construct $\mathcal{P}_{r-1}^{(R)}$ and $\mathcal{P}_{r-1}^{(L)}$ using the description from §2.2 and acting as in the case of the reflections $\mathcal{P}_r^{(R)}$ and $\mathcal{P}_r^{(L)}$. Further reflections can be easily constructed.

Consider the diagonal matrix

$$\Theta = \begin{bmatrix} \pm e^{-i\varphi_1} & & & \mathbf{0} \\ & \pm e^{-i\varphi_2} & & \\ & & \ddots & \\ \mathbf{0} & & & \pm e^{-i\varphi_r} \end{bmatrix}$$

and the matrix $\widetilde{Q}^{(R)} = Q^{(R)}\Theta$. It is obvious that $\widetilde{Q}^{(R)}$ is a unitary matrix. Moreover,

$$Q^{(L)}B\widetilde{Q}^{(R)} = \begin{bmatrix} \pm\beta_1 & & & \mathbf{0} \\ & \pm\beta_2 & & \\ & & \ddots & \\ \mathbf{0} & & & \pm\beta_r \end{bmatrix}.$$

The sign $\pm$ is taken so that the diagonal elements $\pm\beta_j$ are strictly positive.

Thus, we have proved the existence of unitary $r \times r$-matrices $Q^{(L)}$ and $\widetilde{Q}^{(R)}$ such that the matrix

$$Q^{(1)} B \widetilde{Q}^{(R)} = \begin{bmatrix} k_1 & & & 0 \\ & k_2 & & \\ & & \ddots & \\ 0 & & & k_r \end{bmatrix}$$

is diagonal with strictly positive diagonal elements.

Consider the block matrix

$$\underbrace{\begin{bmatrix} B \\ \cdots \\ 0 \end{bmatrix}}_{r \text{ columns}} \begin{matrix} \vdots \\ \vdots \\ \vdots \end{matrix} \underbrace{\begin{bmatrix} 0 \\ \cdots \\ 0 \end{bmatrix}}_{N-r \text{ columns}} \begin{matrix} \}\ r \text{ rows} \\ \}\ M-r \text{ rows} \end{matrix}$$

such that the block located in the upper left corner is a nonsingular $r \times r$-matrix, whereas the remaining blocks are zero. We set

$$\widehat{Q}^{(L)} = \begin{bmatrix} Q^{(L)} & \\ & \boxed{\begin{matrix} 1 & & & 0 \\ & 1 & & \\ & & \ddots & \\ 0 & & & 1 \end{matrix}} \end{bmatrix} \Bigg\} M - r \text{ rows},$$

$$\widehat{Q}^{(R)} = \begin{bmatrix} Q^{(R)} & \\ & \boxed{\begin{matrix} 1 & & & 0 \\ & 1 & & \\ & & \ddots & \\ 0 & & & 1 \end{matrix}} \end{bmatrix} \Bigg\} N - r \text{ rows}.$$

It is obvious that

$$\widehat{Q}^{(L)} \left[\underbrace{\begin{bmatrix} B \\ \cdots \\ 0 \end{bmatrix}}_{r \text{ columns}} \begin{matrix} \vdots \\ \vdots \\ \vdots \end{matrix} \underbrace{\begin{bmatrix} 0 \\ \cdots \\ 0 \end{bmatrix}}_{N-r \text{ columns}} \begin{matrix} \}\ r \text{ rows} \\ \}\ M-r \text{ rows} \end{matrix} \right] \widehat{Q}^{(r)}$$

$$= \left[\begin{array}{cc} \boxed{\begin{array}{cccc} k_1 & & & 0 \\ & k_2 & & \\ & & \ddots & \\ 0 & & & k_r \end{array}} & 0 \\ 0 & 0 \end{array} \right],$$

where the diagonal elements k_j, $j = 1, 2, \ldots, r$, are strictly positive. Moreover, the matrices $\widehat{Q}^{(L)}$ and $\widehat{Q}^{(R)}$ are unitary.

The above representation and the possibility of finding unitary transformations $U^{(L)}$ and $U^{(R)}$ such that

$$U^{(L)} A U^{(R)} = \begin{bmatrix} B & 0 \\ 0 & 0 \end{bmatrix}$$

(see §3.1), imply that the matrix A can be represented in the form $A = UKV^*$, where U is a unitary $M \times M$-matrix, V is a unitary $N \times N$-matrix, and K is a "diagonal" matrix, i.e.,

$$K = \left[\begin{array}{c} \boxed{\begin{array}{cccc} \sigma_N & & & 0 \\ & \sigma_{N-1} & & \\ & & \ddots & \\ 0 & & & \sigma_1 \end{array}} \\ 0 \end{array} \right] \qquad \text{for } N = N_0 < M,$$

$$K = \begin{bmatrix} \sigma_N & & & 0 \\ & \sigma_{N-1} & & \\ & & \ddots & \\ 0 & & & \sigma_1 \end{bmatrix} \qquad \text{for } N = M = N_0,$$

$$K = \left[\begin{array}{cc} \boxed{\begin{array}{cccc} \sigma_M & & & 0 \\ & \sigma_{M-1} & & \\ & & \ddots & \\ 0 & & & \sigma_1 \end{array}} & 0 \end{array} \right] \qquad \text{for } N > M = N_0.$$

In addition, $\sigma_j \geq 0$ for $j = 1, 2, \ldots, N_0$. The number of strictly positive σ_j coincides with the rank r ($r \leq N_0 = \min\{M, N\}$) of the matrix A.

3.4. Singular values

Singular value decomposition and singular values of a rectangular matrix. The uniqueness of singular values of a given matrix. The Sturm method for the computation of singular values. The computation of singular values of a rectangular matrix in terms of eigenvalues of Hermitian square matrices.

As we know, a rectangular $M \times N$-matrix A can be represented as $A = U_M K V_N^*$, where U_M and U_N are unitary square matrices of order M and N, respectively, whereas the matrix K is composed of the diagonal square matrix Σ ($N_0 = \min\{N, M\}$) with nonnegative diagonal elements and (if $N \neq M$) zero rows or zero columns, i.e.,

$$K = \begin{cases} \begin{bmatrix} \Sigma \vdots 0 \end{bmatrix} & \text{if } M < N, \\ \begin{bmatrix} \Sigma \\ \dots \\ 0 \end{bmatrix} & \text{if } M > N, \\ \Sigma & \text{if } N = M \end{cases}$$

$$(\sigma_{N_0} \geq \sigma_{N_0-1} \geq \cdots \geq \sigma_1 > 0),$$

where

$$\Sigma = \begin{bmatrix} \sigma_{N_0} & & & 0 \\ & \sigma_{N_0-1} & & \\ & & \ddots & \\ 0 & & & \sigma_1 \end{bmatrix}.$$

The numbers σ_j, $j = 1, 2, \dots, N_0$, are called *singular values* of the matrix A and the representation

$$A = U_M K V_N^*$$

is called the *singular value decomposition* of the matrix A.

We show that singular values are uniquely determined by the matrix A. Consider the matrix

$$\mathcal{A} = \begin{bmatrix} 0 & A^* \\ A & 0 \end{bmatrix}$$

and the characteristic equation

$$0 = \det \left[\ \mathcal{A} - \lambda I_{N+M}\ \right] = \det \begin{bmatrix} -\lambda I_N & A^* \\ A & -\lambda I_M \end{bmatrix}$$

$$= \det \left\{ \begin{bmatrix} V_N & 0 \\ 0 & U_M \end{bmatrix} \begin{bmatrix} -\lambda I_N & K^* \\ K & -\lambda I_M \end{bmatrix} \begin{bmatrix} V_N^* & 0 \\ 0 & U_M^* \end{bmatrix} \right\},$$

which is equivalent to the equation

$$\det\begin{bmatrix} -\lambda I_N & K^* \\ K & -\lambda I_M \end{bmatrix} = 0.$$

We can simplify the last equation. For $N = M$ we have

$$\det\left[\begin{array}{cc}
\boxed{\begin{matrix}-\lambda & & & 0\\ & -\lambda & & \\ & & \ddots & \\ 0 & & & -\lambda\end{matrix}} & \boxed{\begin{matrix}\sigma_N & & & 0\\ & \sigma_{N-1} & & \\ & & \ddots & \\ 0 & & & \sigma_1\end{matrix}} \\
\boxed{\begin{matrix}\sigma_N & & & 0\\ & \sigma_{N-1} & & \\ & & \ddots & \\ 0 & & & \sigma_1\end{matrix}} & \boxed{\begin{matrix}-\lambda & & & 0\\ & -\lambda & & \\ & & \ddots & \\ 0 & & & -\lambda\end{matrix}}
\end{array}\right]$$

$$= (\lambda^2 - \sigma_N^2)(\lambda^2 - \sigma_{N-1}^2)\dots(\lambda^2 - \sigma_1^2) = 0.$$

For $M > N$ we have

$$\det\left[\begin{array}{ccc}
\boxed{\begin{matrix}-\lambda & & 0\\ & \ddots & \\ 0 & & -\lambda\end{matrix}} & \boxed{\begin{matrix}\sigma_N & & 0\\ & \ddots & \\ 0 & & \sigma_1\end{matrix}} & 0 \\
\boxed{\begin{matrix}\sigma_N & & 0\\ & \ddots & \\ 0 & & \sigma_1\end{matrix}} & \boxed{\begin{matrix}-\lambda & & 0\\ & \ddots & \\ 0 & & -\lambda\end{matrix}} & 0 \\
0 & 0 & \boxed{\begin{matrix}-\lambda & & 0\\ & \ddots & \\ 0 & & -\lambda\end{matrix}}
\end{array}\right]$$

$$= -\lambda^{M-N}(\lambda^2 - \sigma_N^2)(\lambda^2 - \sigma_{N-1}^2)\dots(\lambda^2 - \sigma_1^2).$$

For $M < N$ we have

$$\det \begin{bmatrix} \boxed{\begin{matrix} -\lambda & & 0 \\ & \ddots & \\ 0 & & -\lambda \end{matrix}} & 0 & \boxed{\begin{matrix} \sigma_M & & 0 \\ & \ddots & \\ 0 & & \sigma_1 \end{matrix}} \\ 0 & \boxed{\begin{matrix} -\lambda & & 0 \\ & \ddots & \\ 0 & & -\lambda \end{matrix}} & 0 \\ \boxed{\begin{matrix} \sigma_M & & 0 \\ & \ddots & \\ 0 & & \sigma_1 \end{matrix}} & 0 & \boxed{\begin{matrix} -\lambda & & 0 \\ & \ddots & \\ 0 & & -\lambda \end{matrix}} \end{bmatrix}$$

$$= (-\lambda)^{N-M}(\lambda^2 - \sigma_M^2)(\lambda^2 - \sigma_{M-1}^2)\dots(\lambda^2 - \sigma_1^2).$$

Thus, the roots of the equation $\det(\mathcal{A} - \lambda I_{M+N}) = 0$ consist of all diagonal elements of the matrix Σ, the same diagonal elements with the negative sign, and, in addition, $|N - M|$ zeros. Thereby, the singular values $\sigma_{N_0}, \sigma_{N_0-1}, \dots \sigma_1$ can be regarded as $N_0 = \min\{M, N\}$ largest roots of the characteristic equation for $\mathcal{A}$. We have shown that singular values are uniquely determined by the matrix A. However, generally speaking, the unitary matrices U_M and V_N in the singular value decomposition $A = U_M K V_N^*$ are not unique. One of the obvious reasons is that we can simultaneously multiply the columns of the matrices U_M and V_N with the same indices by the same multipliers of the form $e^{i\theta_j}$, $\theta \in \mathbb{R}$. However, there are other reasons.

PROBLEM 3.4.1. Construct an example of a matrix A for which U_M and V_N are not unique.

We note that singular values of the matrix A do not change under the multiplication of A by a unitary $M \times M$-matrix $W_M^{(L)}$ from the left and a unitary $N \times N$-matrix $W_N^{(R)}$ from the right. Indeed, if $A = U_M K V_N^*$ and $B = W_M^{(L)} A W_N^{(R)}$, then $B = \widetilde{U}_M K \widetilde{V}_N^*$, where $\widetilde{U}_M = W_M^{(L)} U_M$ and $\widetilde{V}_N = [W_N^{(R)}]^* V_N$ are unitary matrices. Consequently, there are the singular value decompositions of the matrices A and B with the same matrix K that contains the singular values at one of its diagonals.

If $A = U_M K V_N^*$ establishes the correspondence $x = Ay$ between vectors $y \in H_N$ and $x \in H_M$, then it is possible to choose other orthonormal bases for the Euclidean spaces H_N and H_M so that the components of the vectors x and y in these bases form the vectors $\widetilde{x} = U_M x$ and $\widetilde{y} = V_N y$, and the above correspondence can be expressed by the formula $\widetilde{x} = K\widetilde{y}$, where K consists of a square diagonal block and (if $N \neq M$) zero rows (if $M > N$) or zero columns (if $N > M$). Singular values of A are nonnegative numbers located on the diagonal of the square block in K.

As was shown in §3.2, there is an effective algorithm that allows us to reduce congruently (without changing singular values) an arbitrary matrix to a two-diagonal matrix. Therefore, we can compute singular values of an arbitrary matrix if we know how to compute singular values of square matrices; moreover, it suffices to consider only two-diagonal matrices

$$A = \begin{bmatrix} a_1 & b_2 & & & & \mathbf{0} \\ & a_2 & b_3 & & & \\ & & \ddots & \ddots & & \\ & & & & a_{r-1} & b_r \\ \mathbf{0} & & & & & a_r \end{bmatrix}.$$

In other words, it suffices to know how to compute the eigenvalues λ of the symmetric matrix

$$\begin{bmatrix} 0 & A^* \\ A & 0 \end{bmatrix},$$

which yields the consistency (i.e., the existence of nonzero solutions) of the system of linear equations

$$\lambda \begin{pmatrix} u \\ v \end{pmatrix} = \begin{bmatrix} 0 & A^* \\ A & 0 \end{bmatrix} \begin{pmatrix} u \\ v \end{pmatrix}.$$

We write these equations term-by-term:

$$\begin{aligned} \lambda v_1 &= a_1 u_1 + b_2 u_2, & \lambda u_1 &= a_1 v_1, \\ \lambda v_2 &= a_2 u_2 + b_3 u_3, & \lambda u_2 &= b_2 v_1 + a_2 v_2, \\ &\cdots\cdots & &\cdots\cdots \\ \lambda v_{r-1} &= a_{r-1} u_{r-1} + b_r u_r, & \lambda u_{r-1} &= b_{r-1} v_{r-2} + a_{r-1} v_{r-1}, \\ \lambda v_r &= a_r u_r, & \lambda u_r &= b_r v_{r-1} + a_r v_r. \end{aligned}$$

It is convenient to rewrite them as follows:

$$\begin{aligned} \lambda u_1 &= a_1 v_1, \\ \lambda v_1 &= a_1 u_1 + b_2 u_2, \\ &\cdots\cdots \\ \lambda u_{r-1} &= a_{r-1} u_{r-1} + b_r u_r, \\ \lambda v_r &= b_r v_{r-1} + a_r v_r, \\ \lambda v_r &= a_r u_r. \end{aligned}$$

We note that they are equivalent to the vector equation

$$\lambda \begin{pmatrix} u_1 \\ v_1 \\ u_2 \\ \vdots \\ v_{r-1} \\ u_r \\ v_r \end{pmatrix} = \begin{bmatrix} 0 & a_1 & & & & & \mathbf{0} \\ a_1 & 0 & b_2 & & & & \\ & b_2 & 0 & a_2 & & & \\ & & \ddots & \ddots & \ddots & & \\ & & & b_r & 0 & a_r \\ \mathbf{0} & & & & a_r & 0 \end{bmatrix} \begin{pmatrix} u_1 \\ v_1 \\ u_2 \\ \vdots \\ v_{r-1} \\ u_r \\ v_r \end{pmatrix}.$$

Thus, the computation of singular values of a two-diagonal matrix is reduced to the computation of the spectrum of a three-diagonal symmetric matrix with zero main diagonal. There is a very effective algorithm based on the so-called Sturm method.

For its detailed description we refer the reader to [**15**]. The main idea of the Sturm method is described in §5.7.

To find singular values of a matrix A, one often uses eigenvalues of the Hermitian matrices A^*A and AA^*. If A has M rows and N columns, then A^*A is an $N \times N$-matrix and AA^* is an $M \times M$-matrix. From the equalities

$$A^*A = V_N K^* K V_N^*, \qquad AA^* = U_M K K^* U_M^*,$$

$$K^*K = \begin{cases} \begin{bmatrix} \sigma_M^2 & & & & & & \\ & \sigma_{M-1}^2 & & & & \mathbf{0} & \\ & & \ddots & & & & \\ & & & \sigma_1^2 & & & \\ & & & & 0 & & \\ & \mathbf{0} & & & & \ddots & \\ & & & & & & 0 \end{bmatrix} & \text{for } M < N, \\ \begin{bmatrix} \sigma_N^2 & & & \mathbf{0} \\ & \sigma_{N-1}^2 & & \\ & & \ddots & \\ \mathbf{0} & & & \sigma_1^2 \end{bmatrix} & \text{for } M \geq N, \end{cases}$$

$$KK^* = \begin{cases} \begin{bmatrix} \sigma_M^2 & & & \mathbf{0} \\ & \sigma_{M-1}^2 & & \\ & & \ddots & \\ \mathbf{0} & & & \sigma_1^2 \end{bmatrix} & \text{for } M \leq N, \\ \begin{bmatrix} \sigma_N^2 & & & & & & \\ & \sigma_{N-1}^2 & & & & \mathbf{0} & \\ & & \ddots & & & & \\ & & & \sigma_1^2 & & & \\ & & & & 0 & & \\ & \mathbf{0} & & & & \ddots & \\ & & & & & & 0 \end{bmatrix} & \text{for } M > N, \end{cases}$$

it is obvious that

$$\sigma_{N_0-j}(A) = \sqrt{\lambda_{N-j}(A^*A)} = \sqrt{\lambda_{M-j}(AA^*)}, \quad j = 0, 1, \ldots, N_0 - 1,$$

where $N_0 = \min\,\{M, N\}$. We assume that the eigenvalues $\lambda_k(A^*A)$ and $\lambda_k(AA^*)$ are numbered so that they do not decrease as k increases.

If A is a square matrix, then the matrices A^*A and AA^* have the same eigenvalues which are equal to the squared singular values of the matrix A.

Part 2

Matrices of Operators in the Euclidean Space

CHAPTER 4

Unitary Similar Transformations. The Schur Theorem

4.1. Reduction of a square matrix to the triangular form

The characteristic polynomial and eigenvalues of a square matrix. Annihilation of elements located in the upper (lower) triangle of a matrix. Similar transformations. Spectral exhaustion.

For a square $N \times N$-matrix A the characteristic polynomial

$$\det[\lambda I - A] = \lambda^N + a_1\lambda^{N-1} + a_2\lambda^{N-2} + \cdots + a_{N-1}\lambda + a_N$$

has N roots (multiple roots are allowed). The roots of the characteristic polynomial $\det[\lambda I - A]$ are called the *eigenvalues* of the matrix A.

Let λ_1 be an eigenvalue of a matrix A. Since $\det[\lambda_1 I - A] = 0$, the columns of the matrix $\lambda_1 I - A$ are linearly dependent, i.e., there exists a nonzero vector $x^{(1)}$ ($\|x^{(1)}\| \neq 0$) such that $[\lambda_1 I - A]x^{(1)} = 0$ or $Ax^{(1)} = \lambda_1 x^{(1)}$. The vector $x^{(1)}$ is called an *eigenvector* of the matrix A corresponding to the eigenvalue λ_1.

We take a unitary reflection $\mathcal{P}_1$ ($\mathcal{P}_1^2 = I$, $\mathcal{P}_1^* = \mathcal{P}_1$) that transforms the eigenvector $x^{(1)}$ to a vector collinear to the basis vector $e^{(N)}$, i.e.,

$$\mathcal{P}_1 x^{(1)} = \rho e^{(N)} = (0, 0, \dots, 0, \rho)^T \quad (\rho \neq 0).$$

If $x^{(1)}$ and $e^{(N)}$ are collinear, then for $\mathcal{P}_1$, instead of a reflection, we can take the transformation corresponding to the identity matrix. The realization of the algorithm requires no computation at all. This fact is often used in practice.

Consider the matrix $A^{(1)} = \mathcal{P}_1 A \mathcal{P}_1 = \mathcal{P}_1 A \mathcal{P}_1^*$. The last ($N$th) column of the matrix $A^{(1)}$ is collinear to the basis vector $e^{(N)}$. Indeed,

$$\begin{pmatrix} a_{1N}^{(1)} \\ \vdots \\ a_{NN}^{(1)} \end{pmatrix} = A^{(1)} e^{(N)} = \frac{1}{\rho} A^{(1)} \mathcal{P}_1 x^{(1)} = \frac{1}{\rho} \mathcal{P}_1 A \mathcal{P}_1 \mathcal{P}_1 x^{(1)}$$

$$= \frac{1}{\rho} \mathcal{P}_1 A x^{(1)} = \frac{\lambda_1}{\rho} \mathcal{P}_1 x^{(1)} = \frac{\lambda_1}{\rho} \rho e^{(N)} = \lambda_1 e^{(N)} = \begin{pmatrix} 0 \\ \vdots \\ \lambda_1 \end{pmatrix}.$$

The elements $a_{1N}^{(1)}, \dots, a_{1N-1}^{(1)}$ of the Nth column of the matrix $A^{(1)}$ vanish, whereas the last element $a_{NN}^{(1)}$ is equal to λ_1. Thus,

$$A^{(1)} = \mathcal{P}_1 A \mathcal{P}_1 = \begin{bmatrix} a_{11}^{(1)} & a_{12}^{(1)} & \dots & a_{1\,N-1}^{(1)} & 0 \\ a_{21}^{(1)} & a_{22}^{(1)} & \dots & a_{2\,N-1}^{(1)} & 0 \\ \vdots & \vdots & \ddots & \vdots & \vdots \\ a_{N-1\,1}^{(1)} & a_{N-1\,2}^{(1)} & \dots & a_{N-1\,N-1}^{(1)} & 0 \\ a_{N\,1}^{(1)} & a_{N\,2}^{(1)} & \dots & a_{N\,N-1}^{(1)} & \lambda_1 \end{bmatrix}$$

$$= \begin{bmatrix} & & & & 0 \\ & \widehat{A}^{(1)} & & & 0 \\ & & & & \vdots \\ & & & & 0 \\ a_{N1}^{(1)} & a_{N2}^{(1)} & \dots & a_{N\,N-1}^{(1)} & \lambda_1 \end{bmatrix}.$$

The characteristic polynomial $\det(\lambda I - A)$ can be written in the form

$$\begin{aligned} \det(\lambda I_N - A) &= \det(\mathcal{P}_1(\lambda I_N - A)\mathcal{P}_1^{-1}) = \det(\lambda I_N - A^{(1)}) \\ &= (\lambda - \lambda_1)\det(\lambda I_{N-1} - \widehat{A}^{(1)}). \end{aligned}$$

Therefore, if λ_2 is an eigenvalue of the matrix $\widehat{A}^{(1)}$, then it is simultaneously an eigenvalue of the initial matrix A. We choose a unitary reflection $\widehat{\mathcal{P}}_2$ of the $(N-1)$-dimensional space so that only one (the lowest) element of the last column of the matrix $\widehat{\mathcal{P}}_2 \widehat{A}^{(1)} \widehat{\mathcal{P}}_2$ can differ from zero; namely, it is equal to λ_2. Starting with the reflection $\widehat{\mathcal{P}}_2$, we construct the $N \times N$-matrix

$$\mathcal{P}_2 = \begin{bmatrix} & & & & 0 \\ & & & & 0 \\ & & \widehat{\mathcal{P}}_2 & & \vdots \\ & & & & 0 \\ 0 & 0 & \dots & 0 & 1 \end{bmatrix}.$$

Moreover, we can check that

$$\mathcal{P}_2 \mathcal{P}_1 A \mathcal{P}_1 \mathcal{P}_2 = \mathcal{P}_2 A^{(1)} \mathcal{P}_2 = \begin{bmatrix} & & & & 0 & 0 \\ & & & & 0 & 0 \\ & & \widehat{A} & & \vdots & \vdots \\ & & & & 0 & 0 \\ \times & \times & \dots & \times & \lambda_2 & 0 \\ \times & \times & \dots & \times & \times & \lambda_1 \end{bmatrix}.$$

We can continue the above procedure and, constructing unitary reflections successively, annihilate elements in the upper right triangle of the matrix. We obtain a lower triangular matrix whose diagonal elements are eigenvalues of the matrix A in any chosen order. This procedure is called the *spectral exhaustion.*

Recall that the product of unitary transformations $U = \mathcal{P}_1\mathcal{P}_2\dots\mathcal{P}_{N-1}$ is unitary, i.e., $U^*U = I_N$ $(U^* = \mathcal{P}_{N-1}\dots\mathcal{P}_1)$.

Completing the exhaustion, we obtain

$$\mathcal{P}_{N-1}\dots\mathcal{P}_2\mathcal{P}_1 A\mathcal{P}_1\mathcal{P}_2\dots\mathcal{P}_{N-1} = U^*AU = \begin{bmatrix} \lambda_N & & & & \\ \times & \lambda_{N-1} & & \text{\Large 0} & \\ \times & \times & \lambda_{N-2} & & \\ \dots & \dots & \dots & \ddots & \\ \times & \times & \times & \dots & \lambda_1 \end{bmatrix}.$$

If U is a nondegenerate linear transformation, then the matrices A and $\widetilde{A} = U^{-1}AU$ are said to be *similar.* Such matrices correspond to the same transformation but in different bases. If both bases are orthonormal, then U is unitary $(U^{-1} = U^*)$. Since

$$1 = \det I_N = \det\left[U \cdot U^{-1}\right] = \det U \cdot \det U^{-1},$$

we have

$$\lambda_1\lambda_2\dots\lambda_N = \det\widetilde{A} = \det U^{-1}\cdot\det A\cdot\det U = \det A.$$

4.2. The Schur theorem

The formulation of the Schur theorem on the reduction of a square matrix to the lower (upper) triangular form. The diagonal form in the case of a self-adjoint matrix.

In §4.1 we presented one of the proofs of a very important theorem suggested in 1909. Now we give an exact formulation of the theorem.

THEOREM 4.2.1 (Schur). *For any square matrix A there exists a unitary similar transformation U such that $U^{-1}AU \equiv U^*AU$ is a lower triangular matrix. On the main diagonal of A there are eigenvalues of A in any chosen order.*

PROBLEM 4.2.1. Show that the Schur theorem remains valid if the words "lower triangular matrix" in the formulation of the theorem are replaced by the words "upper triangular matrix".

For a self-adjoint matrix A the Schur theorem can be made more precise. Let U be a unitary matrix such that $U^*AU = R$, where R is an upper triangular matrix. Since $R^* = [U^*AU]^* = U^*A^*U = U^*AU = R$, the matrix R is necessarily diagonal with real elements.

THEOREM 4.2.2. *For any self-adjoint $N \times N$-matrix A ($A = A^*$) there exists a unitary transformation U such that*

$$A = U \begin{bmatrix} \alpha_1 & & & 0 \\ & \alpha_2 & & \\ & & \ddots & \\ 0 & & & \alpha_N \end{bmatrix} U^*.$$

In this representation, α_j are arbitrarily ordered eigenvalues of the matrix A (α_j are real).

PROBLEM 4.2.2. Prove that the eigenvalues of the matrices A^{-1}, A^*, and $q_0 I + q_1 A + \cdots + q_k A^k$ can be expressed in terms of the known eigenvalues $\lambda_j(A)$ of A by the following formulas:

$$\lambda_j(A^{-1}) = \lambda_j^{-1}(A), \quad \lambda_j(A^*) = \overline{\lambda_j(A)},$$
$$\lambda_j(q_0 I + q_1 A + \cdots + q_k A^k) = q_0 + q_1 \lambda_j(A) + \cdots + q_k \lambda_j^k(A).$$

4.3. Criterion for the solvability of a matrix Sylvester equation

The matrix Sylvester equation $AX - XB = G$. The solvability for 2×2-matrices A and B. Criterion for the solvability of a matrix Sylvester equation in the general case. The necessity of the condition that the spectra of the matrices A and B are disjoint. The distance $\operatorname{sep}(A, B)$ between the spectrum of the matrix A and the spectrum of the matrix B.

The Schur theorem asserts that any square $N \times N$-matrix A can be reduced to the triangular form

$$A = U^* \begin{bmatrix} \lambda_1 & \times & \times & \times & \times & \times \\ & \lambda_2 & \times & \times & \times & \times \\ & & \cdot & \times & \times & \times \\ & & & \cdot & \times & \times \\ & 0 & & & \cdot & \times \\ & & & & & \lambda_N \end{bmatrix} U \qquad (U^* U = I)$$

by a similar unitary transformation U, where λ_j are eigenvalues of the matrix A.

We will often use the following simplified form of the Schur theorem:

$$A = U^* \begin{bmatrix} A_{11} & A_{12} \\ 0 & A_{22} \end{bmatrix} U, \tag{4.3.1}$$

under the assumption that the eigenvalues of the matrices A_{11} and A_{22} form the spectrum of the matrix A.

From the proof of the Schur theorem it follows that the eigenvalues (which are located on the main diagonal of the triangular matrix) can be ordered arbitrarily. Therefore, considering the simplified form (4.3.1), we can assume that the eigenvalues are arbitrarily distributed between the blocks A_{11} and A_{22}.

Instead of the upper triangular form, we can use the lower triangular form, i.e., we start with the canonical form

$$B = V^* \begin{bmatrix} B_{11} & 0 \\ B_{21} & B_{22} \end{bmatrix} V \quad (V^*V = I).$$

LEMMA 4.3.1. *If* 2×2-*matrices*

$$\begin{bmatrix} a_{11} & a_{12} \\ 0 & a_{22} \end{bmatrix}, \quad \begin{bmatrix} b_{11} & 0 \\ b_{21} & b_{22} \end{bmatrix}$$

are such that $a_{11} - b_{11} \neq 0$, $a_{11} - b_{22} \neq 0$, $a_{22} - b_{11} \neq 0$, $a_{22} - b_{22} \neq 0$, *then* y_{ij} *are uniquely found from the system*

$$\begin{bmatrix} a_{11} & a_{12} \\ 0 & a_{22} \end{bmatrix} \begin{bmatrix} y_{11} & y_{12} \\ y_{21} & y_{22} \end{bmatrix} - \begin{bmatrix} y_{11} & y_{12} \\ y_{21} & y_{22} \end{bmatrix} \begin{bmatrix} b_{11} & 0 \\ b_{21} & b_{22} \end{bmatrix} = \begin{bmatrix} f_{11} & f_{12} \\ f_{21} & f_{22} \end{bmatrix}.$$

PROOF. The matrix equation in Lemma 4.3.1 can be rewritten as scalar equations

$$\begin{aligned} a_{22}y_{22} - y_{22}b_{22} &= f_{22}, \\ a_{22}y_{21} - y_{21}b_{11} &= y_{22}b_{21} + f_{21}, \\ a_{11}y_{12} - y_{12}b_{22} &= -a_{12}y_{22} + f_{12}, \\ a_{11}y_{11} - y_{11}b_{11} &= -a_{12}y_{21} + f_{11} + y_{12}b_{21}. \end{aligned}$$

Now the assertion of the lemma becomes obvious. □

Let A and B be second order matrices such that no eigenvalue of the matrix A coincides with an eigenvalue of the matrix B. We prove that the matrix equation

$$AX - XB = G \tag{4.3.2}$$

is uniquely solvable for any right-hand side G.

By the Schur theorem, we represent the matrices A and B in the form

$$A = U \begin{bmatrix} a_{11} & a_{12} \\ 0 & a_{22} \end{bmatrix} U^*, \quad B = V \begin{bmatrix} b_{11} & 0 \\ b_{21} & b_{22} \end{bmatrix} V^*,$$

where U and V are unitary matrices, a_{11} and a_{22} are eigenvalues of the matrix A, b_{11} and b_{22} are eigenvalues of the matrix B. Let us multiply equation (4.3.2) from the left by U^* and from the right by V. Set $Y = U^*XV$ and rewrite the result in the form

$$F = U^*GV = \begin{bmatrix} a_{11} & a_{12} \\ 0 & a_{22} \end{bmatrix} Y - Y \begin{bmatrix} b_{11} & 0 \\ b_{21} & b_{22} \end{bmatrix}.$$

By Lemma 4.3.1, the last equation is solvable with respect to Y. Consequently, equation (4.3.2) is solvable with respect to $X = UYV^*$.

Consider a second order matrix A and a first order matrix B, so that B is a number. In this case, X is a matrix consisting of a single column:

$$X = \begin{bmatrix} x_{11} \\ x_{12} \end{bmatrix}.$$

The criterion for the solvability of the equation

$$\begin{bmatrix} a_{11} & a_{12} \\ 0 & a_{22} \end{bmatrix} \begin{pmatrix} y_{11} \\ y_{12} \end{pmatrix} - \begin{pmatrix} y_{11} \\ y_{12} \end{pmatrix} b_{11} = \begin{pmatrix} f_{11} \\ f_{12} \end{pmatrix}$$

or the equivalent triangular system

$$\begin{aligned} a_{11}y_{11} + a_{12}y_{12} - y_{11}b_{11} &= f_{11}, \\ a_{22}y_{12} - y_{12}b_{11} &= f_{12} \end{aligned}$$

implies that the equation $AX - XB = G$ is uniquely solvable if the matrices A and B have no common eigenvalues.

Consider an $n \times n$-matrix A and an $m \times m$-matrix B. Let the unknown matrix X have n rows and m columns. Using the Schur theorem, we will establish a criterion for the solvability of the matrix equation $AX - XB = G$. We represent the matrices A and B in the canonical form

$$A = U \begin{bmatrix} A_{11} & A_{12} \\ 0 & A_{22} \end{bmatrix} U^*, \quad B = V \begin{bmatrix} B_{11} & 0 \\ B_{21} & B_{22} \end{bmatrix} V^*,$$

where A_{11} is a square $n_1 \times n_1$-block, A_{22} is a square $n_2 \times n_2$-block ($n_1 + n_2 = n$), B_{11} is an $m_1 \times m_1$-block, and B_{22} is an $m_2 \times m_2$-block ($m_1 + m_2 = m$). If the matrices A and B have no common eigenvalues, then the blocks A_{ii}, $i = 1, 2$, and B_{jj}, $j = 1, 2$, have no common eigenvalues either. Furthermore, if the solvability is established for matrices of orders n_i and m_i, then we can conclude that the equation under consideration is solvable. The proof of this fact repeats the above arguments concerning unknown 2×2- or 2×1-matrices X. The solvability in the case $1 \le n_i \le 2$, $1 \le m_i \le 2$ was proved earlier.

Thus, we have established the following criterion.

CRITERION (the solvability of the matrix Sylvester equation). *Let the matrices A and B have no common eigenvalues. Then a solution X to the matrix Sylvester equation $AX - XB = G$ exists and is unique for any matrix G.*

We did not discuss the uniqueness of a solution. We only note that the uniqueness is obvious in view of the above constructions.

The above arguments show that the condition imposed on the eigenvalues of A and B is sufficient for the solvability of the equation with an arbitrary right-hand side. We show that this condition is also necessary.

Let λ be a common eigenvalue of matrices A and B. By the Schur theorem, A and B can be reduced to the canonical form

$$A = U \begin{bmatrix} \times & \times & \cdots & \times & \times \\ & \times & \cdots & \times & \times \\ & & \ddots & \vdots & \vdots \\ & \text{\Large 0} & & \times & \times \\ & & & & \lambda \end{bmatrix} U^*, \quad B = V \begin{bmatrix} \times & & & & \\ \times & \times & & \text{\Large 0} & \\ \vdots & \vdots & \ddots & & \\ \times & \times & \cdots & \times & \\ \times & \times & \cdots & \times & \lambda \end{bmatrix} V^*.$$

If the equation $AX - XB = G$ is solvable for any right-hand side G, then the transformed equation ($Y = U^*XV$)

$$\begin{bmatrix} \times & \times & \cdots & \times & \times \\ & \times & \cdots & \times & \times \\ & & \ddots & \vdots & \vdots \\ & \text{\Large 0} & & \times & \times \\ & & & & \lambda \end{bmatrix} \begin{bmatrix} y_{11} & y_{12} & \cdots & y_{1m} \\ y_{21} & y_{22} & \cdots & y_{2m} \\ \vdots & \vdots & \ddots & \vdots \\ y_{n-1,1} & y_{n-1,2} & \cdots & y_{n-1,m} \\ y_{n1} & y_{n2} & \cdots & y_{nm} \end{bmatrix}$$

$$
-\begin{bmatrix} y_{11} & y_{12} & \cdots & y_{1m} \\ y_{21} & y_{22} & \cdots & y_{2m} \\ \vdots & \vdots & \ddots & \vdots \\ y_{n-1,1} & y_{n-1,2} & \cdots & y_{n-1,m} \\ y_{n1} & y_{n2} & \cdots & y_{nm} \end{bmatrix} \begin{bmatrix} \times & & & & \\ \times & \times & & 0 & \\ \vdots & \vdots & \ddots & & \\ \times & \times & \ldots & \times & \\ \times & \times & \ldots & \times & \lambda \end{bmatrix}
$$

$$
= \begin{bmatrix} f_{11} & f_{12} & \cdots & f_{1m-1} & f_{1m} \\ f_{21} & f_{22} & \cdots & f_{2m-1} & f_{2m} \\ \vdots & \vdots & \ddots & \vdots & \vdots \\ f_{n-1,1} & f_{n-1,2} & \cdots & f_{n-1,m-1} & f_{n-1,m} \\ f_{n1} & f_{n2} & \cdots & f_{nm-1} & f_{nm} \end{bmatrix}
$$

is also solvable for any right-hand side $F = U^*GV$. To find y_{nm}, we must solve the equation $\lambda y_{nm} - y_{nm}\lambda = f_{nm}$ which is solvable only if $f_{nm} = 0$; moreover, a solution is not unique.

If A and B have a common eigenvalue, then the homogeneous Sylvester equation $AX - XB = 0$ has a nonzero solution X, whereas the nonhomogeneous equation (4.3.2) is not solvable for an arbitrary right-hand side G.

Equation (4.3.2) defines a linear operator $\mathcal{L}$ that transforms a rectangular matrix X to a matrix G with the same number of rows and columns as X. If A and B have no common eigenvalues, then (see the above arguments) there exists the inverse operator $\mathcal{L}^{-1}$ mapping G to X.

The space of rectangular matrices can be regarded as a Euclidean space equipped with the inner product

$$
(X, Y)_E = \sum_{ij} x_{ij}\overline{y}_{ij} = \operatorname{tr}(XY^*)
$$

and the norm

$$
\|X\|_E = \sqrt{\sum_{ij} |x_{ij}|^2} = \sqrt{\operatorname{tr}(XX^*)}.
$$

Following Stewart, the parameter $\|\mathcal{L}^{-1}\|^{-1}$ is often used to characterize the distance between the spectrum of the matrix A and the spectrum of the matrix B. This parameter is denoted by $\operatorname{sep}(A, B)$, i.e.,

$$
\operatorname{sep}(A, B) = \begin{cases} \|\mathcal{L}^{-1}\|^{-1} & \text{if } \mathcal{L}^{-1} \text{ exists, i.e., if } A \text{ and } B \\ & \text{have no common eigenvalues,} \\ 0 & \text{otherwise.} \end{cases}
$$

To introduce $\operatorname{sep}(A, B)$, one can use other norms in the space of rectangular matrices. We note that the (nonzero) value of this characteristic depends on the norm.

4.4. Applications of the criterion

The case of a matrix A with eigenvalues in the left half-plane. The Lyapunov equation as a special case of the Sylvester equation. The solvability theorem for the Lyapunov equation. The case of a matrix

A with eigenvalues in the unit disk. The case of a nilpotent matrix. The solution is a Hermitian matrix if the right-hand side is a Hermitian matrix.

We give a special case of the criterion for the solvability of the Sylvester equation (see §4.3) which will be used later. Suppose that the eigenvalues of a matrix A lie in the left half-plane ($\operatorname{Re}\lambda_j(A) < 0$). Then the eigenvalues of the matrix $B = -A^*$ lie in the right half-plane. Consequently, the spectra of the matrices A and $B = -A^*$ are disjoint. In this case, the Sylvester equation $-A^*X - XA \equiv BX - XA = C$ is uniquely solvable for any matrix C. We assume that $C = C^*$ and consider the equations

$$A^*X + XA + C = 0,$$
$$A^*Y + YA + C^* = 0.$$

Passing to adjoint matrices in the last equality, we find

$$Y^*A + A^*Y^* + C = 0,$$

i.e., Y^* satisfies the same equation as X. Since a solution to the Sylvester equation is unique, we have $Y^* = X$. Since $C = C^*$, the equations for Y and for X coincide. Consequently, $Y = X$. (We again use the uniqueness of a solution.) Thus, we have proved that $X = X^*$.

THEOREM 4.4.1. *If the eigenvalues of a matrix A lie in the left half-plane and $C = C^*$, then the matrix equation*

$$XA + A^*X + C = 0 \tag{4.4.1}$$

has a unique solution X; moreover, $X = X^*$.

Equation (4.4.1) is called the *Lyapunov equation.*

If A and B are square $N \times N$-matrices and the matrix B is nonsingular, then the matrix equation $X - BXA = D$ is equivalent to the Sylvester equation $B^{-1}X - XA = B^{-1}D$. Consequently, if the eigenvalues of a matrix A lie in the unit disk and the matrix A is nonsingular, then the matrix equation

$$X - A^*XA = C \tag{4.4.2}$$

is uniquely solvable for any matrix C.

In fact, in (4.4.2) the assumption about the nonsingularity of A is not necessary. First we consider a matrix A that has only zero eigenvalues. By the Schur theorem, there exists a unitary matrix U such that

$$A = U^{-1}\begin{bmatrix} 0 & a_{12} & a_{13} & \cdots & a_{1N} \\ & 0 & a_{23} & \cdots & a_{2N} \\ & & & \ddots & \vdots \\ & \mathbf{0} & & & 0 \end{bmatrix} U \quad (U^* = U^{-1}).$$

It is easy to see that the matrix A is nilpotent, i.e., there exists p such that $A^p = 0$. In any case, $A^N = 0$ and $(A^*)^N = 0$. For any right-hand side C the matrix

$$X = C + A^*CA + (A^*)^2CA^2 + \cdots + (A^*)^{N-1}CA^{N-1}$$

is a solution to equation (4.4.2). Indeed, in the representation $A^*XA = A^*CA + (A^*)^2CA^2 + \cdots + (A^*)^2CA^2 + \cdots + (A^*)^{N-1}CA^{N-1} + (A^*)^NCA^N$ the last nonzero term vanishes and the formula $X - A^*XA = C$ becomes obvious. If the matrix A has only zero eigenvalues, then the equation $X - A^*XA = C$ is solvable for any right-hand side C. Equation (4.4.2) can be regarded as a system of N^2 linear equations with respect to N^2 variables x_{ij}. The solvability of the system for any right-hand side C means that a solution X can be uniquely determined from C.

Let a matrix A have the block triangular form

$$A = \begin{bmatrix} A_{11} & A_{12} \\ 0 & A_{22} \end{bmatrix},$$

where the diagonal block A_{11} is nonsingular and every eigenvalue of A_{11} is less than 1, whereas the diagonal block A_{22} is nilpotent. The matrix equation

$$\begin{bmatrix} X_{11} & X_{12} \\ X_{21} & X_{22} \end{bmatrix} - \begin{bmatrix} A_{11} & A_{12} \\ 0 & A_{22} \end{bmatrix}^* \begin{bmatrix} X_{11} & X_{12} \\ X_{21} & X_{22} \end{bmatrix} \begin{bmatrix} A_{11} & A_{12} \\ 0 & A_{22} \end{bmatrix} = \begin{bmatrix} C_{11} & C_{12} \\ C_{21} & C_{22} \end{bmatrix}$$

can be rewritten as follows:

$$\begin{gathered} X_{11} - A_{11}^* X_{11} A_{11} = C_{11}, \\ (A_{11}^*)^{-1} X_{12} - X_{12} A_{22} = (A_{11}^*)^{-1} [C_{12} + A_{11}^* X_{11} A_{12}], \\ X_{21} A_{11}^{-1} - A_{22}^* X_{21} = (C_{21} + A_{12}^* X_{11} A_{11}) A_{11}^{-1}, \\ X_{22} - A_{22}^* X_{22} A_{22} = C_{22} + A_{12}^* X_{11} A_{12} + A_{12}^* X_{12} A_{22} + A_{22}^* X_{21} A_{12}. \end{gathered}$$

From the first equation we can find X_{11}. After that we can solve the second and third equations with respect to X_{12} and X_{21}. Having X_{11}, X_{12}, and X_{21}, from the fourth equation we find X_{22}. Each of these four matrices is uniquely determined.

By the Schur theorem, any matrix with eigenvalues of modulus less than 1 can be unitarily transformed to the block triangular form with the same properties of the blocks A_{11} and A_{22} as above. To solve the equation $X - A^*XA = C$, we must replace it by the equivalent equation

$$\begin{bmatrix} \widehat{X}_{11} & \widehat{X}_{12} \\ \widehat{X}_{21} & \widehat{X}_{22} \end{bmatrix} - \begin{bmatrix} A_{11} & A_{12} \\ 0 & A_{22} \end{bmatrix}^* \begin{bmatrix} \widehat{X}_{11} & \widehat{X}_{12} \\ \widehat{X}_{21} & \widehat{X}_{22} \end{bmatrix} \begin{bmatrix} A_{11} & A_{12} \\ 0 & A_{22} \end{bmatrix} = \begin{bmatrix} \widehat{C}_{11} & \widehat{C}_{12} \\ \widehat{C}_{21} & \widehat{C}_{22} \end{bmatrix}$$

setting

$$\widehat{X} = \begin{bmatrix} \widehat{X}_{11} & \widehat{X}_{12} \\ \widehat{X}_{21} & \widehat{X}_{22} \end{bmatrix} = U \begin{bmatrix} X_{11} & X_{12} \\ X_{21} & X_{22} \end{bmatrix} U^*, \quad \widehat{C} = \begin{bmatrix} \widehat{C}_{11} & \widehat{C}_{12} \\ \widehat{C}_{21} & \widehat{C}_{22} \end{bmatrix} = UCU^*.$$

As was shown, this equation is uniquely solvable. Then we find $X = U^*\widehat{X}U$. The fact that for any right-hand side C a system of N^2 linear equations (written in the matrix form as $X - A^*XA = C$) is solvable with respect to N^2 elements of the matrix X implies that the solution is unique.

Thus, we have proved the following theorem.

THEOREM 4.4.2. *If the module of every eigenvalue of a square $N \times N$-matrix A is strictly less than 1, i.e., $|\lambda_j(A)| < 1$, $j = 1, 2, \dots, N$, then the matrix equation $X - A^*XA = C$ is uniquely solvable for any matrix C.*

COROLLARY 4.4.1. *Let the assumptions of Theorem 4.4.2 hold. If $C = C^*$, then the solution X is a Hermitian matrix, i.e., $X = X^*$.*

4.5. Invariant subspaces

The reduction of a matrix A to a special block diagonal form depending on the spectrum. The maximal invariant subspace corresponding to an eigenvalue of a matrix. Matrix polynomials. The Cayley–Hamilton theorem.

We use the matrix Sylvester equation to obtain an important consequence of the Schur theorem. Suppose that a matrix A has at most two different eigenvalues. Consider the canonical form

$$A = U \begin{bmatrix} A_{11} & A_{12} \\ 0 & A_{22} \end{bmatrix} U^*,$$

where the blocks A_{11} and A_{22} have no common eigenvalues. We can construct the block triangular matrix

$$\begin{bmatrix} I & -M \\ 0 & I \end{bmatrix}$$

and the inverse matrix

$$\begin{bmatrix} I & -M \\ 0 & I \end{bmatrix}^{-1} = \begin{bmatrix} I & M \\ 0 & I \end{bmatrix}$$

so that

$$\begin{bmatrix} I & M \\ 0 & I \end{bmatrix} \begin{bmatrix} A_{11} & A_{12} \\ 0 & A_{22} \end{bmatrix} \begin{bmatrix} I & -M \\ 0 & I \end{bmatrix} = \begin{bmatrix} A_{11} & 0 \\ 0 & A_{22} \end{bmatrix}.$$

For that, it suffices to use the equality

$$\begin{bmatrix} I & M \\ 0 & I \end{bmatrix} \begin{bmatrix} A_{11} & A_{12} \\ 0 & A_{22} \end{bmatrix} \begin{bmatrix} I & -M \\ 0 & I \end{bmatrix} = \begin{bmatrix} A_{11} & A_{12} + MA_{22} - A_{11}M \\ 0 & A_{22} \end{bmatrix}$$

and find the matrix M from the Sylvester equation $A_{11}M - MA_{22} - A_{12} = 0$, which is possible under the assumptions on A_{11} and A_{22}. Setting

$$T = U \begin{bmatrix} I & -M \\ 0 & I \end{bmatrix} \qquad (|\det T| = |\det U| = 1 \neq 0),$$

we can represent A in the form

$$A = T \begin{bmatrix} A_{11} & 0 \\ 0 & A_{22} \end{bmatrix} T^{-1}.$$

It is clear that the procedure of splitting the matrix A into blocks containing different eigenvalues can be continued. Thus, the matrix A can be written as follows:

$$A = T \begin{bmatrix} A_{11} & & & 0 \\ & A_{22} & & \\ & & \ddots & \\ 0 & & & A_{ll} \end{bmatrix} T^{-1}, \tag{4.5.1}$$

where in every diagonal block the eigenvalues coincide, whereas different blocks have different eigenvalues. By the Schur theorem, we can assume that all blocks are triangular (upper or lower). For example,

$$A_{jj} = \begin{bmatrix} \lambda_j & \times & \times & \times \\ & \lambda_j & \times & \times \\ & & \ddots & \vdots \\ \mathbf{0} & & & \lambda_j \end{bmatrix}, \quad j = 1, 2, \ldots, l.$$

It is obvious that the order ν_j of the square matrix A_{jj} is determined by the multiplicity ν_j of the eigenvalue λ_j of the matrix A.

We divide the matrix T into groups of columns as follows:

$$T = [T_1 \vdots T_2 \vdots \cdots \vdots T_l] \tag{4.5.2}$$

so that the number of the columns in the jth group coincides with the order ν_j of A_{jj}. Since $\det T \neq 0$, all columns are linearly independent. It is easy to verify that linear combinations of columns of T_j are transformed to linear combinations of columns of the same group T_j under the transformation A. It is obvious that linear combinations of columns of T_j form an invariant subspace of A, called the *maximal invariant subspace corresponding to the eigenvalue* λ_j.

In the representation of A, the blocks A_{jj} can be written in the form $A_{jj} = \lambda_j I + R_j$, where R_j is an upper triangular $\nu_j \times \nu_j$-matrix with zero main diagonal:

$$R_j = \begin{bmatrix} 0 & \times & \times & \ldots & \times & \times \\ & 0 & \times & \ldots & \times & \times \\ & & 0 & \ldots & \times & \times \\ & & & \ddots & \vdots & \vdots \\ & \mathbf{0} & & & 0 & \times \\ & & & & & 0 \end{bmatrix}.$$

It is easy to check that R_j^2 is also an upper triangular matrix with zero main diagonal. Moreover, the diagonal lying above the main diagonal also consists of the zero elements. For the upper triangular matrix R_j^3 not only main diagonal, but two diagonals above it consists of the zero elements. As m increases, the number of zero diagonals of R_j^m increases so that $R_j^{\nu_j}$ consists only of zero elements, i.e., $R_j^{\nu_j} = 0$. Thus, if ν_j is the multiplicity of λ_j, then $[A_{jj} - \lambda_j I_{\nu_j}]^{\nu_j} = 0$.

On the other hand, if $k \neq j$, then $A_{kk} - \lambda_j I_{\nu_k}$ is an upper triangular matrix; moreover, every diagonal element is equal to $\lambda_k - \lambda_j$. The matrix $[A_{kk} - \lambda_j I_{\nu k}]^p$ also has the upper triangular form and every diagonal element is equal to $(\lambda_k - \lambda_j)^p$. Therefore,

$$\det(A_{kk} - \lambda_j I_{\nu_k})^p = (\lambda_k - \lambda_j)^p \neq 0, \quad k \neq j.$$

Let $P(A)$ be a polynomial in A. The representation (4.5.1) implies that

$$P(A) = T \begin{bmatrix} P(A_{11}) & & & \mathbf{0} \\ & P(A_{22}) & & \\ & & \ddots & \\ \mathbf{0} & & & P(A_{ll}) \end{bmatrix} T^{-1}.$$

If the polynomial $P(\lambda)$ is divisible by $(\lambda - \lambda_j)^{\nu_j}$, i.e., $P(\lambda) = q(\lambda) \cdot (\lambda - \lambda_j)^{\nu_j}$, then the matrix polynomials $P(A)$ and $P(A_{ij})$ are represented in the form

$$P(A) = q(A)(A - \lambda_j I_N)^{\nu_j}, \quad j = 1, 2, \dots, l,$$
$$P(A_{jj}) = q(A_{jj})(A_{jj} - \lambda_j I_{\nu_j})^{\nu_j} = 0.$$

Taking for $P(\lambda)$ the characteristic polynomial of the matrix A, i.e.,

$$P(\lambda) = \det[A - \lambda I] = \prod_{j=1}^{l} \det[A_{jj} - \lambda I_{\nu_j}] = \prod_{j=1}^{l} (\lambda_j - \lambda)^{\nu_j},$$

we obtain $P(A_{jj}) = 0$ for all j and, consequently, $P(A) = 0$. This assertion is known as the *Cayley–Hamilton theorem.*

THEOREM 4.5.1. *Any square matrix A satisfies the matrix equation $P(A) = 0$, where P is the characteristic polynomial of A.*

The matrix $(A - \lambda_j I)^{\nu_j}$ is represented in the form

$$T \begin{bmatrix} (A_{11} - \lambda_j I_{\nu_1})^{\nu_j} & & & 0 \\ & (A_{22} - \lambda_j I_{\nu_2})^{\nu_j} & & \\ & & \ddots & \\ 0 & & & (A_{ll} - \lambda_j I_{\nu_l})^{\nu_j} \end{bmatrix} T^{-1}.$$

Moreover, only one diagonal block $(A_{jj} - \lambda_j I_{\nu_j})^{\nu_j}$ is zero, whereas the remaining blocks are nonsingular.

We recall that the columns of the matrix T is divided into the groups $T_1, T_2, \dots,$ T_l (see (4.5.2)). Any solution x to the equation $(A - \lambda_j I_N)^{\nu_j} x = 0$ is represented as a linear combination of columns of T_j. In other words, the maximal invariant subspace corresponding to the eigenvalue λ_j consists of exactly those vectors that are annihilated by the matrix $(A - \lambda_j I)^{\nu_j}$. It is convenient to use the following almost obvious results.

Vectors of the maximal invariant subspace corresponding to the eigenvalue λ_j (and only these vectors) are solutions to the equation $(A - \lambda_j I_N)^{\nu_j} x = 0$, where ν_j is the multiplicity of the root λ_j of the characteristic polynomial $P(\lambda) = \det(A - \lambda I)$.

CHAPTER 5

Alternation Theorems

5.1. Formulation of alternation theorems

Alternation of the eigenvalues of a Hermitian $(m+1)\times(m+1)$-matrix A_{m+1} and the eigenvalues of an $m\times m$-matrix A_m obtained from A_{m+1} by elimination of a row and a column. Alternation of singular values. One-sided borderings. The existence of a matrix that borders a given matrix with given alternating sequences of singular values.

The location of eigenvalues of symmetric matrices and their submatrices is not arbitrary. The so-called alternation conditions must be satisfied. Similar conditions connect singular values of a rectangular matrix with singular values of its submatrices. We formulate the main alternation theorems.

THEOREM 5.1.1. *If A_{m+1} is a Hermitian $(m+1)\times(m+1)$-matrix ($A_{m+1}=A^*_{m+1}$, $a_{ij}=\bar{a}_{ji}$) of the form*

$$A_{m+1}=\begin{bmatrix} a_{11} & a_{12} & \cdots & a_{1m} & a_{1\,m+1} \\ a_{21} & a_{22} & \cdots & a_{2m} & a_{2\,m+1} \\ \vdots & \vdots & & \vdots & \vdots \\ a_{m1} & a_{m2} & \cdots & a_{mm} & a_{m\,m+1} \\ a_{m+1\,1} & a_{m+1\,2} & \cdots & a_{m+1\,m} & a_{m+1\,m+1} \end{bmatrix}$$

and an $m\times m$-matrix A_m is obtained from A_{m+1} by eliminating the last column and the last row, i.e.,

$$A_m=\begin{bmatrix} a_{11} & a_{12} & \cdots & a_{1m} \\ a_{21} & a_{22} & \cdots & a_{2m} \\ \vdots & \vdots & & \vdots \\ a_{m1} & a_{m2} & \cdots & a_{mm} \end{bmatrix},$$

then the eigenvalues $\lambda_1,\lambda_2,\dots,\lambda_{m+1}$ of A_{m+1} and the eigenvalues $\mu_1,\mu_2,\dots,\mu_m$ of A_m ordered so that

$$\lambda_{j+1}\geq\lambda_j,\quad \mu_{j+1}\geq\mu_j,$$

form alternating sequences, i.e.,

$$\lambda_1\leq\mu_1\leq\lambda_2\leq\mu_2\leq\cdots\leq\lambda_m\leq\mu_m\leq\lambda_{m+1}.$$

The following theorem is a converse to Theorem 5.1.1.

THEOREM 5.1.2. *For a Hermitian $m\times m$-matrix A_m with eigenvalues $\mu_1,\mu_2,\dots,\mu_m$, ($\mu_{j+1}\geq\mu_j$, $j=1,2,\dots,m$) and a sequence $\lambda_1\,\lambda_2,\dots,\lambda_{m+1}$ alternating with $\mu_1,\mu_2,\dots,\mu_m$ there exists a Hermitian $(m+1)\times(m+1)$-matrix A_{m+1} obtained* *from A_m by the bordering, i.e., adding a last column and a last row, such that $\lambda_1\,\lambda_2,\dots,\lambda_{m+1}$ are the eigenvalues of A_{m+1}.*

It suffices to prove Theorems 5.1.1 and 5.1.2 for a diagonal $m\times m$-matrix A_m. Indeed, for an $m\times m$-matrix A_m it is possible to find a unitary $m\times m$-matrix U_m ($U_m^*U_m = I_m$) such that

$$U_m^* A_m U_m = \begin{bmatrix} \alpha_1 & & & \mathbf{0} \\ & \alpha_2 & & \\ & & \ddots & \\ \mathbf{0} & & & \alpha_m \end{bmatrix},$$

where a_i are real numbers. Denote by U_{m+1} a unitary $(m+1)\times(m+1)$-matrix obtained from U_m as follows:

$$U_{m+1} = \begin{bmatrix} & & & & 0 \\ & U_m & & & 0 \\ & & & & \vdots \\ & & & & 0 \\ 0 & 0 & \dots & 0 & 1 \end{bmatrix}.$$

It is obvious that

$$U_{m+1}^* A_{m+1} U_{m+1} = U_{m+1}^* \begin{bmatrix} & & & a_{1m+1} \\ & A_m & & a_{2m+1} \\ & & & \vdots \\ a_{m+11} & a_{m+12} & \dots & a_{m+1m+1} \end{bmatrix} U_{m+1}$$

$$= \begin{bmatrix} a_1 & & \mathbf{0} & & b_1 \\ & a_2 & & & b_2 \\ & & \ddots & & \vdots \\ \mathbf{0} & & & a_m & b_m \\ \bar b_1 & \bar b_2 & \dots & \bar b_m & b_{m+1} \end{bmatrix},$$

where

$$\begin{pmatrix} b_1 \\ b_2 \\ \vdots \\ b_m \end{pmatrix} = U_{m+1}^* \begin{pmatrix} a_{1m+1} \\ a_{2m+1} \\ \vdots \\ a_{mm+1} \end{pmatrix}.$$

It is obvious that Theorems 5.1.1 and 5.1.2 remain valid for an arbitrary Hermitian $m\times m$-matrix A_m. Choosing U_m in a suitable way, we can assume that $a_1,\ a_2\ \dots, a_m$ are located on the main diagonal in a given order.

THEOREM 5.1.3. *The singular values $\sigma_1 \le \sigma_2 \le \dots \le \sigma_m$ of an $m\times m$-matrix A_m and the singular values $\delta_1 \le \delta_2 \le \dots \le \delta_m \le \delta_{m+1}$ of a one-sided bordering,* *the* $(m+1) \times (m+1)$*-matrix*

$$A_{m+1} = \begin{bmatrix} a_{11} & a_{12} & \dots & a_{1m} & a_{1\,m+1} \\ a_{21} & a_{22} & \dots & a_{2m} & a_{2\,m+1} \\ \dots & \dots & \dots & \dots & \dots \\ a_{m\,1} & a_{m\,2} & \dots & a_{m\,m} & a_{m\,m+1} \\ a_{m+1\,1} & a_{m+1\,2} & \dots & a_{m+1\,m} & a_{m+1\,m+1} \end{bmatrix}$$

$$\equiv \begin{bmatrix} & & & & a_{1\,m+1} \\ & A_m & & & a_{2\,m+1} \\ & & & & \dots \\ & & & & a_{m\,m+1} \\ 0 & 0 & 0 & 0 & a_{m+1\,m+1} \end{bmatrix},$$

satisfy the following alternation condition:

$$0 \le \delta_1 \le \sigma_1 \le \delta_2 \le \sigma_2 \le \dots \le \delta_m \le \sigma_m \le \delta_{m+1}. \tag{5.1.1}$$

The proof of Theorem 5.1.3 is given in §§5.2 and 5.3.

The next theorem shows that except for (5.1.1), there is no condition on the location of singular values of the matrices A_m and A_{m+1}.

THEOREM 5.1.4. *For any nonnegative numbers* $\delta_1, \delta_2, \dots, \delta_{m+1}$ *alternate with the singular values* $\sigma_1, \sigma_2, \dots, \sigma_m$ *of an* $m \times m$*-matrix* A_m*, i.e.,*

$$0 \le \delta_1 \le \sigma_1 \le \delta_2 \le \sigma_2 \le \dots \le \delta_m \le \sigma_m \le \delta_{m+1}, \tag{5.1.2}$$

there exists a matrix

$$A_{m+1} = \begin{bmatrix} & & & & a_{1\,m+1} \\ & A_m & & & a_{2\,m+1} \\ & & & & \dots \\ & & & & a_{m\,m+1} \\ 0 & 0 & 0 & 0 & a_{m+1\,m+1} \end{bmatrix}$$

for which $\delta_1, \delta_2, \dots, \delta_{m+1}$ *are the singular values.*

It suffices to prove Theorems 5.1.3 and 5.1.4 for a diagonal $m \times m$-matrix $\widetilde{A}_m$ with nonnegative diagonal elements $a_1, \dots, a_m$ which coincide with the singular values $\sigma_1, \sigma_2, \dots, \sigma_m$ of the $m \times m$-matrix A_m. Indeed, if the singular value decomposition of A_m has the form

$$A_m \equiv U_m \widetilde{A}_m V_m^* \quad (U_m^* U_m = V_m^* V_m = I_m),$$

$$\widetilde{A}_{m+1} = \begin{bmatrix} a_1 & & \mathbf{0} & & d_1 \\ & a_2 & & & d_2 \\ & & \ddots & & \vdots \\ \mathbf{0} & & & a_m & d_m \\ 0 & 0 & \dots & 0 & d_{m+1} \end{bmatrix},$$

then, using the unitary matrices

$$U_{m+1} = \begin{bmatrix} & & U_m & & 0 \\ & & & & 0 \\ & & & & \vdots \\ & & & & 0 \\ 0 & 0 & \dots & 0 & 1 \end{bmatrix}, \quad V_{m+1} = \begin{bmatrix} & & V_m & & 0 \\ & & & & 0 \\ & & & & \vdots \\ & & & & 0 \\ 0 & 0 & \dots & 0 & 1 \end{bmatrix},$$

from $\widetilde{A}_{m+1}$ we can construct the matrix

$$A_{m+1} = U_{m+1}\widetilde{A}_{m+1}V_{m+1}^* = \begin{bmatrix} & & A_m & & a_{1\,m+1} \\ & & & & a_{2\,m+1} \\ & & & & \dots \\ & & & & a_{m\,m+1} \\ 0 & 0 & 0 & 0 & a_{m+1\,m+1} \end{bmatrix}$$

with the same singular values as $\widetilde{A}_{m+1}$; moreover, A_{m+1} is a one-sided bordering of A_m. In addition, the last column of A_{m+1} is obtained from the last column of $\widetilde{A}_{m+1}$ by a unitary transformation:

$$\begin{pmatrix} a_{1\,m+1} \\ a_{2\,m+1} \\ \vdots \\ a_{m\,m+1} \\ a_{m+1\,m+1} \end{pmatrix} = U_{m+1} \begin{pmatrix} d_1 \\ d_2 \\ \vdots \\ d_m \\ d_{m+1} \end{pmatrix}.$$

Having proved Theorems 5.1.3 and 5.1.4 for a diagonal $m \times m$-matrix A_m, we may conclude that they remain valid in the general case. In view of this remark, together with the similar remark after Theorems 5.1.1 and 5.1.2, we can study only a diagonal $m \times m$-matrix A_m (see §§5.2 and 5.3).

5.2. The proof of the simplified alternation theorems

The case of nonzero bordering elements and different eigenvalues. The existence of a bordering in the case of strictly alternating sequences. The case of a diagonal matrix.

THEOREM 5.2.1 (simplified variant of Theorem 5.1.1). *Let $a_1 < a_2 < \cdots < a_m$ be any real numbers, $b_1, b_2, \dots, b_m$ nonzero numbers, b_{m+1} a real number. Consider the $(m+1) \times (m+1)$-matrix*

$$A_{m+1} = \begin{bmatrix} a_1 & & \mathbf{0} & & b_1 \\ & a_2 & & & b_2 \\ & & \ddots & & \vdots \\ \mathbf{0} & & & a_m & b_m \\ \bar{b}_1 & \bar{b}_2 & \dots & \bar{b}_m & b_{m+1} \end{bmatrix},$$

which borders the diagonal matrix

$$A_m = \begin{bmatrix} a_1 & & & 0 \\ & a_2 & & \\ & & \ddots & \\ 0 & & & a_m \end{bmatrix}$$

by $b_1, b_2, \ldots, b_m, b_{m+1}$. *Then the eigenvalues* $\lambda_1, \lambda_2, \ldots, \lambda_{m+1}$ *of* A_{m+1} *strictly alternate with* $a_1, a_2, \ldots, a_m$, *i.e.,*

$$\lambda_1 < a_1 < \lambda_2 < a_2 < \cdots < \lambda_m < a_m < \lambda_{m+1}.$$

PROOF. Expanding the determinant of the matrix $A_{m+1} - \lambda I_{m+1}$ in the form

$$\det(A_{m+1} - \lambda I_{m+1}) = \det \begin{bmatrix} a_1 - \lambda & & & 0 & b_1 \\ & a_2 - \lambda & & & b_2 \\ & & \ddots & & \vdots \\ 0 & & & a_m - \lambda & b_m \\ \bar{b}_1 & \bar{b}_2 & \ldots & \bar{b}_m & b_{m+1} - \lambda \end{bmatrix}$$

with respect to the last row (or the last column), we find that

$$\begin{aligned} \det[A_{m+1} - \lambda I_{m+1}] &= (b_{m+1} - \lambda) \prod_{k=1}^{m} (a_k - \lambda) - \sum_{j=1}^{m} |b_j|^2 \prod_{l \neq j, 1 \leq l \leq m} (a_l - \lambda) \\ &= \left[\prod_{j=1}^{m} (a_j - \lambda) \right] \left[b_{m+1} - \lambda - \sum_{j=1}^{m} \frac{|b_j|^2}{a_j - \lambda} \right]. \end{aligned}$$

The rational function

$$\varphi(\lambda) = b_{m+1} - \lambda - \sum_{j=1}^{m} \frac{|b_j|^2}{a_j - \lambda}$$

is continuous and monotonically decreasing on the real axis except the points $a_1, a_2, \ldots, a_m$ at which this function is not defined. In neighborhoods of the points $\lambda = a_j$, $j = 1, 2, \ldots, m$, the function $\varphi(\lambda)$ is unbounded and tends to $-\infty$ from the left and to $+\infty$ from the right, i.e.,

$$\lim_{\lambda \to a_j - 0} \varphi(\lambda) = -\infty, \qquad \lim_{\lambda \to a_j + 0} \varphi(\lambda) = +\infty.$$

Furthermore,

$$\lim_{\lambda \to -\infty} \varphi(\lambda) = +\infty, \qquad \lim_{\lambda \to +\infty} \varphi(\lambda) = -\infty.$$

Each of $m+1$ intervals $-\infty < \lambda < a_1$, $a_1 < \lambda < a_2$, $a_2 < \lambda < a_3$, $\cdots$, $a_{m-1} < \lambda < a_m$, $a_m < \lambda < +\infty$, contains one (and only one) interior point at which the function $\varphi(\lambda)$ changes the sign passing through the zero (see Figure 5.2.1). The number of all such points is equal to $m + 1$. We enumerate these points so that their numbers increase as λ increases. We obtain the sequence $\lambda_1, \lambda_2, \ldots, \lambda_{m+1}$, which strictly alternates with the sequence $a_1, a_2, \ldots, a_m$. All λ_j's are different,

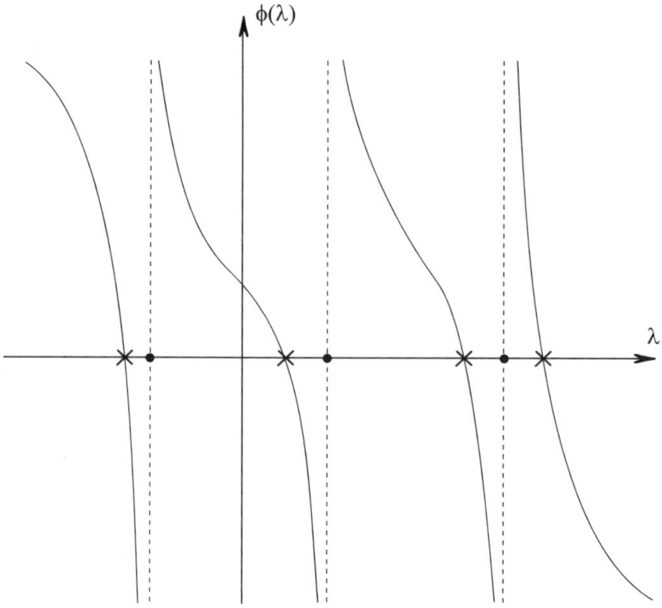

FIGURE 5.2.1. The graph of the function $\varphi(\lambda)$ for $m = 3$.

and the polynomial $\det(A_{m+1} - \lambda I_{m+1})$ vanishes at these points. Since the leading coefficient of the polynomial

$$\det(A_{m+1} - \lambda I_{m+1}) = (-1)^{m+1}\lambda^{m+1} + p_1\lambda^m + p_2\lambda^{m-1} + \cdots + p_0$$

differs from zero, it must have exactly $m+1$ roots, taking into account their multiplicity. Consequently, $\lambda_1, \lambda_2, \ldots, \lambda_{m+1}$ are all the roots of this polynomial. Moreover, they alternate with $a_1, a_2, \ldots, a_m$. □

THEOREM 5.2.2 (simplified variant of Theorem 5.1.2). *For any real numbers $a_1 < a_2 < \cdots < a_m$ and real numbers $\lambda_1, \lambda_2, \ldots, \lambda_{m+1}$ such that*

$$\lambda_1 < a_1 < \lambda_2 < a_2 < \cdots < \lambda_m < a_m < \lambda_{m+1},$$

there exist nonzero numbers $b_1, b_2, \ldots, b_m$ and a real number b_{m+1} such that $\lambda_1, \lambda_2, \ldots, \lambda_{m+1}$ are the eigenvalues of the $(m+1) \times (m+1)$-matrix

$$A_{m+1} = \begin{bmatrix} a_1 & & 0 & & b_1 \\ & a_2 & & & b_2 \\ & & \ddots & & \vdots \\ 0 & & & a_m & b_m \\ \bar{b}_1 & \bar{b}_2 & \ldots & \bar{b}_m & b_{m+1} \end{bmatrix},$$

which borders the diagonal matrix

$$A_m = \begin{bmatrix} a_1 & & & 0 \\ & a_2 & & \\ & & \ddots & \\ 0 & & & a_m \end{bmatrix}.$$

PROOF. We show how, by choosing b_j appropriately, to construct the matrix A_{m+1} with a given determinant

$$\det(A_{m+1} - \lambda I_{m+1}) = (\lambda_1 - \lambda)(\lambda_2 - \lambda)\dots(\lambda_{m+1} - \lambda).$$

Consider

$$\varphi(\lambda) = \frac{\det(A_{m+1} - \lambda I_{m+1})}{\det(A_m - \lambda I_m)} = \frac{(\lambda_1 - \lambda)(\lambda_2 - \lambda)\dots(\lambda_m - \lambda)(\lambda_{m+1} - \lambda)}{(a_1 - \lambda)(a_2 - \lambda)\dots(a_m - \lambda)}. \tag{5.2.1}$$

Represent $\varphi(\lambda)$ as the sum of irreducible fractions:

$$\varphi(\lambda) = -\lambda + f_{m+1} + \sum_{j=1}^{m} \frac{f_j}{a_j - \lambda}.$$

The coefficients $f_1, f_2, \dots, f_m$ can be found from the formulas

$$\begin{aligned} f_j &= \frac{\prod_{k=1}^{m+1} (\lambda_k - a_j)}{\prod_{k\neq j, 1\le k\le m} (a_k - a_j)} \\ &= \left[\frac{\lambda_1 - a_j}{a_1 - a_j}\,\frac{\lambda_2 - a_j}{a_2 - a_j}\cdots\frac{\lambda_{j-1} - a_j}{a_{j-1} - a_j}\right](\lambda_j - a_j)(\lambda_{m+1} - a_j) \\ &\quad\times \left[\frac{\lambda_{j+1} - a_j}{a_{j+1} - a_j}\,\frac{\lambda_{j+2} - a_j}{a_{j+2} - a_j}\cdots\frac{\lambda_m - a_j}{a_m - a_j}\right]. \end{aligned} \tag{5.2.2}$$

Then f_{m+1} can be found from the condition

$$\varphi(\mu) = -\mu + \sum_{j=1}^{m} \frac{f_j}{a_j - \mu} + f_{m+1},$$

where any point (not equal to $a_1, a_2, \dots, a_m$) can be taken for μ.

It is obvious that $f_1, f_2, \dots, f_m, f_{m+1}$ are real and strictly negative. Indeed, the alternation condition for λ_j and a_j imply that the expressions in square brackets in (5.2.2) are positive and the remaining two multipliers on the right-hand sides of (5.2.2) have different signs: $\lambda_{m+1} - a_j > 0$, whereas $\lambda_j - a_j < 0$. Setting

$$b_{m+1} = f_{m+1}, \quad b_j = e^{i\omega_j}\sqrt{-f_j} \neq 0 \quad (1 \le j \le m),$$

where ω_j are real, we can show that the function $\varphi(\lambda)$ is represented in the form

$$\varphi(\lambda) = -\lambda + b_{m+1} - \sum_{j=1}^{m} \frac{|b_j|^2}{a_j - \lambda}.$$

As was shown in the proof of Theorem 5.2.1, this means that $\varphi(\lambda)$ is the ratio of the characteristic determinant $\det(A_{m+1} - \lambda I)/\det(A_m - \lambda I)$ of A_{m+1} and the characteristic determinant of A_m. On the other hand,

$$\varphi(\lambda) = \prod_{j=1}^{m+1} (\lambda_j - \lambda)/\det(A_m - \lambda I_m).$$

Thereby, we get the rule for computing the elements $b_j, \bar{b}_j$ such that $\lambda_1, \lambda_2, \ldots, \lambda_{m+1}$ are the eigenvalues of A_{m+1}. □

From Theorems 5.2.1 and 5.2.2 we easily obtain the following theorem.

THEOREM 5.2.3 (simplified variant of Theorem 5.1.3). *Assume that the diagonal elements $a_1, a_2, \ldots, a_m$ of the matrix*

$$A_m = \begin{bmatrix} a_1 & & \text{\Large 0} \\ & a_2 & \\ & & \ddots & \\ \text{\Large 0} & & & a_m \end{bmatrix}$$

are strictly positive and distinct (it is obvious that $a_1, a_2, \ldots, a_m$ coincide with the singular values $\sigma_1, \sigma_2, \ldots, \sigma_m$ of the matrix A_m) and the elements $d_1, d_2, \ldots, d_{m+1}$ of the last column of the matrix

$$A_{m+1} = \begin{bmatrix} a_1 & & \text{\Large 0} & & d_1 \\ & a_2 & & & d_2 \\ & & \ddots & & \vdots \\ \text{\Large 0} & & & a_m & d_m \\ 0 & 0 & \ldots & 0 & d_{m+1} \end{bmatrix}$$

are nonzero. Then the singular values $\delta_1, \delta_2, \ldots, \delta_{m+1}$ of A_{m+1} are strictly positive and strictly alternate with the singular values $\sigma_1 = a_1, \sigma_2 = a_2, \ldots, \sigma_m = a_m$ of A_m:

$$0 < \delta_1 < \sigma_1 < \delta_2 < \cdots < \delta_m < \sigma_m < \delta_{m+1}. \tag{5.2.3}$$

THEOREM 5.2.4 (simplified variant of Theorem 5.1.4). *Let the singular values $\sigma_1 = a_1, \sigma_2 = a_2, \ldots, \sigma_m = a_m$ of the diagonal $m \times m$-matrix*

$$A_m = \begin{bmatrix} a_1 & & \text{\Large 0} \\ & a_2 & \\ & & \ddots & \\ \text{\Large 0} & & & a_m \end{bmatrix}$$

strictly alternate with positive numbers $\delta_1, \delta_2, \ldots, \delta_{m+1}$, i.e.,

$$0 < \delta_1 < \sigma_1 < \delta_2 < \cdots < \delta_m < \sigma_m < \delta_{m+1}. \tag{5.2.4}$$

Then there exists the matrix

$$A_{m+1} = \begin{bmatrix} a_1 & & \mathbf{0} & & d_1 \\ & a_2 & & & d_2 \\ & & \ddots & & \vdots \\ \mathbf{0} & & & a_m & d_m \\ 0 & 0 & \dots & 0 & d_{m+1} \end{bmatrix},$$

which borders the matrix A_m and has singular values $\delta_1, \delta_2, \dots, \delta_{m+1}$.

PROOF OF THEOREM 5.2.3. We note that $\delta_1^2, \delta_2^2, \dots, \delta_{m+1}^2$ are the eigenvalues of the Hermitian matrix

$$\begin{aligned} H &= A_{m+1}^* A_{m+1} \\ &= \begin{bmatrix} a_1 & & \mathbf{0} & & 0 \\ & a_2 & & & 0 \\ & & \ddots & & \vdots \\ \mathbf{0} & & & a_m & 0 \\ \bar{d}_1 & \bar{d}_2 & \dots & \bar{d}_m & d_{m+1} \end{bmatrix} \begin{bmatrix} a_1 & & \mathbf{0} & & d_1 \\ & a_2 & & & d_2 \\ & & \ddots & & \vdots \\ \mathbf{0} & & & a_m & d_m \\ 0 & 0 & \dots & 0 & d_{m+1} \end{bmatrix} \\ &= \begin{bmatrix} a_1^2 & & \mathbf{0} & & b_1 \\ & a_2^2 & & & b_2 \\ & & \ddots & & \vdots \\ \mathbf{0} & & & a_m^2 & b_m \\ \bar{b}_1 & \bar{b}_2 & \dots & \bar{b}_m & b_{m+1} \end{bmatrix} \end{aligned}$$

such that

$$b_j = a_j d_j, \quad j = 1, 2, \dots, m, \qquad b_{m+1} = \sum_{j=1}^{m+1} |d_j|^2.$$

Under the assumptions of the above theorems, the numbers $b_1, b_2, \dots, b_{m+1}$ differ from zero and the matrix H is nonsingular ($\det H = |\det A_{m+1}|^2 > 0$). By Theorem 5.2.1, we have the relation

$$0 < \delta_1^2 < \sigma_1^2 < \delta_2^2 < \cdots < \delta_m^2 < \sigma_m^2 < \delta_{m+1}^2,$$

which is equivalent to (5.2.3) since δ_j is nonnegative and σ_j is positive. □

PROOF OF THEOREM 5.2.4. Using Theorem 5.2.2, for given $a_1^2 = \sigma_1^2 > 0$, $a_2^2 = \sigma_2^2 > \sigma_1^2, \dots, a_m^2 = \sigma_m^2 > \sigma_{m-1}^2$ and the diagonal $m \times m$-matrix with diagonal elements $a_1^2, a_2^2, \dots, a_m^2$, we construct the matrix

$$H_{m+1} = \begin{bmatrix} a_1^2 & & & \mathbf{0} & b_1 \\ & a_2^2 & & & b_2 \\ & & \ddots & & \vdots \\ \mathbf{0} & & & a_m^2 & b_m \\ \bar{b}_1 & \bar{b}_2 & \dots & \bar{b}_m & b_{m+1} \end{bmatrix}$$

with eigenvalues $\delta_1^2, \delta_2^2, \dots, \delta_{m+1}^2$. This is possible in view of the assumption (5.2.4), which is equivalent to the inequalities

$$0 < \delta_1^2 < \sigma_1^2 < \delta_2^2 < \sigma_2^2 < \cdots < \delta_m^2 < \sigma_m^2 < \delta_{m+1}^2.$$

It is obvious that $\det H_{m+1} > 0$, i.e.,

$$a_1^2 a_2^2 \dots a_m^2 \left(b_{m+1} - \sum_{j=1}^{m} \frac{|b_j|^2}{a_j^2} \right) > 0.$$

We set

$$d_j = \frac{b_j}{a_j}, \quad j = 1, 2, \dots, m, \quad d_{m+1} = \sqrt{b_{m+1} - \sum_{j=1}^{m} \frac{|b_j|^2}{a_j^2}} > 0$$

and denote

$$A_{m+1} = \begin{bmatrix} a_1^2 & & & 0 & d_1 \\ & a_2^2 & & & d_2 \\ & & \ddots & & \vdots \\ 0 & & & a_m^m & d_m \\ 0 & 0 & \dots & 0 & d_{m+1} \end{bmatrix}.$$

We obtain the equality $A_{m+1}^* A_{m+1} = H_{m+1}$.

The eigenvalues $\delta_1^2, \delta_2^2, \dots, \delta_{m+1}^2$ of the Hermitian matrix H_{m+1} are squares of the singular values $\delta_1, \delta_2, \dots, \delta_{m+1}$ of the matrix A_{m+1}. □

Theorems 5.1.1–5.1.4 can be obtained from the simplified variants (Theorems 5.2.1–5.2.4) by continuity. The detailed proof is left to the reader. We recall that the formulations of the general theorems differ from the simplified variants in that the strict inequalities (5.2.3) and (5.2.4) are replaced with the inequalities (5.1.1), (5.1.2). As was mentioned in §5.1, the assumption that the matrix A_m is diagonal is not essential.

The formal deduction of Theorems 5.1.1–5.1.4 from their simplified variants (Theorems 5.2.1–5.2.4) will be given in the next section. We will not use arguments concerning the continuous dependence of eigenvalues and singular values on elements of a matrix. The presented proof is elementary, but rather cumbersome. The reader may omit it for the first reading.

5.3. The proof of the general alternation theorems

The case of a matrix with equal diagonal elements. The general case. The lemma about extended alternating sequences. The proof of theorems about the existence of a matrix that borders a given matrix in the case where some of singular values might coincide.

In this section, we show how to prove Theorems 5.1.1–5.1.4 on the basis of the simplified variants (Theorems 5.2.1–5.2.4) without continuity arguments. The proof is cumbersome and, at the first glance, is too tedious (moreover, the same assertions can be obtained by using limit arguments; in §5.2, we suggested that the reader verify this independently). However, we pay attention to some interesting details lost in the proofs based on the limit passage. For the first reading, this section can be omitted.

PROOF OF THEOREM 5.1.1 (general case). Let $a_1, a_2, \dots, a_m$ be the diagonal elements of a diagonal matrix A_m. Assume that some of them are equal. For definiteness, let

$$a_1 = a_2 = \cdots = a_q = a, \quad a_j \neq a \quad (q < j \leq m).$$

This order of $a_1, a_2, \dots, a_m$ can be always achieved by the choice of a unitary transformation

$$U_m^* A_m U_m = \begin{bmatrix} a_1 & & 0 \\ & a_2 & \\ & & \ddots \\ 0 & & & a_m \end{bmatrix}$$

at the preliminary stage of the diagonalization of the original Hermitian matrix A_m. We obtain the $(m+1) \times (m+1)$-matrix

$$A_{m+1} = \begin{bmatrix} a & & & & & & b_1 \\ & a & & & & 0 & b_2 \\ & & \ddots & & & & \vdots \\ & & & a & & & b_q \\ & & & & a_{q+1} & & b_{q+1} \\ & & & & & \ddots & \vdots \\ & 0 & & & & a_m & b_m \\ \bar{b}_1 & \bar{b}_2 & \dots & \bar{b}_q & \bar{b}_{q+1} & \dots & \bar{b}_m & b_{m+1} \end{bmatrix},$$

which borders the diagonal $m \times m$-matrix A_m. For any unitary $q \times q$-matrix V_q $(V_q^* V_q = I_q)$ the unitary matrix

$$A_{m+1}^{(1)} = \begin{bmatrix} V_q & 0 \\ 0 & I_{m+1-q} \end{bmatrix}^* A \begin{bmatrix} V_q & 0 \\ 0 & I_{m+1-q} \end{bmatrix}$$

is similar to A_{m+1}, has the same eigenvalues as A_{m+1}, and takes the form

$$A_{m+1}^{(1)} = \begin{bmatrix} a & & & & & & b'_1 \\ & a & & & & 0 & b'_2 \\ & & \ddots & & & & \vdots \\ & & & a & & & b'_q \\ & & & & a_{q+1} & & b_{q+1} \\ & & & & & \ddots & \vdots \\ & 0 & & & & a_m & b_m \\ \bar{b}'_1 & \bar{b}'_2 & \dots & \bar{b}'_q & \bar{b}_{q+1} & \dots & \bar{b}_m & b_{m+1} \end{bmatrix},$$

where

$$\begin{pmatrix} b_1' \\ b_2' \\ \vdots \\ b_q' \end{pmatrix} = V_q^* \begin{pmatrix} b_1 \\ b_2 \\ \vdots \\ b_q \end{pmatrix}.$$

As we already know (see §2.3), it is possible to choose a matrix V_q that polarizes the vector $(b_1, b_2, \ldots, b_q)^T$, i.e., among the components $b_1', b_2', \ldots, b_q'$, only b_q' may be nonzero. Hence in $A_{m+1}^{(1)}$ we can take the diagonal $(q-1)\times(q-1)$-block and represent $A_{m+1}^{(1)}$ in the form

$$A_{m+1}^{(1)} = \left[\begin{array}{c|c} \boxed{\begin{array}{cccc} a_1 & & 0 & \\ & a_2 & & \\ & & \ddots & \\ 0 & & & a_{q-1} \end{array}} & 0 \\ 0 & \boxed{\begin{array}{ccccc} a_q & & & 0 & b_q' \\ & a_{q+1} & & & b_{q+1}' \\ & & \ddots & & \vdots \\ 0 & & & a_m & b_m \\ \overline{b}_q' & \overline{b}_q & \cdots & \overline{b}_m & b_{m+1} \end{array}} \end{array}\right].$$

If all $b_1, b_2, \ldots, b_q$ are zero, then A_{m+1} has the same form as $A_{m+1}^{(1)}$ but with the $q \times q$-block instead of the $(q-1)\times(q-1)$-block. In this case, it is convenient to replace q by $q-1$ and assume that $A_{m+1} = A_{m+1}^{(1)}$.

It can happen that some of the elements $a_q, a_{q+1}, \ldots, a_m$ are equal or for some of the elements $a_q, a_{q+1}, \ldots, a_m$ the corresponding $b_q, b_{q+1}, \ldots, b_m$ are zero. Then, applying unitary similar transformations several times, we can reduce A to the form

$$\widetilde{A}_{m+1} = \left[\begin{array}{c|c} \boxed{\begin{array}{cccc} a_1 & & 0 & \\ & a_2 & & \\ & & \ddots & \\ 0 & & & a_r \end{array}} & 0 \\ 0 & \boxed{\begin{array}{ccccc} a_{r+1} & & & 0 & b_{r+1} \\ & a_{r+2} & & & b_{r+2} \\ & & \ddots & & \vdots \\ 0 & & & a_m & b_m \\ \overline{b}_{r+1}' & \overline{b}_{r+2} & \cdots & \overline{b}_m & b_{m+1} \end{array}} \end{array}\right];$$

here a_j are diagonal elements of the original $(m+1)\times(m+1)$-matrix A_{m+1} renumbered so that

$$a_{r+1} < a_{r+2} < \cdots < a_{m-1} < a_m,$$
$$b_{r+1} \neq 0, b_{r+2} \neq 0, \ldots, b_m \neq 0, \quad \operatorname{Im} b_{m+1} = 0.$$

Some of $a_1, a_2, \ldots, a_r$ can coincide or be equal to some of a_j with $j \geq r+1$. The eigenvalues of the $(m+1)\times(m+1)$-matrix $\widetilde{A}_{m+1}$ (recall that they coincide with the eigenvalues of the original $(m+1)\times(m+1)$-matrix A_{m+1}) are $a_1, a_2, \ldots, a_r$ and the eigenvalues $\widetilde{\lambda}_{r+1}, \widetilde{\lambda}_{r+2}, \ldots, \widetilde{\lambda}_{m+1}$ of the right lower $(m-r)\times(m-r)$-block of A_{m+1}. By Theorem 5.2.1, all $\widetilde{\lambda}_{r+1}, \widetilde{\lambda}_{r+2}, \ldots, \widetilde{\lambda}_{m+1}$ are different and alternate with $a_{r+1}, a_{r+2}, \ldots, a_m$ under a suitable order, i.e.,

$$\widetilde{\lambda}_{r+1} < a_{r+1} < \widetilde{\lambda}_{r+2} < a_{r+2} < \cdots < \widetilde{\lambda}_m < a_m < \widetilde{\lambda}_{m+1},$$
$$\widetilde{\lambda}_1 = a_1, \widetilde{\lambda}_2 = a_2, \ldots, \widetilde{\lambda}_r = a_r.$$

If we add several pairs of equal numbers $\widetilde{\lambda}_j$, a_j ($\widetilde{\lambda}_j = a_j$) to the alternating sequences $\widetilde{\lambda}_{r+1}, \ldots, \widetilde{\lambda}_m$ and $a_{r+1}, \ldots, a_{m+1}$, then the extended sequences also alternate after a suitable renumbering of elements. It suffices to justify this assertion in the case of one pair. Lemma 5.3.1 provides this justification and completes the proof of Theorem 5.1.1. □

LEMMA 5.3.1 (extended alternating sequences). *Let*

$$x_1 \leq y_1 \leq x_2 \leq y_2 \leq \cdots \leq y_k \leq x_{k+1}, \qquad x = y.$$

Including x in the sequence $x_1, x_2, \ldots, x_k$ and y in the sequence $y_1, y_2, \ldots, y_k$ and renumbering in a suitable way, we obtain alternating sequences.

PROOF. Consider all possible variants of the location of $x = y$ with respect to the elements of the initial sequences.

Case 1. $x = y \leq x_1$.
Case 2. $y_j \leq x = y \leq x_{j+1}$, $1 \leq j \leq k$.
Case 3. $x_j \leq y = x \leq y_j$, $1 \leq j \leq k$.
Case 4. $y_{k+1} \leq x = y$.

If $x = y$ can be inserted in several ways, we choose any of them. Extended sequences must be ordered as follows.

Case 1. $x, x_1, x_2, \ldots, x_k, x_{k+1}$; $y, y_1, y_2, \ldots, y_k$.
Case 2. $x_1, \ldots, x_j, x, x_{j+1}, \ldots, x_{k+1}$; $y_1, \ldots, y_j, y, y_{j+1}, \ldots, y_k$.
Case 3. $x_1, \ldots, x_j, x, x_{j+1}, \ldots, x_{k+1}$; $y_1, \ldots, y_{j-1}, y, y_j, \ldots, y_k$.
Case 4. $x_1, \ldots, x_k, x_{k+1}, x$; $y_1, \ldots, y_k, y$.

After a natural renumbering of the sequences, the assertion that the sequences are alternating becomes obvious. □

Later in this volume, we need an assertion that is in a sense converse to Lemma 5.3.1.

LEMMA 5.3.2 (about reduced alternating sequences). *Let*

$$x_1 \leq y_1 \leq x_2 \leq y_2 \leq \cdots \leq y_k \leq x_{k+1}$$

and $y_j = x_j$ or $y_j = x_{j+1}$ for some j. Eliminating the coinciding pair from the sequences

$$x_1, x_2, \ldots, x_{k+1}; y_1, y_2, \ldots, y_k$$

and re-enumerating in a natural way, we again obtain alternating sequences.

The proof of this lemma is obvious.

Eliminating the pairs of equal elements from the sequences $x_1, x_2, \ldots, x_{k+1}$ and $y_1, y_2, \ldots, y_k$, we obtain strictly alternating sequences. This fact is used in the proof of Theorem 5.1.2.

PROOF OF THEOREM 5.1.2. The diagonal elements $a_1, a_2, \ldots, a_m$ of the $m \times m$-matrix A_m alternate with the eigenvalues $\lambda_1, \lambda_2, \ldots, \lambda_{m+1}$ of the $(m+1)\times(m+1)$-matrix A_{m+1} constructed from λ_j, $j = 1, \ldots, m+1$. First we take the sequences so that the elements a_j and λ_j coincide for $j = 1, 2, \ldots, r$ and must be eliminated. The remaining elements alternate strictly. We take them so that

$$\lambda_{r+1} < a_{r+1} < \lambda_{r+2} < a_{r+2} < \cdots < \lambda_m < a_m < \lambda_{m+1} \tag{5.3.1}$$

and $\lambda_1 = a_1, \lambda_2 = a_2, \ldots, \lambda_r = a_r$. In fact, it is not necessary to eliminate λ_j and a_j. We only need a suitable numbering of these sequences. We will look for an $(m+1) \times (m+1)$-matrix A_{m+1} with eigenvalues $\lambda_1, \lambda_2, \ldots, \lambda_{m+1}$ in the form

$$A_{m+1} = \left[\begin{array}{cc} \begin{array}{|cccc|} \hline a_1 & & 0 & \\ & a_2 & & \\ & & \ddots & \\ 0 & & & a_r \\ \hline \end{array} & 0 \\ 0 & \begin{array}{|ccccc|} \hline a_{r+1} & & & 0 & b_{r+1} \\ & a_{r+2} & & & b_{r+2} \\ & & \ddots & & \vdots \\ 0 & & & a_m & b_m \\ \bar{b}_{r+1} & \bar{b}_{r+2} & \cdots & \bar{b}_m & b_{m+1} \\ \hline \end{array} \end{array}\right],$$

i.e., we assume that $b_1, b_2, \ldots, b_r$ are zero. The existence of such a matrix with given $\lambda_{r+1}, \lambda_{r+2}, \ldots, \lambda_m$ satisfying (5.3.1) follows from Theorem 5.2.2. □

The proof of Theorem 5.1.3 is based on Theorem 5.1.1 and almost literally repeats the arguments used in deducing Theorem 5.2.3 from Theorem 5.2.1 (see §5.2). The only difference is that we do not need the conditions $0 \neq \sigma_i = a_i \neq a_k = \sigma_k$, $d_j \neq 0$, $b_j \neq 0$, because these conditions were used only in the simplified variant (see Theorem 5.2.1). Since Theorem 5.1.1 does not contain these conditions, we cannot guarantee that the singular values of A_{m+1} are nonzero and alternate with the singular values of A_m. We recall that in Theorem 5.1.3 the inequalities are nonstrict, $0 \leq \delta_1 \leq \sigma_1 \leq \delta_2 \leq \sigma_2 \leq \cdots \leq \delta_m \leq \sigma_m \leq \delta_{m+1}$, whereas the assumptions of Theorem 5.2.3 contains the strict inequalities $0 < \delta_1 < \sigma_1 < \delta_2 < \cdots < \delta_m < \sigma_m < \delta_{m+1}$.

Theorem 5.1.4 is proved by the same arguments as in the proof of Theorem 5.1.2. Let the diagonal elements $a_1^2, a_2^2, \dots, a_m^2$ of the Hermitian matrix

$$H = \left[\begin{array}{c|c} \begin{matrix} a_1^2 & & & 0 \\ & a_2^2 & & \\ & & \ddots & \\ 0 & & & a_r^2 \end{matrix} & 0 \\ \hline 0 & \begin{matrix} a_{r+1}^2 & & & 0 & b_{r+1} \\ & a_{r+2}^2 & & & b_{r+2} \\ & & \ddots & & \vdots \\ 0 & & & a_m^2 & b_m \\ \overline{b}_{r+1} & \overline{b}_{r+2} & \cdots & \overline{b}_m & b_{m+1} \end{matrix} \end{array}\right]$$

coincide with $\sigma_1^2, \sigma_2^2, \dots, \sigma_m^2$. Assume that they are ordered somehow and alternate with given $\delta_1^2, \delta_2^2, \dots, \delta_{m+1}^2$, i.e.,

$$0 \le \delta_1^2 \le \sigma_1^2 \le \delta_2^2 \le \cdots \le \delta_m^2 \le \sigma_m^2 \le \delta_{m+1}^2.$$

The proof of Theorem 5.2.2 shows that σ_j^2 can be ordered so that $a_1^2, a_2^2, \dots, a_r^2$ coincide with some δ_j^2's; moreover, there is no δ_j^2 that is equal to any element of the sequence $a_{r+1}^2, a_{r+2}^2, \dots, a_m^2$. In addition $a_{r+1}^2, a_{r+2}^2, \dots, a_m^2$ are strictly positive and different. As was shown in the proof of Theorem 5.2.2, the bordering $b_1 = b_2 = \cdots = b_r = 0, b_{r+1}, \dots, b_m, b_{m+1}$ can be chosen so that the Hermitian matrix H has the eigenvalues $\delta_1^2, \delta_2^2, \dots, \delta_m^2, \delta_{m+1}^2$. It is obvious that for such a choice of b_j the determinant

$$\det \begin{bmatrix} a_{r+1}^2 & & & 0 & b_{r+1} \\ & a_{r+2}^2 & & & b_{r+2} \\ & & \ddots & & \vdots \\ 0 & & & a_m^2 & b_m \\ \overline{b}_{r+1} & \overline{b}_{r+2} & \cdots & \overline{b}_m & b_{m+1} \end{bmatrix} = \left[b_{m+1} - \sum_{j=r+1}^{m} \frac{|b_j|^2}{a_j^2} \right] a_{r+1}^2 a_{r+2}^2 \cdots a_m^2$$

is nonnegative (it coincides with the product of several nonnegative δ_j^2's). Therefore, we have the representation

$$b_{m+1} = d_{m+1} + \sum_{j=r+1}^{m} \frac{|b_j|^2}{a_j^2}, \qquad d_{m+1}^2 \ge 0, d_{m+1} \ge 0.$$

Setting

$$d_j = \begin{cases} 0 & \text{for } 1 \le j \le r, \\ b_j / a_j & \text{for } r+1 \le j \le m \end{cases}$$

and introducing the notation

$$A_{m+1} = \begin{bmatrix} a_1 & & & & & & & d_1 \\ & a_2 & & & \mathbf{0} & & & d_2 \\ & & \ddots & & & & & \vdots \\ & & & a_r & & & & \vdots \\ & & & & a_{r+1} & & & \vdots \\ & & & & & \ddots & & \vdots \\ & \mathbf{0} & & & & & a_m & d_m \\ & & & & & & & d_{m+1} \end{bmatrix},$$

we obtain the equality

$$A^*_{m+1} A_{m+1} = H.$$

We have shown that for a diagonal matrix with diagonal elements $\sigma_1, \sigma_2, \ldots, \sigma_m$ and any sequence of nonnegative numbers $\delta_1, \delta_2, \ldots, \delta_{m+1}$ that alternate with $\sigma_1, \sigma_2, \ldots, \sigma_m$, i.e., satisfy the conditions

$$0 \le \delta_1 \le \sigma_1 \le \delta_2 \le \cdots \le \delta_m \le \sigma_m \le \delta_{m+1},$$

we can choose a matrix A_{m+1} such that it borders the diagonal matrix and the eigenvalues of $A^*_{m+1} A_{m+1}$ are δ_j^2's. In this case, $\delta_1, \delta_2, \ldots, \delta_{m+1}$ are singular values of A_{m+1}.

5.4. Corollaries of alternation theorems

Some variants of the alternation theorems. Alternation of singular values of a matrix and singular values of its submatrix. An estimate for singular values of the product of square matrices.

We give several variants of Theorem 5.1.3 proved in §5.3. They are useful in applications.

THEOREM 5.4.1 (Version 1). *Let an $N \times (N+1)$-matrix B be obtained from an $N \times N$-matrix A_N by adding from the right a column of N elements. Then the singular values $\sigma_1(B), \sigma_2(B), \ldots, \sigma_N(B)$ of B alternate with the singular values $\sigma_1(A), \sigma_2(A), \ldots, \sigma_N(A)$ of A as follows:*

$$\sigma_N(B) \ge \sigma_N(A_N) \ge \sigma_{N-1}(B) \ge \cdots \ge \sigma_1(B) \ge \sigma_1(A_N).$$

PROOF. Indeed, the singular values

$$\sigma_1(B), \sigma_2(B), \sigma_3(B), \ldots, \sigma_N(B)$$

of the $N \times (N+1)$-matrix B coincide with the singular values

$$\sigma_1(A_{N+1}), \sigma_2(A_{N+1}), \ldots, \sigma_N(A_{N+1}), \sigma_{N+1}(A_{N+1})$$

of the $(N+1) \times (N+1)$-matrix A_{N+1}, which differs from B by the row of $N+1$ zeros added from below (it is obvious that $\sigma_1(A_{N+1}) = 0$). Therefore, we can refer to the assertion proved in Theorem 5.1.3. □

THEOREM 5.4.2 (Version 2). *Let $M < N$ and let an $M \times (N+1)$-matrix B be obtained from an $M \times N$-matrix A by adding at the right a column of M elements. Then the singular values $\sigma_1(B), \sigma_2(B), \dots, \sigma_M(B)$ of B alternate with the singular values $\sigma_1(A), \sigma_2(A), \dots, \sigma_M(A)$ of A as follows:*

$$\sigma_M(B) \geq \sigma_M(A) \geq \sigma_{M-1}(B) \geq \dots \geq \sigma_1(B) \geq \sigma_1(A) \geq 0. \tag{5.4.1}$$

PROOF. The $M \times N$-matrix A and the $M \times (N+1)$-matrix B can be transformed to the $N \times N$-matrix $\widetilde{A}$ and the $N \times (N+1)$-matrix $\widetilde{B}$, respectively, by adding the zero rows from below. Then the sets of singular values of the matrices A and B are complemented by the corresponding number of zeros, so that

$$\sigma_N(\widetilde{A}) = \sigma_M(A), \quad \sigma_{N-1}(A) = \sigma_{M-1}(A), \dots, \sigma_{N-M+1}(\widetilde{A}) = \sigma_1(A),$$
$$\sigma_{N-M}(\widetilde{A}) = \sigma_{N-M-1}(\widetilde{A}) = \dots = \sigma_1(\widetilde{A}) = 0,$$
$$\sigma_N(\widetilde{B}) = \sigma_M(B), \sigma_{N-1}(\widetilde{B}) = \sigma_{M-1}(B), \dots, \sigma_{N-M}(\widetilde{B}) = \sigma_1(B),$$
$$\sigma_{N-M}(\widetilde{B}) = \sigma_{N-M-1}(\widetilde{B}) = \dots = \sigma_1(\widetilde{B}) = 0.$$

Since $\widetilde{A}$ is a square matrix, we can apply Theorem 5.4.1. We obtain

$$\sigma_N(\widetilde{B}) \geq \sigma_N(\widetilde{A}) \geq \sigma_{N-1}(\widetilde{B}) \geq \sigma_{N-1}(\widetilde{A}) \geq \dots \geq \sigma_1(\widetilde{B}) \geq \sigma_1(\widetilde{A}). \tag{5.4.2}$$

The alternation condition (5.4.2) can be rewritten in the form (5.4.1) if we use the rule that gives the correspondence between the singular values of the matrices $\widetilde{A}$, A and $\widetilde{B}$, B. □

THEOREM 5.4.3 (Version 3). *Let $M > N$. Let an $M \times (N+1)$-matrix B be obtained from an $M \times N$-matrix A by adding from the right by a column of M elements. Then the singular values $\sigma_1(B), \sigma_2(B), \dots, \sigma_N(B)$ of B alternate with the singular values $\sigma_1(A), \sigma_2(A), \dots, \sigma_N(A)$ of A as follows:*

$$\sigma_N(A) \geq \sigma_N(B) \geq \sigma_{N-1}(A) \geq \dots \geq \sigma_1(B) \geq \sigma_1(A) \geq 0. \tag{5.4.3}$$

PROOF. We can find an orthogonal transformation $\mathcal{P}$ ($\mathcal{P} \times \mathcal{P} = I_M$) such that

$$\mathcal{P}A = \begin{bmatrix} A_0 \\ 0 \end{bmatrix}, \quad \mathcal{P}a = \begin{pmatrix} a_0 \\ \alpha \\ 0 \end{pmatrix} \begin{matrix} \} \, N \text{ components} \\ \} \, 1 \text{ component} \\ \} \, M-N-1 \text{ components,} \end{matrix}$$

where a denotes the column which is added to A, i.e., $B = [A \vdots a]$. It is clear that

$$\mathcal{P}B = \begin{bmatrix} A_0 & a_0 \\ 0 \dots 0 & \alpha \\ 0 & 0 \end{bmatrix} \begin{matrix} \} \, N \text{ rows} \\ \} \, 1 \text{ row} \\ \} \, M-N-1 \text{ rows.} \end{matrix}$$

The singular values of A coincide with the singular values of A_0. The singular values of the matrix B coincide with the corresponding singular values of the matrix

$$\begin{bmatrix} A_0 & a_0 \\ 0 \dots 0 & \alpha \end{bmatrix} \begin{matrix} \} \, N \text{ rows} \\ \} \, 1 \text{ row,} \end{matrix}$$

which borders the matrix A_0. By Theorem 5.1.3, the singular values of A_0 alternate with the singular values of a matrix that borders A_0. Since $\sigma_j(\mathcal{P}B) = \sigma_j(B)$ and $\sigma_j(\mathcal{P}A) = \sigma_j(A)$, we obtain the required condition (5.4.3). □

Thus, we have proved the following proposition.

PROPOSITION 5.4.1. *Adding from the right a column to any matrix (rectangular or square) we obtain a matrix whose singular values alternate with the singular values of the original matrix. The same assertion is true if we add a row from below (this case is reduced to the first assertion by transposition).*

Let a_{ij}, $1 \le i \le M$, $1 \le j \le N$, be elements of a matrix A and let $\widetilde{A}$ be an $M_1 \times N_1$-submatrix ($1 \le M_1 < M$, $1 \le N_1 < N$) composed of $a_{i_k j_l}$, $1 \le i_{k_1} \le i_{k_2} \le \cdots \le i_{k_{M_1}} \le M$, $1 \le j_{l_1} \le j_{l_2} \le \cdots \le j_{l_{N_1}} \le N$. The singular values $\sigma_j(A)$ and $\sigma_i(\widetilde{A})$ do not change under any permutations of rows and columns. Therefore we can assume that all elements of the matrix $\widetilde{A}$ are located in the upper left corner of the matrix A, i.e., $i_{k_1} = 1$, $i_{k_2} = 2, \ldots, i_{k_{M_1}} = M_1$, $j_{l_1} = 1$, $j_{l_2} = 2, \ldots, j_{l_{N_1}} = N_1$. Then we can pass from the matrix $\widetilde{A}$ to the matrix A by adding a column or a row. Let each of matrices

$$A^{(1)},\ A^{(2)},\ A^{(3)}, \ldots, A^{(n)} \quad (\widetilde{A} = A^{(0)} = A^{(1)},\ A^{(n)} = A),$$

be obtained from the previous matrix by adding a row or a column. Then the singular values of neighboring matrices alternate.

This means that $j+1$ singular values of the matrices A and $\widetilde{A}$

$$\sigma_L(A), \sigma_{L-1}(A), \ldots, \sigma_{L-j}(A),$$
$$\sigma_l(\widetilde{A}), \sigma_{l-1}(\widetilde{A}), \ldots, \sigma_{l-j}(\widetilde{A}),$$

where $j+1 \le l$, $l = \min\{M_1, N_1\}$, $L = \min\{M, N\}$, $M_1 \le M$, $N_1 \le N$, satisfy the conditions

$$\sigma_L(A) \ge \sigma_l(\widetilde{A}), \sigma_{L-1}(A) \ge \sigma_{l-1}(\widetilde{A}), \ldots, \sigma_{L-j}(A) \ge \sigma_{l-j}(\widetilde{A}).$$

We have just proved the following theorem.

THEOREM 5.4.4. *The singular values of an $M \times N$-matrix A and the singular values of the $M_1 \times N_1$-submatrix $\widetilde{A}$ satisfy the inequalities ($p > 0$)*

$$[\sigma_L(A)]^p + [\sigma_{L-1}(A)]^p + \cdots + [\sigma_{L-j}(A)]^p \ge [\sigma_l(\widetilde{A})]^p + [\sigma_{l-1}(\widetilde{A})]^p + \cdots + [\sigma_{l-j}(\widetilde{A})]^p,$$
$$\sigma_L(A)\sigma_{L-1}(A)\ldots\sigma_{L-j}(A) \ge \sigma_l(\widetilde{A})\sigma_{l-1}(\widetilde{A})\ldots\sigma_{l-j}(\widetilde{A})$$

for $0 \le j \le l-1$, $L = \min\{M, N\}$, $l = \min\{M_1, N_1\}$, where $M_1 \le M$, $N_1 \le N$.

Let A be a square $N \times N$-matrix. Let P and Q be unitary $N \times N$-matrices ($P^*P = I_N$ and $Q^*Q = I_N$). As is known, the singular values of A coincide with the singular values of P^*AQ. Take the upper principal $n \times n$-minor ($1 \le n < N$) D of the matrix P^*AQ.

By Theorem 5.4.4

$$\sigma_n(D) \le \sigma_N(A),$$
$$\sigma_n(D)\sigma_{n-1}(D)\ldots\sigma_{n-j+1}(D) \le \sigma_N(A)\sigma_{N-1}(A)\ldots\sigma_{N-j+1}(A),$$
$$[\sigma_n(D)]^p + [\sigma_{n-1}(D)]^p + \cdots + [\sigma_{n-j+1}(D)]^p$$
$$\le [\sigma_N(A)]^p + [\sigma_{N-1}(A)]^p + \cdots + [\sigma_{N-j+1}(A)]^p,$$

where $j = 1, 2, 3, \ldots, n$. To compute elements of D, we use only the first n columns of the unitary matrices P and Q. Hereinafter, we also use the notation P and Q for the rectangular submatrices of the original matrices P and Q that are composed of those columns that $P^*P = Q^*Q = I_n$. We can obtain unitary matrices by adding $N - n$ columns to arbitrary matrices P and Q whose columns satisfy these orthonormality conditions. Therefore,

$$(5.4.4)\qquad \begin{gathered} \sum_{j=1}^{k} \sigma_{n-j+1}^{p}(P^*AQ) \le \sum_{j=1}^{k} \sigma_{N-j+1}^{p}(A), \quad k = 1, 2, \dots n, \\ \prod_{j=1}^{k} \sigma_{n-j+1}(P^*AQ) \le \prod_{j=1}^{k} \sigma_{N-j+1}(A). \end{gathered}$$

In particular, replacing A with A^*A in the last inequality and assuming that $P = Q$ and $k = n$, we conclude that

$$\prod_{j=1}^{n} \sigma_{n-j+1}(Q^*A^*AQ) \le \prod_{j=1}^{n} \sigma_{N-j+1}(A^*A) = \prod_{j=1}^{n} \sigma_{N-j+1}^2(A).$$

The left-hand side of this estimate is the determinant of Q^*A^*AQ (equal to the product of the singular values). Thus, $n \le N$,

$$\det(Q^*A^*AQ) \le \left[\prod_{j=1}^{n} \sigma_{N-j+1}(A)\right]^2.$$

An arbitrary $N \times n$-matrix X ($n \le N$) can be represented as the product of a matrix Q with orthonormal columns and a square $n \times n$-matrix R (see the QR decomposition in §3.1):

$$X = QR \quad (Q^*Q = I_n);$$

in addition,

$$X^*A^*AX = R^*Q^*A^*AQR, \quad X^*X = R^*R,$$
$$\det(X^*A^*AX) = (\det R)^2 \det(Q^*A^*AQ) = \det(Q^*A^*AQ)\det(X^*X).$$

Consequently,

$$\det(X^*A^*AX) \le \left[\prod_{j=1}^{n} \sigma_{N-j+1}(A)\right]^2 \det(X^*X).$$

Let $X = BY$, where B is an $N \times N$-matrix. Then

$$\begin{aligned} \det(Y^*B^*A^*ABY) &\le \left[\prod_{j=1}^{n} \sigma_{N-j+1}(A)\right]^2 \det(Y^*B^*BY) \\ &\le \left[\prod_{j=1}^{n} \sigma_{N-j+1}(A)\right]^2 \left[\prod_{j=1}^{n} \sigma_{N-j+1}(B)\right]^2 \det(Y^*Y). \end{aligned}$$

We take a matrix Y such that its columns are orthonormal and coincide with the singular vectors of the product AB corresponding to the n largest singular values. Moreover,

$$Y^*Y = I_n, \quad \det(Y^*Y) = 1,$$

$$ABY = W \begin{bmatrix} \sigma_N(AB) & & & 0 \\ & \sigma_{N-1}(AB) & & \\ & & \ddots & \\ 0 & & & \sigma_{N-n+1}(AB) \end{bmatrix}, \quad W^*W = I_n,$$

$$Y^*B^*A^*ABY = \begin{bmatrix} \sigma_N^2(AB) & & & 0 \\ & \sigma_{N-1}^2(AB) & & \\ & & \ddots & \\ 0 & & & \sigma_{N-n+1}^2(AB) \end{bmatrix},$$

$$\det\,(Y^*B^*A^*ABY) = \prod_{j=1}^{n} \sigma_{N-j+1}^2(AB).$$

Thus, we have proved the following theorem.

THEOREM 5.4.5. *If A and B are square $N \times N$-matrices and $1 \le n \le N$, then*

$$\prod_{j=1}^{n} \sigma_{N-j+1}(AB) \le \prod_{j=1}^{n} \sigma_{N-j+1}(A) \prod_{j=1}^{n} \sigma_{N-j+1}(B). \tag{5.4.5}$$

5.5. Useful inequalities for convex functions

Integral representations of convex functions. Some properties of non-negative piecewise linear functions and their consequences.

First we impose conditions on convex functions in the inequalities studied below.

Let a function $\Phi(x)$ be twice continuously differentiable for $x > x_0$. Suppose that the second order derivative $\Phi''(x)$ is strictly positive for $x > x_0$. We also suppose that $\Phi(x)$, $\Phi'(x)$, and $x\Phi''(x)$ tend to zero as $x \to x_0$. It is natural to assume that $\Phi(x_0) = 0$. In view of the relation

$$\begin{aligned}\Phi(x) - \Phi(y) - (x-y)\Phi'(y) &= \int_y^x (x-\xi)\Phi''(\xi)\,d\xi \\ &= (x-\xi)\Phi'(\xi)\Big|_{\xi=y}^{\xi=x} + \int_y^x \Phi'(\xi)\,d\xi, \quad x > y > x_0,\end{aligned}$$

we have the equality

$$\Phi(x) = \int_{x_0}^{x} (x-\xi)\Phi''(\xi)\,d\xi. \tag{5.5.1}$$

The integral on the right-hand side of (5.5.1) is convergent.

Hereinafter, we assume that either $x_0 = -\infty$ or $x_0 = 0$. For admissible functions $\Phi(x)$ we take

$$\begin{gathered}\Phi(x) = e^{kx}, \quad k > 0,\ x > x_0 = -\infty, \\ \Phi(x) = \log(1 + re^{x}), \quad x > x_0 = -\infty,\ r > 0, \\ \Phi(x) = x^p, \quad p > 1,\ x \geq x_0 = 0.\end{gathered} \tag{5.5.2}$$

PROBLEM 5.5.1. Verify that the functions (5.5.2) satisfy the required conditions.

Using the notation

$$(x-\xi)_+ = \begin{cases} x - \xi & \text{if } \xi < x, \\ 0 & \text{if } \xi \geq x, \end{cases}$$

we rewrite (5.5.1) in the following more convenient form:

$$\Phi(x) = \int_{x_0}^{\infty} (x-\xi)_+ \Phi''(\xi)\, d\xi. \tag{5.5.3}$$

THEOREM 5.5.1. *If*

$$a_N \geq a_{N-1} \geq \cdots \geq a_1 \geq x_0, \quad b_n \geq b_{N-1} \geq \cdots \geq b_1 \geq x_0,$$
$$\sum_{j=k}^{N} a_j \leq \sum_{j=k}^{N} b_j, \quad k = 1, 2, \ldots, N,$$

then

$$\sum_{j=k}^{N} \Phi(a_j) \leq \sum_{j=k}^{N} \Phi(b_j), \quad k = 1, 2, \ldots, N; \tag{5.5.4}$$

moreover, the equality

$$\sum_{j=1}^{N} \Phi(a_j) = \sum_{j=1}^{N} \Phi(b_j)$$

in (5.5.4) *holds if and only if* $a_j = b_j$ *for all* $j = 1, 2, \ldots, N$.

PROOF. It is convenient to use the notation

$$A_k(\xi) = \sum_{j=k}^{N} (a_j - \xi)_+, \quad B_k(\xi) = \sum_{j=k}^{N} (b_j - \xi)_+$$

in the representations

$$\sum_{j=k}^{N} \Phi(a_j) = \int_{x_0}^{\infty} A_k(\xi)\Phi''(\xi)\, d\xi, \quad \sum_{j=k}^{N} \Phi(b_j) = \int_{x_0}^{\infty} B_k(\xi)\Phi''(\xi)\, d\xi.$$

These formulas follow from (5.5.3).

We show that the conditions on $a_1, a_2, \ldots, a_N$ and $b_1, b_2, \ldots, b_N$ yield the inequalities $A_k(\xi) \leq B_k(\xi)$, $k = 1, 2, \ldots, N$, for $\xi \geq x_0$. Furthermore, $A_1(\xi) = B_1(\xi)$ implies $a_j = b_j$ for all $j = 1, 2, \ldots, N$. If we prove this assertion, the theorem follows from the representation of the sums $\sum \Phi(a_j)$ and $\sum \Phi(b_j)$.

Let us prove the inequality $A_k(\xi) \le B_k(\xi)$. Since $A_N(\xi) = (a_N - \xi)_{+1}$, $B_N(\xi) = (b_N - \xi)_+$, and $b_N \ge a_N$, we have

$$A_N(\xi) - B_N(\xi) = \begin{cases} (a_N - \xi) - (b_N - \xi) = a_N - b_N \le 0, & \xi < a_N, \\ -(b_N - \xi)_+ \le 0, & \xi \ge a_N, \end{cases}$$

i.e., $A_N(\xi) \le B_N(\xi)$. We proceed by induction. Assuming that the inequality $A_k(\xi) \le B_k(\xi)$ is already proved, we show that $A_{k+1}(\xi) \le B_{k+1}(\xi)$, $k = 1, 2, \dots, N-1$.

If a point ξ lies outside the segment with endpoints a_k and b_k ($\xi_0 \le \xi < \min\{a_k, b_k\}$ or $\xi > \max\{a_k, b_k\}$) or $b_k \ge \xi \ge a_k$, it is not hard to check that

$$\Delta_k(\xi) = [A_k(\xi) - B_k(\xi)] - [A_{k+1}(\xi) - B_{k+1}(\xi)] = (a_k - \xi)_+ - (b_k - \xi)_+.$$

Then we have the inequality $A_k(\xi) - B_k(\xi) \le 0$. Indeed,

$$\Delta_k(\xi) = (a_k - \xi)_+ - (b_k - \xi)_+ = \begin{cases} a_k - b_k \le 0, & \xi < \min\{a_k, b_k\}, \\ 0, & \xi > \max\{a_k, b_k\}, \\ -(b_k - \xi) \le 0, & b_k \ge \xi \ge a_k. \end{cases}$$

It remains to prove the inequality $A_k(\xi) \le B_k(\xi)$ for $a_k \ge \xi$. In this case, we have $(a_j - \xi)_+ = a_j - \xi$ for $j \ge k$ since $a_j \ge a_k \ge \xi$ and, consequently,

$$\begin{aligned} A_k(\xi) - B_k(\xi) &= \sum_{j=k}^{N} (a_j - \xi)_+ - \sum_{j=k}^{N} (b_j - \xi)_+ = \sum_{j=k}^{N} (a_j - \xi) - \sum_{j=k}^{N} (b_j - \xi)_+ \\ &= \sum_{j=k}^{N} [(a_j - \xi) - (b_j - \xi)] + \sum_{j=k}^{N} [(b_j - \xi) - (b_j - \xi)_+] \\ &\le \sum_{j=k}^{N} [(a_j - \xi) - (b_j - \xi)] = \sum_{j=k}^{N} a_j - \sum_{j=k}^{N} b_j \le 0. \end{aligned}$$

Thus,

$$A_k(\xi) \le B_k(\xi), \quad k = N, N-1, \dots, 1$$

for all $\xi_0 \le \xi < \infty$.

The graphs of the functions $A_k(\xi)$ and $B_k(\xi)$ are convex polygonal lines with vertices at the points $\xi = a_k$ and $\xi = b_k$, respectively. We note that for

$$a_{j+1} > \xi > a_j, \quad N - 1 \ge j \ge k,$$

the derivative $A_k'(\xi)$ is constant and equal to $j - N$, whereas for $\xi > a_N$ and $\xi < a_k$ it is equal to 0 and $k - N - 1$, respectively. The derivative $B_k'(\xi)$ takes the same values for $b_{j+1} > \xi > b_j$, $N - 1 \ge j \ge k$, for $\xi > b_N$, and for $\xi < b_k$. In other words, the corresponding edges of the graphs of the piecewise linear functions $A_k(\xi)$ and $B_k(\xi)$ are parallel. The equality $A_k(\xi) = B_k(\xi)$ expresses the fact that the graphs of the functions $A_k(\xi)$ and $B_k(\xi)$ coincide. This implies that the x-coordinates of the vertices also coincide, i.e., $\{a_k, a_{k+1}, \dots, a_N\}$ and $\{b_k, b_{k+1}, \dots, b_N\}$ coincide. □

REMARK 5.5.1 (important!). Except for the equality $a_j = b_j$, all the assertions of Theorem 5.5.1 remain valid if for $x \ge x_0$ we take $\Phi(x) = x^p$, $p = 1$, i.e., $\Phi(x) = x$. Indeed, for $\Phi(x) = x$ the required inequalities coincide with one of the assumptions of the theorem.

5.6. Singular values of products of matrices

An estimate for the singular values of the product of two matrices. A generalization to the product of arbitrary number of matrices. An estimate for the singular numbers of a power of a matrix.

The estimates we derive in this section are based on Theorem 5.4.2 and the inequalities for convex functions proved in §5.5. We also use the inequalities (see Theorem 5.4.2)

$$\prod_{j=1}^{n} \sigma_{N-j+1}(AB) \leq \prod_{j=1}^{n} \sigma_{N-j+1}(A) \prod_{j=1}^{n} \sigma_{N-j+1}(B) \tag{5.6.1}$$

for singular values of $N \times N$-matrices A, B, and AB. Put $a_j = \log \sigma_j(AB)$, $b_j = \log \sigma_j(A) + \log \sigma_j(B)$. For $n = 1, 2, \ldots, N$ we have the inequality

$$\sum_{j=1}^{n} a_{N-j+1} \leq \sum_{j=1}^{n} b_{N-j+1},$$

which, by Theorem 5.5.1 with $\Phi(x) = e^{px}$ $(p > 0)$, leads to the inequality

$$\sum_{j=N-n+1}^{N} \sigma_j^p(AB) \leq \sum_{j=N-n+1}^{N} \sigma_j^p(A)\sigma_j^p(B).$$

The inequality (5.6.1) admits a generalization to the case of an arbitrary number of factors; namely,

$$\prod_{j=1}^{n} \sigma_{N-j+1}(A_1 A_2 \ldots A_l) \leq \prod_{k=1}^{l} \left[\prod_{j=1}^{n} \sigma_{N-j+1}(A_k) \right].$$

Assuming that $\Phi(x) = e^{px/l}$ $(p > 0)$ and setting

$$a_j = \log \left[\prod_{j=1}^{n} \sigma_{N-j+1}(A_1 A_2 \ldots A_l) \right], \quad b_j = \sum_{k=1}^{l} \log \prod_{j=1}^{n} \sigma_{N-j+1}(A_k),$$

from Theorem 5.5.1 we obtain the inequality

$$\sum_{j=N-n+1}^{N} \sigma_j^{p/l}(A_1 A_2 \ldots A_l) \leq \sum_{j=N-n+1}^{N} \prod_{k=1}^{l} \sigma_j^{p/l}(A_k), \tag{5.6.2}$$

where $N \geq n \geq 1$. For $A_1 = A_2 = \cdots = A_k = A$ the inequality (5.6.2) takes the form

$$\sum_{j=N-n+1}^{N} \sigma_j^{p/l}(A^l) \leq \sum_{j=N-n+1}^{N} \sigma_j^p(A), \qquad p > 0, \quad l = 1, 2, 3, \ldots,$$

where $N \geq n \geq 1$. In particular, for $l = p$ we have

$$\sum_{j=N-n+1}^{N} \sigma_j(A^l) \leq \sum_{j=N-n+1}^{N} \sigma_j^l(A), \quad N \geq n \geq 1, \ l = 1, 2, \ldots.$$

5.7. Foundations of the Sturm method

The eigenvalues of a symmetric (Hermitian) matrix. The Sturm theorem. Computation of the eigenvalues of Hermitian matrices by the Sturm method. A remark about the realization of the Sturm method. The Sylvester criterion.

We discuss an elementary consequence of the theorem about the alternation of the eigenvalues of a symmetric (generally speaking, Hermitian) matrix and the eigenvalues of the symmetric matrix that borders it. This consequence, known as the Sturm theorem, is the basis of the widely used procedure for computing eigenvalues of symmetric matrices.

As was noted (see §3.4), the computation of the singular values of an arbitrary matrix is reduced to the case of a Hermitian (or real symmetric) matrix. Therefore, we can use the Sturm theorem.

Consider a Hermitian $N \times N$-matrix

$$A_N = \begin{bmatrix} a_{11} & a_{12} & \dots & a_{1N} \\ a_{21} & a_{22} & \dots & a_{2N} \\ \dots & \dots & \dots & \dots \\ a_{N1} & a_{N2} & \dots & a_{NN} \end{bmatrix} \qquad (a_{ij} = \overline{a}_{ji}).$$

Introduce the notation

$$A_k = \begin{bmatrix} a_{11} & a_{12} & \dots & a_{1k} \\ a_{21} & a_{22} & \dots & a_{2k} \\ \dots & \dots & \dots & \dots \\ a_{k1} & a_{k2} & \dots & a_{kk} \end{bmatrix}$$

and

$$\Delta_k = \det A_k.$$

We assume that A_k is nonsingular for every k, i.e., all $\Delta_1, \Delta_2, \dots, \Delta_N$ are nonzero (all these numbers are real: $\Delta_k = \det A_k = \det A_k^* = \overline{\det A_k} = \overline{\Delta_k}$). The sign of Δ_k is determined by the number of negative eigenvalues of A_k. Since

$$\Delta_k = \det A_k = \lambda_1(A_k)\lambda_2(A_k)\dots\lambda_{k-1}(A_k)\lambda_k(A_k),$$

we conclude that $\Delta_k > 0$ if the number of negative $\lambda_j(A_k)$ is even and $\Delta_k < 0$ otherwise. Under the assumption $\Delta_k \neq 0$, all $\lambda_j(A_k)$'s differ from zero. Since A_{k+1} borders A_k, we can apply the alternation theorem and obtain

$$\lambda_1(A_{k+1}) \leq \lambda_1(A_k) \leq \dots \leq \lambda_k(A_k) \leq \lambda_{k+1}(A_{k+1}). \tag{5.7.1}$$

In view of (5.7.1), if A_{k+1} has exactly s negative eigenvalues, i.e.,

$$\lambda_1(A_{k+1}) \leq \lambda_2(A_{k+1}) \leq \dots \leq \lambda_s(A_{k+1}) < 0,$$
$$0 < \lambda_{s+1}(A_{k+1}) \leq \lambda_{s+2}(A_{k+1}) \leq \dots \leq \lambda_k(A_{k+1}),$$

then

$$\lambda_1(A_k) \leq \lambda_2(A_k) \leq \dots \leq \lambda_{s-1}(A_k) \leq \lambda_s(A_{k+1}) < 0,$$

i.e., the number of negative $\lambda_j(A_k)$'s is at least $s-1$ and

$$0 < \lambda_{s+1}(A_{k+1}) \leq \lambda_{s+1}(A_k) \leq \lambda_{s+2}(A_k) \leq \dots \leq \lambda_k(A_k). \tag{5.7.2}$$

The inequalities (5.7.2) are equivalent to the assertion that the number of negative $\lambda_j(A_k)$'s is at most s. In other words, the number of negative eigenvalues of A_k

either coincides with the number of negative eigenvalues of A_{k+1} or is less than it by 1. The signs of Δ_k and Δ_{k+1} coincide in the first case and are opposite in the second case. Consider the sequence

$$1 = \Delta_0, \Delta_1, \ldots, \Delta_k, \Delta_{k+1}, \ldots, \Delta_n. \tag{5.7.3}$$

Moving along this sequence from right to left, we see that if the signs of Δ_{k+1} and Δ_k coincide, then A_{k+1} and A_k have the same number of negative eigenvalues; otherwise, the number of negative eigenvalues of A_k is less by 1.

In other words, the number of the sign changes in the sequence $\Delta_1, \Delta_2, \ldots, \Delta_N$ is equal to the difference between the number of negative eigenvalues of A_N and the number of negative eigenvalues of $A_1 = a_{11}$. If $\lambda_1(A_1) = a_{11} > 0$, then $\Delta_1 > 0$, and if $\lambda_1(A_1) = a_{11} < 0$, then $\Delta_1 < 0$. It is obvious that between $\Delta_0 = 1$ and $\det D_1$ the sign changes only if A_1 has one negative eigenvalue, i.e., $a_{11} < 0$.

Thus, the number of the sign alternations in (5.7.3) coincides with the number of negative eigenvalues of the matrix A_N.

It is convenient to use the following notation:

$$P_1 = \frac{\Delta_1}{\Delta_0} = \Delta_1 = \det A_1 \equiv a_{11},$$
$$P_j = \frac{\Delta_j}{\Delta_{j-1}} = \frac{\det A_j}{\det A_{j-1}}, \quad 2 \leq j \leq N.$$

The number of negative $P_1, P_2, \ldots, P_N$ coincides with the number of negative eigenvalues of A_N.

It is clear that $\lambda_j(A_N) - \lambda$ are the eigenvalues of the matrix $A_N - \lambda I_N$. Let $D_k(\lambda)$ denote the determinant of the upper principal minor $A_k - \lambda I_k$ of $A_N - \lambda I_N$, i.e., $D_k(\lambda) = \det(A_k - \lambda I_k)$. Applying the above assertion about the signs of the ratio $P_j = \Delta_j / \Delta_{j-1}$ to the relations

$$P_1(\lambda) = D_1(\lambda) = a_{11} - \lambda,$$
$$P_k(\lambda) = \frac{D_k(\lambda)}{D_{k-1}(\lambda)}, \quad k = 2, 3, \ldots, N,$$

we arrive at the following theorem.

THEOREM 5.7.1 (Sturm). *If all $P_1(\lambda), P_2(\lambda), \ldots, P_N(\lambda)$ are different from zero, then the number of negative $P_k(\lambda)$'s coincides with the number of negative eigenvalues of the matrix $A_N - \lambda I_N$ or, equivalently, with the number of those eigenvalues of the matrix A_N that are less than λ.*

The Sturm theorem is convenient in computing eigenvalues of Hermitian or real symmetric matrices. For example, compute $\lambda_3(A_N)$ under the assumption that the eigenvalues are ordered in ascending order, i.e.,

$$\lambda_1(A_N) \leq \lambda_2(A_N) \leq \lambda_3(A_N) \leq \lambda_4(A_N) \leq \cdots \leq \lambda_N(A_N).$$

First, we find the endpoints of the segment $[\lambda', \lambda'']$ containing $\lambda_3(A_N)$ ($\lambda' \leq \lambda_3(A_N) \leq \lambda''$). We can put

$$\lambda' = -\max_i \sum_i |a_{ij}|, \quad \lambda'' = +\max_i \sum_j |a_{ij}|,$$

because any eigenvalue $\lambda_j(A_N)$ of the matrix A_N satisfies the inequality

$$\lambda' \leq \lambda_j(A_N) \leq \lambda''.$$

Indeed, if a vector $x \neq 0$ with components $x_1, x_2, \dots, x_N$ is an eigenvector corresponding to the eigenvalue λ_j, $|x_m| = \max_k |x_k| > 0$, then

$$\lambda_j x_m = \sum_{k=1}^{N} a_{mk} x_k,$$

$$|\lambda_j||x_m| \leq \sum_{k=1}^{N} |a_{mk}||x_m| \leq (\max_i \sum_{k=1}^{N} |a_{ik}|)|x_m|,$$

which implies the relations

$$|\lambda_j| \leq \max_i \sum_{k=1}^{N} |a_{ik}|,$$

$$\lambda' = -\max_i \sum_{k=1}^{N} |a_{ik}| \leq \lambda_j \leq \max_i \sum_{k=1}^{N} |a_{ik}| = \lambda''.$$

Setting $\lambda = (\lambda' + \lambda'')/2$, we compute $P_1(\lambda), P_2(\lambda), \dots, P_N(\lambda)$ (the Sturm sequence) and find out the number of negative elements in this sequence. If there are three or more such elements, then the eigenvalue λ_3 lies in the left half-interval $[\lambda', \lambda)$ of the segment $[\lambda', \lambda'']$. If there are one or two such elements, then $\lambda_3 \in (\lambda, \lambda'']$. Selecting the appropriate half-interval, we can divide it in halves in a similar way and find the half (i.e., the quarter of the initial interval $[\lambda', \lambda'']$) that contains λ_3.

Continuing the division of the intervals containing λ_3, we close the bounds for this eigenvalue. The process is complete when we find λ_3 with the required accuracy.

At the first glance, the described process is too cumbersome. However, the above procedure is now widely used.

In practice, significant modifications and simplifications of the procedure are introduced. In order to facilitate the computation of the determinants $D_j(\lambda)$ and their ratios $P_j(\lambda)$, a symmetric real matrix A is reduced to the symmetric three-diagonal form by a unitary similar transformation (see Problem 3.2.3). The eigenvalues do not change, but the computation of the determinants becomes simpler. Indeed, the successive minors

$$D_1(\lambda) = d_1 - \lambda,$$

$$D_2(\lambda) = \det \begin{bmatrix} d_1 - \lambda & b_2 \\ b_2 & d_2 - \lambda \end{bmatrix},$$

$$D_3(\lambda) = \det \begin{bmatrix} d_1 - \lambda & b_2 & 0 \\ b_2 & d_2 - \lambda & b_3 \\ 0 & b_3 & d_3 - \lambda \end{bmatrix}$$

....

of the three-diagonal matrix

$$P^*AP - \lambda I_N = \begin{bmatrix} d_1-\lambda & b_2 & & & & & \\ b_2 & d_2-\lambda & b_3 & & & \mathbf{0} & \\ & b_3 & d_3-\lambda & b_4 & & & \\ & & \ddots & \ddots & \ddots & & \\ & & & b_{N-2} & d_{N-2}-\lambda & b_{N-1} & \\ & \mathbf{0} & & & b_{N-1} & d_{N-1}-\lambda & b_N \\ & & & & & b_N & d_N-\lambda \end{bmatrix}$$

are connected by the recurrence relations

$$\begin{aligned} &D_0(\lambda) = 1, \\ &D_1(\lambda) = d_1 - \lambda, \\ &D_2(\lambda) = (d_2-\lambda)D_1(\lambda) - b_2^2 D_0(\lambda), \\ &D_3(\lambda) = (d_3-\lambda)D_2(\lambda) - b_3^2 D_1(\lambda), \\ &\dots\dots\dots\dots\dots\dots\dots\dots\dots\dots\dots\dots \\ &D_k(\lambda) = (d_k-\lambda)D_{k-1}(\lambda) - b_k^2 D_k(\lambda), \\ &\dots\dots\dots\dots\dots\dots\dots\dots\dots\dots\dots\dots \end{aligned}$$

which leads to the following computation rule for $P_j(\lambda) = D_j(\lambda)/D_{j-1}(\lambda)$:

$$\begin{aligned} &P_0(\lambda) = 1, \\ &P_1(\lambda) = d_1 - \lambda, \\ &P_j(\lambda) = d_j - \lambda - \frac{b_j^2}{P_{j-1}(\lambda)}, \quad j = 2, 3, \dots, N. \end{aligned}$$

These formulas are very simple. However, in practice, it is necessary to pay attention to round-off errors and to try to avoid division by zero or a very small number.

We refer the reader to [**15**] for details. We recall that the computation of the singular values of an arbitrary matrix is reduced to the computation of the eigenvalues of a symmetric (or Hermitian) matrix. Such a reduction was described in §3.4.

PROBLEM 5.7.1. Using Theorem 5.1.1, establish the following Sylvester criterion.

CRITERION (Sylvester). The eigenvalues of a Hermitian $N \times N$-matrix $A = A^*$ are positive if and only if the determinants of all its successive upper principal minors are strictly positive.

CHAPTER 6

The Weyl Inequalities

6.1. The Weyl inequalities and the Horn theorem

Eigenvalues and singular values. The Weyl inequalities: the estimate of the product of modules of eigenvalues in terms of the product of singular values. The Horn theorem as the converse to the Weyl inequalities. The Mirsky lemma. The proof of the Horn theorem.

Let A be a square $N \times N$-matrix. We choose a unitary transformation U such that the matrix $\widetilde{A} = U^*AU = U^{-1}AU$, which is similar to the matrix A, takes the upper triangular form

$$\widetilde{A} = U^*AU = \begin{bmatrix} \lambda_N & \widetilde{a}_{N\,N-1} & \widetilde{a}_{N\,N-2} & \cdots & \widetilde{a}_{N\,1} \\ & \lambda_{N-1} & \widetilde{a}_{N-1\,N-2} & \cdots & \widetilde{a}_{N-1\,1} \\ & & \ddots & & \vdots \\ & 0 & & \lambda_2 & \widetilde{a}_{21} \\ & & & & \lambda_1 \end{bmatrix},$$

where the eigenvalues on the diagonal are ordered by module

$$|\lambda_N| \geq |\lambda_{N-1}| \geq \cdots \geq |\lambda_2| \geq |\lambda_1|.$$

It is obvious that the eigenvalues $\lambda_j(A)$ and $\lambda_j(\widetilde{A})$, as well as the singular values $\sigma_j(A)$ and $\sigma_j(\widetilde{A})$, of the matrices A and $\widetilde{A}$ coincide, i.e., $\lambda_j(\widetilde{A}) = \lambda_j(A)$ and $\sigma_j(\widetilde{A}) = \sigma_j(A)$ ($|\lambda_j| \geq |\lambda_{j-1}|$ and $\sigma_j \geq \sigma_{j-1}$).

THEOREM 6.1.1. *The following relations, called Weyl's inequalities, hold:*

$$\begin{gathered} |\lambda_N| \leq \sigma_N, \\ |\lambda_N|\,|\lambda_{N-1}| \leq \sigma_N\sigma_{N-1}, \\ |\lambda_N|\,|\lambda_{N-1}|\,|\lambda_{N-2}| \leq \sigma_N\sigma_{N-1}\sigma_{N-2}, \\ \cdots\cdots\cdots\cdots\cdots\cdots \\ |\lambda_N|\,|\lambda_{N-1}|\ldots|\lambda_3|\,|\lambda_2| = \sigma_N\sigma_{N-1}\ldots\sigma_3\sigma_2, \\ |\lambda_N|\,|\lambda_{N-1}|\ldots|\lambda_3|\,|\lambda_2|\,|\lambda_1| = \sigma_N\sigma_{N-1}\ldots\sigma_3\sigma_2\sigma_1. \end{gathered} \tag{6.1.1}$$

PROOF. The last equality in (6.1.1) is obvious since the left-hand side and the right-hand side are equal to $|\det \widetilde{A}|$.

We begin to prove (6.1.1). Together with the triangular matrix $\widetilde{A}$, we consider the upper principal minors of $\widetilde{A}$, i.e., a sequence of triangular matrices

$$\widetilde{A}_N = \lambda_N,$$

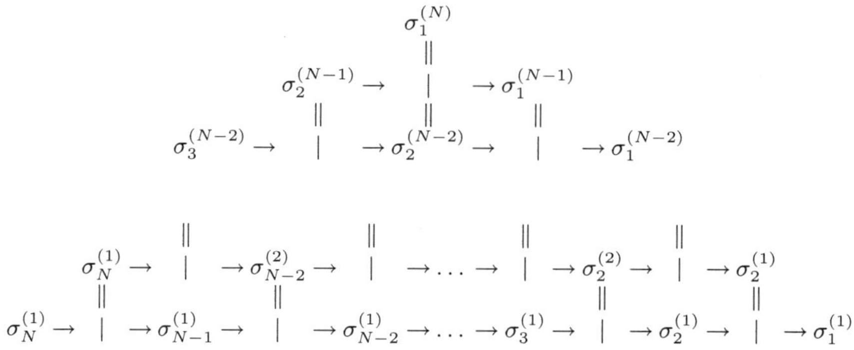

FIGURE 6.1.1

$$\widetilde{A}_{N-1} = \begin{bmatrix} \lambda_N & \widetilde{a}_{N\,N-1} \\ 0 & \lambda_{N-1} \end{bmatrix},$$

$$\widetilde{A}_{N-2} = \begin{bmatrix} \lambda_N & \widetilde{a}_{N\,N-1} & \widetilde{a}_{N\,N-2} \\ & \lambda_{N-1} & \widetilde{a}_{N-1\,N-2} \\ \text{\Large 0} & & \lambda_{N-2} \end{bmatrix},$$

....

$$\widetilde{A}_2 = \begin{bmatrix} \lambda_N & \widetilde{a}_{N\,N-1} & \dots & \widetilde{a}_{N\,1} \\ & \lambda_{N-1} & & \widetilde{a}_{N-1\,2} \\ & & \ddots & \vdots \\ & \text{\Large 0} & & \lambda_2 \end{bmatrix},$$

$$\widetilde{A}_1 = \widetilde{A}.$$

The singular values of the $(N-j+1)\times(N-j+1)$-matrix $\widetilde{A}_j$ are denoted by $\sigma^{(j)}_{N-j+1}, \sigma^{(j)}_{N-j}, \dots, \sigma^{(j)}_1$ and, as usual, are ordered in ascending order:

$$\sigma^{(j)}_{N-j+1} \geq \sigma^{(j)}_{N-j} \geq \dots \geq \sigma^{(j)}_1 \geq 0. \tag{6.1.2}$$

By Theorem 5.1.3, the singular values $\sigma^{(j-1)}_{N-j+2}, \sigma^{(j-1)}_{N-j+1}, \sigma^{(j-1)}_{N-j}, \dots, \sigma^{(j-1)}_1$ and $\sigma^{(j)}_{N-j+1}, \sigma^{(j)}_{N-j}, \dots, \sigma^{(j)}_1$ of the matrices $\widetilde{A}_{j-1}$ and $\widetilde{A}_j$ alternate:

$$\sigma^{(j-1)}_{N-j+2} \geq \sigma^{(j)}_{N-j+1} \geq \sigma^{(j-1)}_{N-j+1} \geq \sigma^{(j)}_{N-j} \geq \dots \geq \sigma^{(j-1)}_2 \geq \sigma^{(j)}_1 \geq \sigma^{(j-1)}_1.$$

It is convenient to represent the set of all $\sigma^{(j)}_k$ as in Figure 6.1.1. The horizontal arrows indicate the direction of decreasing of singular values of $\widetilde{A}_j$. The vertical line located below $\sigma^{(j)}_{k-1}$ and above the vertical line between $\sigma^{(j-1)}_k$ and $\sigma^{(j-1)}_{k-1}$ means that the inequality $\sigma^{(j-1)}_k \geq \sigma^{(j)}_{k-1} \geq \sigma^{(j-1)}_{k-1}$ holds. Figure 6.1.1 shows that for

$N \geq j + k$ we have

$$\sigma^{(j)}_{N-j+1} \leq \sigma^{(j-1)}_{N-j+2},$$
$$\sigma^{(j)}_{N-j} \leq \sigma^{(j-1)}_{N-j+1},$$
$$\cdots\cdots\cdots\cdots$$
$$\sigma^{(j)}_{N-j-k} \leq \sigma^{(j-1)}_{N-j-k+1}.$$

Therefore,

$$\sigma^{(j)}_{N-j+1}\sigma^{(j)}_{N-j}\cdots\sigma^{(j)}_{N-j-k} \leq \sigma^{(j-1)}_{N-j+2}\sigma^{(j-1)}_{N-j-1}\cdots\sigma^{(j-1)}_{N-j-k+1}.$$

By induction, we conclude that

$$\sigma^{(j)}_{N-j+1}\cdots\sigma^{(j)}_{N-j-k} \leq \sigma^{(1)}_{N}\cdots\sigma^{(1)}_{N-k-1} \equiv \sigma_N\sigma_{N-1}\cdots\sigma_{N-k-1}.$$

This relation makes sense if $N \geq j \geq 1$ and $N - j - k \geq 1$. For $k = N - j - 1$ it leads to the inequality

$$\sigma^{(j)}_{N-j+1}\sigma^{(j)}_{N-j}\cdots\sigma^{(j)}_{1} \leq \sigma_N\sigma_{N-1}\cdots\sigma_j.$$

The product of singular values of $\widetilde{A}_j$ coincides with the product of the modules of eigenvalues of $\widetilde{A}_j$, i.e.,

$$\sigma^{(j)}_{N-j+1}\sigma^{(j)}_{N-j}\cdots\sigma^{(j)}_{1} = |\lambda_N|\,|\lambda_{N-1}|\cdots|\lambda_j|.$$

Therefore,

$$|\lambda_N|\,|\lambda_{N-1}|\cdots|\lambda_j| \leq \sigma_N\sigma_{N-1}\cdots\sigma_j$$

and the Weyl inequalities are proved. □

Horn proved the converse assertion to Theorem 6.1.1.

THEOREM 6.1.2 (Horn). *If the relations*

$$\sigma_N \geq |\lambda_N|,$$
$$\sigma_N\sigma_{N-1} \geq |\lambda_N|\,|\lambda_{N-1}|,$$
$$\cdots\cdots\cdots\cdots$$
$$\sigma_N\sigma_{N-1}\cdots\sigma_2 \geq |\lambda_N|\,|\lambda_{N-1}|\cdots|\lambda_2|,$$
$$\sigma_N\sigma_{N-1}\cdots\sigma_2\sigma_1 = |\lambda_N|\,|\lambda_{N-1}|\cdots|\lambda_2|\,|\lambda_1|$$

hold for sequences $\sigma_N, \sigma_{N-1}, \ldots, \sigma_1$ *and* $\lambda_N, \lambda_{N-1}, \ldots, \lambda_2, \lambda_1$ *such that*

$$\sigma_N \geq \sigma_{N-1} \geq \cdots \geq \sigma_1 \geq 0, \qquad |\lambda_N| \geq |\lambda_{N-1}| \geq \cdots \geq |\lambda_2| \geq |\lambda_1| \geq 0,$$

then there exists an $N \times N$*-matrix* A *such that* $\lambda_1, \lambda_2, \ldots, \lambda_N$ *are the eigenvalues and* $\sigma_1, \sigma_2, \ldots, \sigma_N$ *are the singular values of* A.

We first formulate an elementary lemma which will be used in the proof of the Horn theorem.

LEMMA 6.1.1 (Mirsky). *If real nonnegative numbers*

$$x_1, x_2, \ldots, x_m, \qquad y_1, y_2, \ldots, y_m$$

are connected by the conditions

$$\begin{gathered}
x_m \geq x_{m-1} \geq \cdots \geq x_1 \geq 0, \\
y_m \geq y_{m-1} \geq \cdots \geq y_1 \geq 0, \\
x_m \geq y_m, \\
x_m x_{m-1} \geq y_m y_{m-1}, \\
\cdots\cdots\cdots\cdots\cdots\cdots\cdots\cdots \\
x_m x_{m-1} \ldots x_2 \geq y_m y_{m-1} \ldots y_2, \\
x_m x_{m-1} \ldots x_2 x_1 = y_m y_{m-1} \ldots y_2 y_1,
\end{gathered}$$

then there exists a sequence $c_{m-1}, c_{m-2}, \ldots, c_1$ alternating with $x_m, x_{m-1}, \ldots, x_1$, i.e.,

$$x_m \geq c_{m-1} \geq x_{m-1} \geq c_{m-2} \geq \cdots \geq x_2 \geq c_1 \geq x_1,$$

and satisfing the inequalities

$$\begin{gathered}
c_{m-1} \geq y_m, \\
c_{m-1} c_{m-2} \geq y_m y_{m-1}, \\
\cdots\cdots\cdots\cdots\cdots\cdots\cdots\cdots \\
c_{m-1} c_{m-2} \ldots c_2 \geq y_m y_{m-1} \ldots y_3
\end{gathered}$$

and the equality

$$c_{m-1} c_{m-2} \ldots c_2 c_1 = y_m y_{m-1} \ldots y_3 y_2.$$

The proof of Lemma 6.1.1 is rather cumbersome but is not too complicated. The lemma will be proved in §6.2. In this section, we use this lemma to prove the Horn theorem.

PROOF OF THEOREM 6.1.2. Starting with the sequence

$$\sigma_N, \sigma_{N-1}, \ldots, \sigma_2, \sigma_1,$$

denoted, for convenience, by

$$\sigma_N^{(1)}, \sigma_{N-1}^{(1)}, \ldots, \sigma_2^{(1)}, \sigma_1^{(1)} \quad (\sigma_j^{(1)} \equiv \sigma_j),$$

we construct pairwise alternating sequences

$$\sigma_{N-k+1}^{(k)}, \sigma_{N-k}^{(k)}, \ldots, \sigma_1^{(k)}$$

so that

$$\sigma_{N-k+2}^{(k-1)} \geq \sigma_{N-k+1}^{(k)} \geq \sigma_{N-k+1}^{(k-1)} \geq \sigma_{N-k}^{(k)} \geq \cdots \geq \sigma_1^{(k)} \geq \sigma_1^{(k-1)}. \tag{6.1.3}$$

Then the set of sequences $\sigma_j^{(k)}$ can be represented as in Figure 6.1.1. We will show that the equalities

$$\sigma_{N-k+1}^{(k)} \sigma_{N-k}^{(k)} \ldots \sigma_1^{(k)} = |\lambda_N|\,|\lambda_{N-1}| \ldots |\lambda_k|, \quad k = 1, 2, \ldots, N, \tag{6.1.4}$$

and the inequalities

(6.1.5)

$$\sigma_{N-k+1}^{(k)} \sigma_{N-k}^{(k)} \ldots \sigma_j^{(k)} \geq |\lambda_N|\,|\lambda_{N-1}| \ldots |\lambda_{k+j-1}|, \quad j = 2, 3, \ldots, N-k+1,$$

are valid for a suitable (generally speaking, not unique) choice of $\sigma_j^{(k)}$. For $k = 1$ (6.1.4) and (6.1.5) follow from the conditions imposed on $\sigma_j \equiv \sigma_j^{(1)}$ and λ_j in the

Horn theorem. If for some k we construct $\sigma_j^{(k)}$ satisfying (6.1.4) and (6.1.5), then the choice of the required singular values $\sigma_j^{(k+1)}$ is possible in view of the Mirsky lemma. Indeed, introduce the notation

$$N-k+1=m, \quad \sigma_{N-k-j+1}^{(k)}=x_{m-j}, \quad |\lambda_{N-j}|=y_{m-j}, \quad 0\le j\le m-1.$$

Then (6.1.4) and (6.1.5) become conditions on x_j and y_j that are exactly the assumptions of Lemma 6.1.1, i.e.,

$$\begin{gathered} x_m x_{m-1}\dots x_2 x_1 = y_m y_{m-1}\dots y_2 y_1, \\ x_m x_{m-1}\dots x_{2+k} x_{1+k} \ge y_m y_{m-1}\dots y_{2+k} y_{1+k}, \quad m-2\ge k\ge 1. \end{gathered}$$

Using this lemma, we choose $c_m, c_{m-1}, \dots, c_1$ so as to satisfy the inequalities

$$x_m \ge c_{m-1} \ge x_{m-1} \ge c_{m-2} \ge \dots \ge x_2 \ge c_1 \ge x_1$$

and the relations

$$\begin{aligned} c_{m-1} &\ge y_m = |\lambda_N|, \\ c_{m-1}c_{m-2} &\ge y_m y_{m-1} = |\lambda_N|\,|\lambda_{N-1}|, \\ &\dots\dots\dots\dots\dots\dots \\ c_{m-1}c_{m-2}\dots c_2 &\ge y_m y_{m-1}\dots y_3 = |\lambda_N|\,|\lambda_{N-1}|\dots|\lambda_{k+2}|, \\ c_{m-1}c_{m-2}\dots c_2 c_1 &= y_m y_{m-1}\dots y_3 y_2 = |\lambda_N|\,|\lambda_{N-1}|\dots|\lambda_{k+1}|. \end{aligned} \tag{6.1.6}$$

We set $\sigma_{N-k-l+1}^{(k+1)} = c_{m-l}$, $l=1,2,\dots,m-1$. Then

$$\begin{gathered} \sigma_{N-k-l+1}^{(k)} = x_{N-k-l+1} = x_{m-l} \ge c_{m-l} = \sigma_{N-k-l}^{(k+1)} \ge x_{m-l-1} = \sigma_{N-k-l}^{(k)}, \\ \sigma_{N-k-l+1}^{(k)} \ge \sigma_{N-k-l+1}^{(k+1)} \ge \sigma_{N-k-l}^{(k)}, \end{gathered}$$

which means that the sequences $\sigma_j^{(k)}$ and $\sigma_j^{(k+1)}$ alternate and that (6.1.6) is equivalent to (6.1.4) and (6.1.5) with k replaced by $k+1$.

Starting with $\sigma_j^{(1)}$, we construct $\sigma_j^{(2)}$, pass to $\sigma_j^{(3)}$ and so on, until we obtain the sequence $\sigma_j^{(N)}$ consisting of the single element $\sigma_1^{(N)}$. The fact that the constructed $\sigma_j^{(k)}$ satisfy (6.1.3) and (6.1.4) is essential later, whereas the inequality (6.1.5) will not be used in the further consideration.

Now we construct the matrix A from the Horn theorem. We begin by constructing the triangular matrices

$$A_N = a_{NN},$$

$$A_{N-1} = \begin{bmatrix} a_{NN} & a_{N\,N-1} \\ 0 & a_{N-1\,N-1} \end{bmatrix},$$

$$\cdots\cdots\cdots\cdots\cdots\cdots\cdots\cdots\cdots$$

$$A_{N-k} = \begin{bmatrix} a_{NN} & \times & \times \\ & \ddots & \\ 0 & & a_{N-k\,N-k} \end{bmatrix},$$

$$A_1 = \begin{bmatrix} a_{NN} & \times & \times & \times \\ & a_{N-1\,N-1} & & \\ & & \ddots & \\ 0 & & & a_{11} \end{bmatrix}.$$

Each of these matrices is obtained from the preceding matrix by adding one (right) column and one (bottom) row, all elements of which are zero except for the last element. In other words, A_{N-k-1} is a one-sided bordering of A_{N-k}.

Assume that the matrix A_{N-k} with singular values $\sigma_{N-k+1}^{(k)}, \sigma_{N-k}^{(k)}, \dots, \sigma_2^{(k)}, \sigma_1^{(k)}$ is already constructed. By Theorem 5.1.4, we can assert that there exists a matrix A_{N-k-1} that borders A_{N-k} from one side and has

$$\sigma_{N-k}^{(k-1)}, \sigma_{N-k-1}^{(k-1)}, \dots, \sigma_2^{(k-1)}, \sigma_1^{(k-1)}$$

as the singular values. We recall that the choice of such a matrix A_{N-k-1} is possible provided that the sequences $\sigma_j^{(k)}$ and $\sigma_j^{(k-1)}$ alternate.

If we have the matrix A_N consisting of the single element $a_{NN} = \sigma_1^{(N)}$, then we can construct the successive borderings $A_{N-1}, A_{N-2}, \dots$. Finally, we arrive at a triangular $N \times N$-matrix A_1 such that the singular values $\sigma_j^{(1)}$ of A_1 and the singular values $\sigma_j^{(k)}$ of the upper principal minors A_{N-k+1} of A_1 are the required values.

It is obvious that

$$\det A_N = |a_{NN}| = |\sigma_1^{(N)}| = |\lambda_N|,$$

$$\det A_{N-1} = \det \begin{vmatrix} a_{NN} & a_{N\,N-1} \\ 0 & a_{N-1\,N-1} \end{vmatrix} = |\sigma_2^{(N-1)} \sigma_1^{(N-1)}| = |\lambda_N|\,|\lambda_{N-1}|,$$

$$\cdots\cdots\cdots\cdots\cdots\cdots\cdots\cdots\cdots\cdots\cdots\cdots$$

$$\det A_1 = |\sigma_N^{(1)} \sigma_{N-1}^{(1)} \cdots \sigma_1^{(1)}| = |\sigma_N \sigma_{N-1} \cdots \sigma_1| = |\lambda_N|\,|\lambda_{N-1}| \cdots |\lambda_1|.$$

Therefore,

$$|a_{NN}| = |\lambda_N|, \quad |a_{N-1\,N-1}| = |\lambda_{N-1}|, \dots, |a_{11}| = |\lambda_1|.$$

Thus, the constructed matrix A_1 has given singular values σ_j ($\sigma_j^{(1)} = \sigma_j$) and given modules of eigenvalues $|\lambda_j|$. This matrix has the following upper triangular form:

$$A_1 = \begin{bmatrix} e^{i\omega_N}\lambda_{n-1} & \times & \times & \times \\ & e^{i\omega_{N-1}}\lambda_1 & \times & \times \\ & & \ddots & \vdots \\ \text{0} & & & e^{i\omega_1}\lambda_1 \end{bmatrix};$$

moreover, the diagonal elements $a_{ll} = e^{i\omega_l}\lambda_l$, $l = 1, 2, \ldots, N$, differ from λ_l only by the factor $e^{i\omega_l}$ whose module is 1.

The matrix

$$A = \begin{bmatrix} e^{-i\omega_N} & & \text{0} & \\ & e^{-i\omega_{N-1}} & & \\ & & \ddots & \\ \text{0} & & & e^{-i\omega_1} \end{bmatrix} A_1$$

is obtained from A_1 by the multiplication from the left by a diagonal unitary matrix. It has the same singular values as A_1, and $\lambda_N, \lambda_{N-1}, \ldots, \lambda_1$ are exactly the eigenvalues of A_1.

We have shown that under certain conditions on σ_j and λ_j (see Theorem 6.1.2) it is possible to construct a matrix such that σ_j are the singular values and λ_j are the eigenvalues. □

6.2. The proof of the Mirsky lemma

An auxiliary lemma about numerical monotone sequences. A weak variant of the Mirsky lemma.

The proof of the Mirsky lemma (see Lemma 6.1.1) given in this section is rather long. The reader can omit this section for a first reading.

We begin with the following auxiliary lemma.

LEMMA 6.2.1. *If* $m \geq 2$ *and*

$$\begin{gathered} \xi_m \geq \xi_{m-1} \geq \cdots \geq \xi_1, \\ \eta_m \geq \eta_{m-1} \geq \cdots \geq \eta_1, \\ \xi_m \geq \eta_m, \\ \xi_m + \xi_{m-1} \geq \eta_m + \eta_{m-1}, \\ \xi_m + \xi_{m-1} + \cdots + \xi_2 \geq \eta_m + \eta_{m-1} + \cdots + \eta_2, \\ \xi_m + \xi_{m-1} + \cdots + \xi_2 + \xi_1 = \eta_m + \eta_{m-1} + \cdots + \eta_2 + \eta_1, \end{gathered} \tag{6.2.1}$$

then there exist $\gamma_{m-1}, \gamma_{m-2}, \ldots, \gamma_2, \gamma_1$ *such that*

$$\xi_m \geq \gamma_{m-1} \geq \xi_{m-1} \geq \gamma_{m-2} \geq \xi_{m-2} \geq \cdots \geq \xi_2 \geq \gamma_1 \geq \xi_1;$$

moreover,

$$\begin{gathered}\gamma_{m-1} \geq \eta_m,\\ \gamma_{m-1}+\gamma_{m-2} \geq \eta_m+\eta_{m-1},\\ \dots\dots\dots\dots\dots\dots\\ \gamma_{m-1}+\gamma_{m-2}+\cdots+\gamma_2 \geq \eta_m+\eta_{m-1}+\cdots+\eta_3,\\ \gamma_{m-1}+\gamma_{m-2}+\cdots+\gamma_2+\gamma_1 = \eta_m+\eta_{m-1}+\cdots+\eta_3+\eta_2.\end{gathered}$$

PROOF. The set Δ of points ζ with real coordinates ζ_j such that

$$\begin{gathered}\xi_m \geq \zeta_{m-1} \geq \xi_{m-1} \geq \zeta_{m-2} \geq \cdots \geq \xi_2 \geq \zeta_1 \geq \xi_1,\\ \zeta_{m-1}+\zeta_{m-2}+\cdots+\zeta_2+\zeta_1 \geq \eta_m+\eta_{m-1}+\cdots+\eta_2,\\ \zeta_{m-1}+\zeta_{m-2}+\cdots+\zeta_2 \geq \eta_m+\eta_{m-1}+\cdots+\eta_3,\\ \dots\dots\dots\dots\dots\dots\\ \zeta_{m-1}+\zeta_{m-2} \geq \eta_m+\eta_{m-1},\\ \zeta_{m-1} \geq \eta_m\end{gathered} \tag{6.2.2}$$

is nonempty. Indeed, it contains the point $\zeta^{(0)}$ with coordinates $\zeta_1^{(0)} = \xi_2$, $\zeta_2^{(0)} = \xi_3, \dots, \zeta_{m-1}^{(0)} = \xi_m$. It is obvious that Δ is a bounded convex closed polyhedron. The function $f(\zeta) = \zeta_1+\zeta_2+\cdots+\zeta_{m-1}$ on Δ is bounded and continuous. The minimal and maximal values of f on Δ are denoted by μ and M, respectively. It is obvious that

$$M \geq f(\zeta^{(0)}) = \xi_m+\xi_{m-1}+\cdots+\xi_2 \geq \eta_m+\eta_{m-1}+\cdots+\eta_2.$$

If we are able to prove that

$$\mu \leq \eta_m+\eta_{m-1}+\cdots+\eta_2, \tag{6.2.3}$$

then we can prove the existence of a point $\zeta=\gamma$ of Δ at which

$$f(\gamma) = \eta_m+\eta_{m-1}+\cdots+\eta_2,$$

i.e., we can prove the existence of the sequence $\gamma_{m-1}, \gamma_{m-2}, \dots, \gamma_1$, satisfying all the inequalities in the formulation of the lemma.

We consider three possible cases.

Case 1. Let $\mu = f(\zeta^{(\mu)}) = f(\zeta_1^{(\mu)}, \zeta_2^{(\mu)}, \dots, \zeta_{m-1}^{(\mu)})$ and let all inequalities in (6.2.2) be strict except for the first row. This is possible only if

$$\zeta_{m-1}^{(\mu)} = \xi_{m-1},\ \zeta_{m-2}^{(\mu)} = \xi_{m-2}, \dots, \zeta_1^{(\mu)} = \xi_1.$$

Otherwise, we can decrease $f(\zeta^{(\mu)})$ by decreasing those $\zeta_j^{(\mu)}$ for which $\zeta_j^{(\mu)} > \xi_j$. In this case,

$$\begin{aligned}\mu &= \xi_{m-1}+\xi_{m-2}+\cdots+\xi_1 = (\xi_m+\xi_{m-1}+\cdots+\xi_2+\xi_1)-\xi_m\\ &= (\eta_m+\eta_{m-1}+\cdots+\eta_2+\eta_1)-\xi_m = (\eta_m+\eta_{m-1}+\cdots+\eta_2)-(\xi_m-\eta_1).\end{aligned}$$

Since

$$\xi_m-\eta_1 = (\xi_m-\eta_m)+(\eta_m-\eta_1) \geq 0,$$

the inequality (6.2.3) is obvious.

Case 2. Let $\mu = f(\zeta^{(\mu)})$ and let the second row in (6.2.2) be the equality, i.e.,

$$\zeta_{m-1}^{(\mu)}+\zeta_{m-2}^{(\mu)}+\cdots+\zeta_1^{(\mu)} = \eta_m+\eta_{m-1}+\cdots+\eta_2.$$

Then $\mu = \eta_m + \eta_{m-1} + \cdots + \eta_2$.

Case 3. Let $\zeta^{(\mu)}$ be such that there exists s $(2 \le s \le m-1)$ such that the inequalities

$$\zeta^{(\mu)}_{m-1} + \zeta^{(\mu)}_{m-2} + \cdots + \zeta^{(\mu)}_1 > \eta_m + \eta_{m-1} + \cdots + \eta_2,$$
$$\zeta^{(\mu)}_{m-1} + \zeta^{(\mu)}_{m-2} + \cdots + \zeta^{(\mu)}_2 > \eta_m + \eta_{m-1} + \cdots + \eta_3,$$
$$\dots\dots\dots\dots\dots\dots\dots\dots$$
$$\zeta^{(\mu)}_{m-1} + \zeta^{(\mu)}_{m-2} + \cdots + \zeta^{(\mu)}_{s-1} > \eta_m + \eta_{m-1} + \cdots + \eta_s$$

from (6.2.2) are strict, whereas

$$\zeta^{(\mu)}_{m-1} + \zeta^{(\mu)}_{m-2} + \cdots + \zeta^{(\mu)}_s = \eta_m + \eta_{m-1} + \cdots + \eta_{s+1}.$$

Since the remaining inequalities in (6.2.2) do not contain $\zeta^{(\mu)}_1, \zeta^{(\mu)}_2, \ldots, \zeta^{(\mu)}_s$ and (see (6.2.2))

$$\zeta^{(\mu)}_1 \ge \xi_1,\ \zeta^{(\mu)}_2 \ge \xi_2, \ldots, \zeta^{(\mu)}_{s-1} \ge \xi_{s-1},$$

we conclude that $f(\xi)$ attains the minimal value μ at the point $\zeta^{(\mu)}$ with coordinates

$$\zeta^{(\mu)}_1 = \xi_1,\ \zeta^{(\mu)}_2 = \xi_2,\ \ldots,\ \zeta^{(\mu)}_{s-1} = \xi_{s-1};$$

moreover,

$$\begin{aligned}\mu = f(\zeta^{(\mu)}) &= (\zeta^{(\mu)}_{m-1} + \zeta^{(\mu)}_{m-1} + \cdots + \zeta^{(\mu)}_s) + \zeta^{(\mu)}_{s-1} + \cdots + \zeta^{(\mu)}_1 \\ &= (\eta_m + \eta_{m-1} + \cdots + \eta_{s+1}) + (\xi_{s-1} + \xi_{s-2} + \cdots + \xi_1).\end{aligned}$$

The equality (see (6.2.1))

$$(\eta_m + \cdots + \eta_{s+1}) + (\eta_s + \cdots + \eta_1) = (\xi_m + \cdots + \xi_{s+1}) + (\xi_s + \cdots + \xi_1)$$

and the inequalities (see again (6.2.1))

$$\xi_m + \xi_{m-1} + \cdots + \xi_s \ge \eta_m + \eta_{m+1} + \cdots + \eta_s, \quad \eta_s \ge \eta_1$$

imply the inequality

$$\xi_{s-1} + \xi_{s-2} + \cdots + \xi_1 \le \eta_{s-1} + \eta_{s-2} + \cdots + \eta_1.$$

Hence

$$\begin{aligned}\mu = f(\zeta^{(\mu)}) &= (\eta_m + \eta_{m-1} + \cdots + \eta_{s+1}) + (\xi_{s-1} + \xi_{s-2} + \cdots + \xi_1) \\ &= (\eta_m + \cdots + \eta_{s+1}) + (\eta_{s-1} + \cdots + \eta_2) = (\eta_m + \cdots + \eta_2) + (\eta_1 - \eta_s) \\ &\le (\eta_m + \eta_{m-1} + \cdots + \eta_2).\end{aligned}$$

Thus, we have established the inequality $\mu \le \eta_m + \eta_{m-1} + \cdots + \eta_1$ in all possible cases. □

Introduce the notation

$$\xi_m = \log x_m, \quad \eta_m = \log y_m, \quad \gamma_m = \log c_m.$$

From Lemma 6.2.1 we obtain the following assertion which can be regarded as a variant of the Mirsky lemma under some stronger assumptions.

LEMMA 6.2.2 (a weak variant of the Mirsky lemma). *If*

$$\begin{gathered}
x_m \geq x_{m-1} \geq \cdots \geq x_1 > 0, \\
y_m \geq y_{m-1} \geq \cdots \geq y_1 > 0, \\
x_m \geq y_m, \\
x_m x_{m-1} \geq y_m y_{m-1}, \\
\cdots\cdots\cdots\cdots\cdots\cdots \\
x_m x_{m-1} \ldots x_2 \geq y_m y_{m-1} \ldots y_2, \\
x_m x_{m-1} \ldots x_2 x_1 = y_m y_{m-1} \ldots y_2 y_1,
\end{gathered} \tag{6.2.4}$$

then there exists a sequence $c_{m-1}, c_{m-2}, \ldots, c_1$ alternating with $x_m, x_{m-1}, \ldots, x_1$, i.e.,

$$x_m \geq c_{m-1} \geq x_{m-1} \geq c_{m-2} \geq \cdots \geq x_2 \geq c_1 \geq x_1,$$

and satisfies the conditions

$$\begin{gathered}
c_{m-1} \geq y_m, \\
c_{m-1} c_{m-2} \geq y_m y_{m-1}, \\
\cdots\cdots\cdots\cdots\cdots\cdots \\
c_{m-1} c_{m-2} \ldots c_2 \geq y_m y_{m-1} \ldots y_3, \\
c_{m-1} c_{m-2} \ldots c_2 c_1 = y_m y_{m-1} \ldots y_3 y_2.
\end{gathered}$$

PROOF OF LEMMA 6.1.1. It suffices to show that strict inequalities $x_1 > 0$ and $y_1 > 0$ can be replaced by nonstrict inequalities $x_1 \geq 0$ and $y_1 \geq 0$. We note that $y_1 = 0$ if $x_1 = 0$; the converse assertion is also valid. This follows from the equality

$$x_m x_{m-1} \ldots x_1 = y_m y_{m-1} \ldots y_2 y_1.$$

By (6.2.4), the number of j's with $y_j = 0$ is not less than the number of j's with $x_j = 0$. Let

$$\begin{gathered}
x_m \geq x_{m-1} \geq \cdots \geq x_1 = 0, \\
y_m \geq y_{m-1} \geq \cdots \geq y_{p+1} > y_p = y_{p-1} = \cdots = y_1 = 0.
\end{gathered}$$

Setting

$$\begin{gathered}
c_{p-1} = x_{p-1}, \quad c_{p-2} = x_{p-2}, \ldots, c_1 = x_1 = 0, \\
c_p = x_{p+1}, \quad c_{p+1} = x_{p+2}, \ldots, c_{m-1} = x_m,
\end{gathered}$$

we obtain the alternation conditions

$$x_m \geq c_{m-1} \geq x_{m-1} \geq c_{m-1} \geq \cdots \geq x_2 \geq c_1 \geq x_1.$$

Since

$$\begin{aligned} c_{m-1} &= x_m \geq y_m, \\ c_{m-1}c_{m-2} &= x_m x_{m-1} \geq y_m y_{m-1}, \\ c_{m-1}c_{m-2}\dots c_p &= x_m x_{m-1}\dots x_{p+1} \geq y_m y_{m-1}\dots y_{p+1}, \\ c_{m-1}c_{m-2}\dots c_p c_{p-1} &\geq 0 = y_m y_{m-1}\dots y_{p+1} y_p, \\ c_{m-1}c_{m-2}\dots c_p c_{p-1} c_{p-2} &\geq 0 = y_m y_{m-1}\dots y_{p+1} y_p y_{p-1}, \\ &\dots\dots\dots\dots\dots\dots \\ c_{m-1}c_{m-2}\dots c_2 &\geq 0 = y_m y_{m-1}\dots y_3, \\ c_{m-1}c_{m-2}\dots c_2 c_1 &= 0 = y_m y_{m-1}\dots y_2, \end{aligned}$$

the remaining requirements on c_j in the Mirsky lemma are satisfied. Lemma 6.1.1 is proved for all x_j and y_j satisfying the inequalities

$$x_m \geq x_{m-1} \geq \dots \geq x_1 \geq 0, \quad x_m \geq y_{m-1} \geq \dots \geq y_1 \geq 0$$

and the remaining assumptions of the lemma. □

6.3. Corollaries of the Weyl inequalities

An estimate for the sum of powers of the modules of eigenvalues in terms of the sum of powers of singular values. Normal matrices. The logarithmic form of inequalities. A variant of the Weyl inequalities under a transformation of a matrix. The equality of the sum of eigenvalues and the sum of diagonal elements.

Using the Weyl inequalities

$$\prod_{j=N-n+1}^{N} |\lambda_j(A)| \leq \prod_{j=N-n+1}^{N} \sigma_j(A), \quad n = 1, 2, \dots, N,$$

and setting $a_j = \log|\lambda_j(A)|$, $b_j = \log\sigma_j(A)$, we find that

$$\sum_{j=N-n+1}^{N} a_j \leq \sum_{j=N-n+1}^{N} b_j, \quad n = 1, 2, \dots, N.$$

We recall that the eigenvalues λ_j and the singular values σ_j of A are numbered so that

$$|\lambda_1| \leq |\lambda_2| \leq \dots \leq |\lambda_N|, \quad \sigma_1 \leq \sigma_2 \leq \dots \leq \sigma_N.$$

Therefore,

$$a_N \geq a_{N-1} \geq \dots \geq a_1, \quad b_N \geq b_{N-1} \geq \dots \geq b_1.$$

Using Theorem 5.5.1 and setting $\Phi(x) = e^{px}$ (which is admissible as was shown in §5.6), we conclude that

$$\sum_{j=N-n+1}^{N} |\lambda_j(A)|^p \leq \sum_{j=N-n+1}^{N} \sigma_j^p(A), \quad n = 1, 2, \dots, N.$$

PROBLEM 6.3.1. Prove that the equality

$$\sum_{j=1}^{N} |\lambda_j(A)|^p = \sum_{j=1}^{N} \sigma_j^p(A) \tag{6.3.1}$$

holds if and only if the matrix A is unitarily similar to the diagonal matrix Λ with eigenvalues $\lambda_j(A)$:

$$A = U\Lambda U^*, \quad U^*U = I_N,$$

$$\Lambda = \begin{bmatrix} \lambda_1 & & & 0 \\ & \lambda_2 & & \\ & & \ddots & \\ 0 & & & \lambda_N \end{bmatrix}.$$

In other words, the equality in (6.3.1) takes place only if the $N \times N$-matrix A has N mutually orthogonal eigenvectors. Such matrices are said to be *normal.* Again, using the inequality

$$\sum_{j=N-n+1}^{N} \Phi(a_j) \leq \sum_{j=N-n+1}^{N} \Phi(b_j) \tag{6.3.2}$$

from Theorem 5.5.1 and setting $\Phi(x) = \log(1 + re^x)$, where $r > 0$, we obtain the inequality

$$\sum_{j=N-n+1}^{N} \log[1 + r|\lambda_j(A)|] \leq \sum_{j=N-n+1}^{N} \log[1 + r\sigma_j(A)]$$

or the equivalent inequality

$$\prod_{j=N-n+1}^{N} [1 + r|\lambda_j(A)|] \leq \prod_{j=N-n+1}^{N} [1 + r\sigma_j(A)]. \tag{6.3.3}$$

The inequalities (6.3.2) and (6.3.3) will be used often later.

We consider another series of inequalities connecting eigenvalues and singular values of a matrix. In turn, from these inequalities we will derive new estimates for singular values of the sum of matrices.

Let P_n and Q_n be rectangular matrices with orthonormal columns ($P_n^*P_n = I_n$ and $Q_n^*Q_n = I_n$). We will use (5.4.4) with $p = 1$,

$$\sum_{j=1}^{n} \sigma_{n-j+1}(P_n^*AQ_n) \leq \sum_{j=1}^{n} \sigma_{N-j+1}(A)$$

and the inequalities (6.3.1) (again, with $p = 1$), which imply

$$\left|\sum_{j=1}^{n} \lambda_{n-j+1}(P_n^*AQ_n)\right| \leq \sum_{j=1}^{n} |\lambda_{n-j+1}(P_n^*AQ_n)| \leq \sum_{j=1}^{n} \sigma_{N-j+1}(A). \tag{6.3.4}$$

It turns out that there exist P_n and Q_n ($P_n^*P_n = I_n$ and $Q_n^*Q_n = I_n$) such that $\lambda_{n-j+1}(P_n^*AQ_n) = \sigma_{n-j+1}(A)$, which, in view of (6.3.4), leads to the equality

$$\max_{P_n, Q_n} \sum_{j=1}^{n} |\lambda_{n-j+1}(P_n^*AQ_n)| = \sum_{j=1}^{n} \sigma_{N-j+1}(A).$$

To prove the above assertion, we use the singular value decomposition of A:

$$A = P_N \begin{bmatrix} \sigma_N(A) & & 0 \\ & \ddots & \\ 0 & & \sigma_1(A) \end{bmatrix} Q_N^* \quad (P_N^* P_N = I_N,\ Q_N^* Q_N = I_N),$$

where $P_N^* A Q_N$ is the diagonal matrix with eigenvalues $\sigma_N(A), \sigma_{N-1}(A), \dots, \sigma_1(A)$.

For P_n and Q_n we take the first n columns of the matrices P_N and Q_N. It is obvious that

$$P_n^* A Q_n = \begin{bmatrix} \sigma_N(A) & & 0 \\ & \ddots & \\ 0 & & \sigma_{N-n+1}(A) \end{bmatrix},$$

$$\begin{aligned} \sum_{j=1}^{n} \lambda_{n-j+1}(P_n^* A Q_n) &= \left| \sum_{j=1}^{n} \lambda_{n-j+1}(P_n^* A Q_n) \right| \\ &= \sum_{j=1}^{n} |\lambda_{n-j+1}(P_n^* A Q_n)| = \sum_{j=1}^{n} \sigma_{N-j+1}(A). \end{aligned}$$

Thus, we have proved (6.3.4).

We note that the sum of the eigenvalues of $P_n^* A Q_n$ is equal to the sum of the diagonal elements of $P_n^* A Q_n$, i.e.,

$$\sum_{j=1}^{n} \lambda_{n-j+1}(P_n^* A Q_n) = \sum_{j=1}^{n} [P_n^* A Q_n]_{jj}.$$

Let A and B be $N \times N$-matrices and let P_n and Q_n be chosen so that

$$\sum_{j=1}^{n} [P_n^*(A+B)Q_n]_{jj} = \sum_{j=1}^{n} \sigma_{N-j+1}(A+B);$$

moreover,

$$\begin{aligned} \sum_{j=1}^{n} \sigma_{N-j+1}(A+B) &= \sum_{j=1}^{n} [P_n^* A Q_n]_{jj} + \sum_{j=1}^{n} [P_n^* B Q_n]_{jj} \\ &= \left| \sum_{j=1}^{n} \lambda_{n-j+1}(P_n^* A Q_n) + \sum_{j=1}^{n} \lambda_{n-j+1}(P_n^* B Q_n) \right| \\ &\le \sum_{j=1}^{n} |\lambda_{n-j+1}(P_n^* A Q_n)| + \sum_{j=1}^{n} |\lambda_{n-j+1}(P_n^* B Q_n)| \\ &\le \sum_{j=1}^{n} \sigma_{N-j+1}(A) + \sum_{j=1}^{n} \sigma_{N-j+1}(B). \end{aligned}$$

Thus, we have established one more series of important inequalities:

$$\sum_{j=1}^{n} \sigma_{N-j+1}(A+B) \le \sum_{j=1}^{n} \sigma_{N-j+1}(A) + \sum_{j=1}^{n} \sigma_{N-j+1}(B), \quad 1 \le n \le N. \tag{6.3.5}$$

Choose $p \geq 1$ and introduce

$$\Phi(x) = \begin{cases} x^p, & x \geq 0, \\ 0, & x < 0. \end{cases}$$

Applying Theorem 5.5.1, from (6.3.5) we deduce the estimate

$$\sum_{j=1}^{n} \sigma_{N-j+1}^p(A+B) \leq \sum_{j=1}^{n} [\sigma_{N-j+1}(A) + \sigma_{N-j+1}(B)]^p, \quad 1 \leq n \leq N.$$

Since for $x_j \geq 0$ and $y_j \geq 0$ the Minkowski inequality

$$\left(\sum_{j=1}^{n} (x_j + y_j)^p\right)^{1/p} \leq \left(\sum_{j=1}^{n} x_j^p\right)^{1/p} + \left(\sum_{j=1}^{n} y_j^p\right)^{1/p}$$

holds for all $1 \leq n \leq N$ and $p \geq 1$, we have

(6.3.6)

$$\left(\sum_{j=1}^{n} \sigma_{N-j+1}^p(A+B)\right)^{1/p} \leq \left(\sum_{j=1}^{n} \sigma_{N-j+1}^p(A)\right)^{1/p} + \left(\sum_{j=1}^{n} \sigma_{N-j+1}^p(B)\right)^{1/p}.$$

We have completed the proof of the most important corollaries of the Weyl inequalities.

CHAPTER 7

Variational Principles

7.1. Stationary values of a Hermitian form on the unit sphere

Stationary points of a Hermitian form on the unit sphere and eigenvalues.

Suppose that $x = x(t)$ is a differentiable vector-valued function of t such that the curve defined by the equation $x = x(t)$ lies on the sphere $\|x(t)\| = 1$ and passes through the point $x^{(0)}$ for $t = 0$.

Considera Hermitian matrix A ($A = A^*$). Compute the derivative of the scalar function $\psi(t) = (Ax(t), x(t))$ at $t = 0$:

$$\begin{aligned}
\left.\frac{d\psi}{dt}\right|_{t=0} &= \left(Ax^{(0)}, \left.\frac{dx}{dt}\right|_{t=0}\right) + \left(A\left.\frac{dx}{dt}\right|_{t=0}, x^{(0)}\right) \\
&= \left(Ax^{(0)}, \left.\frac{dx}{dt}\right|_{t=0}\right) + \overline{\left(x^{(0)}, A\left.\frac{dx}{dt}\right|_{t=0}\right)} \\
&= \left(Ax^{(0)}, \left.\frac{dx}{dt}\right|_{t=0}\right) + \overline{\left(A^*x^{(0)}, \left.\frac{dx}{dt}\right|_{t=0}\right)} \\
&= \left(Ax^{(0)}, \left.\frac{dx}{dt}\right|_{t=0}\right) + \overline{\left(Ax^{(0)}, \left.\frac{dx}{dt}\right|_{t=0}\right)} = 2\mathrm{Re}\left(Ax^{(0)}, \left.\frac{dx}{dt}\right|_{t=0}\right).
\end{aligned}$$

Fixing a point $x^{(0)}$ on the unit sphere ($\|x^{(0)}\| = 1$) and choosing different curves on the sphere $(x, x) = 1$ that pass through $x^{(0)}$, we obtain different vectors $z = \left.\frac{dx}{dt}\right|_{t=0}$. For such vectors z we have

$$\left.\frac{d}{dt}(x, x)\right|_{t=0} \equiv 2\mathrm{Re}\left.\left(x, \frac{dx}{dt}\right)\right|_{t=0} \equiv 2\mathrm{Re}\,(x^{(0)}, z) = 0.$$

Any vector z such that $2\mathrm{Re}\,(x^{(0)}, z) = 0$ can be regarded as a tangent vector to some curve at the point $x^{(0)}$ on the sphere $\|x\| = 1$. Indeed, it is easy to check that for such a curve we can take, for example, one of the following curves:

$$x(t) = \frac{1}{\|x^{(0)} + tz\|}[x^{(0)} + tz],$$

$$x(t) = \frac{1}{\sqrt{\|x^{(0)}\|^2 + 2t\mathrm{Re}\,(x^{(0)}, z) + t^2\|z\|^2}}[x^{(0)} + tz],$$

$$x(t) = \frac{1}{\sqrt{1+t^2\|z\|^2}}[x^{(0)} + tz].$$

In other words, $\psi'(t)$ has those and only those values that $(Ax^{(0)}, z)$ has for z such that $\operatorname{Re}(x^{(0)}, z) = 0$.

Using a unitary transformation U, we reduce a Hermitian matrix A to the diagonal form

$$U^*AU = \begin{bmatrix} \lambda_1 & & & \text{\Large 0} \\ & \lambda_2 & & \\ & & \ddots & \\ \text{\Large 0} & & & \lambda_N \end{bmatrix}.$$

Setting

$$x^{(0)} = U\begin{pmatrix} y_1 \\ y_2 \\ \vdots \\ y_N \end{pmatrix}, \qquad z = U\begin{pmatrix} h_1 \\ h_2 \\ \vdots \\ h_N \end{pmatrix},$$

we obtain the following relations:

$$\begin{aligned}
(x^{(0)}, x^{(0)}) &= \sum_{j=1}^N y_j\overline{y}_j = 1, \\
(Ax^{(0)}, x^{(0)}) &= \sum_{j=1}^N \lambda_j y_j\overline{y}_j, \\
(Ax^{(0)}, z) &= \sum_{j=1}^N \lambda_j(\operatorname{Re} y_j \operatorname{Re} h_j + \operatorname{Im} y_j \operatorname{Im} h_j), \\
(x^{(0)}, z) &= \sum_{j=1}^N (\operatorname{Re} y_j \operatorname{Re} h_j + \operatorname{Im} y_j \operatorname{Im} h_j).
\end{aligned}$$

A point on the sphere $(x, x) = 1$ is a stationary point of the Hermitian form (Ax, x) if $\psi'(t)\Big|_{t=0} = 0$ along any curve passing through this point for $t = 0$. Such points have coordinates $y_1, y_2, \dots, y_N$ satisfying the condition

$$\sum_{j=1}^N \lambda_j(\operatorname{Re} y_j \operatorname{Re} h_j + \operatorname{Im} y_j \operatorname{Im} h_j) = 0$$

for any $h_1, h_2, \dots, h_N$ such that

$$\sum_{j=1}^N (\operatorname{Re} y_j \operatorname{Re} h_j + \operatorname{Im} y_j \operatorname{Im} h_j) = 0.$$

Assume that $y_{j1} \neq 0$, $y_{j2} \neq 0$, and $\lambda_{j1} \neq \lambda_{j2}$ at some point. Setting

$$h_j = 0 \text{ for } j \neq j_1, j_2, \quad h_{j_1} = \frac{1}{|y_{j_1}|^2} y_{j_1}, \quad h_{j_2} = -\frac{1}{|y_{j_2}|^2} y_{j_2},$$

we obtain

$$\sum_{j=1}^{N}(\operatorname{Re} y_j \operatorname{Re} h_j + \operatorname{Im} y_j \operatorname{Im} h_j) = 0,$$

$$\sum_{j=1}^{N}\lambda_j(\operatorname{Re} y_j \operatorname{Re} h_j + \operatorname{Im} y_j \operatorname{Im} h_j) = \lambda_{j_1} - \lambda_{j_2} \neq 0.$$

We see that the point under consideration is not stationary. At a stationary point, the only nonvanishing y_j's can be those that correspond to λ_j. Therefore, on the unit sphere $\|x\| = 1$, only the endpoints of eigenvectors of A are stationary points of (Ax, x) and only eigenvalues of A are stationary values of (Ax, x). We show that the quadratic form (Ax, x) attains the stationary value at any normalized eigenvector. In view of the choice of a unitary similar transformation U reducing A to the diagonal form, we can assume that the eigenvector $x^{(0)}$ has the form

$$x^{(0)} = U\begin{pmatrix}1\\0\\\vdots\\0\end{pmatrix} \qquad (y_1 = 1, y_2 = y_3 = \cdots = y_N = 0).$$

The corresponding eigenvalue is denoted by λ_1.

Let $h_1, h_2, \ldots, h_N$ satisfy the condition

$$\sum_{j=1}^{N}(\operatorname{Re} y_j \operatorname{Re} h_j + \operatorname{Im} y_j \operatorname{Im} h_j) = \operatorname{Re} h_1 = 0.$$

Then

$$\sum_{j=1}^{N}\lambda_j(\operatorname{Re} y_j \operatorname{Re} h_j + \operatorname{Im} \operatorname{Im} h_j) = \lambda_1 \operatorname{Re} h_1 = 0,$$

which means that $x^{(0)}$ is a stationary point.

Thus, we have shown that only eigenvalues of A are stationary values and only eigenvectors of A are stationary points of the Hermitian form (Ax, x) on the unit sphere $\|x\| = 1$.

7.2. Stationary values of a Hermitian form

Eigenvalues of a matrix and stationary points of the Rayleigh quotient. Alternation of stationary points of the Rayleigh quotient on the N-dimensional space and on its $(N-1)$-dimensional subspace. Estimates for eigenvalues in terms of the minimal (maximal) values of the Rayleigh quotient on subspaces of smaller dimension.

Consider a Hermitian form (Ax, x), $x \in H_N$. As was shown in §7.1, the roots $\lambda_1, \lambda_2, \ldots, \lambda_N$ of the characteristic equation $\det[A - \lambda I_N] = 0$ are the stationary values of the Hermitian form (Ax, x) on the unit sphere $\|x\| = 1$. Assume that $\lambda_1, \lambda_2, \ldots, \lambda_N$ are numbered so that $\lambda_N \geq \lambda_{N-1} \geq \cdots \geq \lambda_2 \geq \lambda_1$.

For an arbitrary unit vector q ($\|q\| = 1$), consider the Hermitian form (Ax, x) not on the entire N-dimensional sphere $\|x\| = 1$ but only on the cut of the sphere by the hyperplane $(q, x) = 0$.

First we choose a unitary reflection W ($W^*W = I_N$) polarizing the vector q, i.e.,

$$Wq = \begin{pmatrix} 0 \\ 0 \\ \vdots \\ 0 \\ e^{i\alpha} \end{pmatrix}.$$

Setting $y = Wx$ ($x = W^*y$), we can write

$$(Ax, x) = (A^{(1)}y, y), \quad A^{(1)} = WAW^*.$$

The stationary values of (Ax, x) on the $(N-1)$-dimensional sphere $\|x\| = 1$, $(x, q) = 0$, coincide with the stationary values of $(A^{(1)}y, y)$ on the intersection of the sphere $\sum_{j=1}^{n} y_j \overline{y}_j = 1$ and the hyperplane $e^{-i\alpha} y_N = 0$. Note that this hyperplane can be defined by the equality $y_N = 0$. Let

$$A^{(1)} = \begin{bmatrix} a_{11}^{(1)} & a_{12}^{(1)} & \dots & a_{1\,N-1}^{(1)} & a_{1\,N}^{(1)} \\ a_{21}^{(1)} & a_{22}^{(1)} & \dots & a_{2\,N-1}^{(1)} & a_{2\,N}^{(1)} \\ \vdots & \vdots & & \vdots & \vdots \\ a_{N-1\,1}^{(1)} & a_{N-1\,2}^{(1)} & \dots & a_{N-1\,N-1}^{(1)} & a_{N-1\,N}^{(1)} \\ a_{N\,1}^{(1)} & a_{N\,2}^{(1)} & \dots & a_{N\,N-1}^{(1)} & a_{N\,N}^{(1)} \end{bmatrix}.$$

The values of $(A^{(1)}y, y)$ on the hyperplane $y_N = 0$ can be found with the help of the truncated matrix

$$\widehat{A}^{(1)} = \begin{bmatrix} a_{11}^{(1)} & a_{12}^{(1)} & \dots & a_{1\,N-1}^{(1)} \\ a_{21}^{(1)} & a_{22}^{(1)} & \dots & a_{2\,N-1}^{(1)} \\ \vdots & \vdots & & \vdots \\ a_{N-1\,1}^{(1)} & a_{N-1\,2}^{(1)} & \dots & a_{N-1\,N-1}^{(1)} \end{bmatrix},$$

which differs from $A^{(1)}$ by the absence of the right column and the lowest row. It is clear that the stationary values of $(A^{(1)}y, y)$ on the $(N-1)$-dimensional sphere $\|y\| = 1$, $y^{(N)} = 0$ can be found as the eigenvalues $\lambda_{N-1}^{(1)}, \lambda_{N-2}^{(1)} \dots, \lambda_1^{(1)}$ of $\widehat{A}^{(1)}$. Enumerate the eigenvalues of $\widehat{A}^{(1)}$ in ascending order:

$$\lambda_{N-1}^{(1)} \geq \lambda_{N-2}^{(1)} \geq \dots \geq \lambda_1^{(1)}.$$

By Theorem 5.1.1, the eigenvalues of the original matrix $A^{(1)}$ and the truncated matrix $\widehat{A}^{(1)}$ satisfy the Rayleigh inequalities

$$\lambda_N \geq \lambda_{N-1}^{(1)} \geq \lambda_{N-1} \geq \lambda_{N-2}^{(1)} \geq \dots \geq \lambda_2 \geq \lambda_1^{(1)} \geq \lambda_1.$$

We recall that λ_j are the stationary values of the quadratic form (Ax, x) on the unit sphere $\|x\| = 1$ or, equivalently, the stationary values of the Rayleigh quotient $\dfrac{(Ax, x)}{(x, x)}$. Therefore, $\lambda_j^{(1)}$ are the stationary values of the same Rayleigh quotient $\dfrac{(Ax, x)}{(x, x)}$, but for $x \in H_N$ lying on the hyperplane $(x, q) = 0$. We could assume that

$\|q\| = 1$ as above. However, it suffices to assume that $q \neq 0$. We have proved the following important lemma.

LEMMA 7.2.1. *Consider a Hermitian form (Ax, x), the stationary values $\lambda_1^{(1)}$, $\lambda_2^{(1)}, \dots, \lambda_{N-1}^{(1)}$ of the Rayleigh quotient $\dfrac{(Ax,x)}{(x,x)}$ on the $(N-1)$-dimensional linear subspace $(q, x) = 0$ $(q \neq 0)$, and the stationary values $\lambda_1, \lambda_2, \dots, \lambda_N$ of the Rayleigh quotient $\dfrac{(Ax,x)}{(x,x)}$ on the entire N-dimensional space. The following inequalities hold:*

$$\lambda_1 \leq \lambda_1^{(1)} \leq \lambda_2 \leq \lambda_2^{(1)} \leq \lambda_3 \leq \cdots \leq \lambda_{N-1}^{(1)} \leq \lambda_N.$$

By Lemma 7.2.1, the stationary values $\lambda_1^{(2)}, \lambda_2^{(2)}, \dots, \lambda_{N-2}^{(2)}$ of the Rayleigh quotient $\dfrac{(Ax,x)}{(x,x)}$ on the $(N-2)$-dimensional plane $(q^{(1)}, x) = 0$, $(q^{(2)}, x) = 0$ satisfy the inequalities[1]

$$\lambda_1^{(1)} \leq \lambda_1^{(2)} \leq \lambda_2^{(1)} \leq \lambda_2^{(2)} \leq \lambda_3^{(1)} \leq \cdots \leq \lambda_{N-2}^{(2)} \leq \lambda_{N-1}^{(1)}.$$

We can continue this procedure and distinguish subspaces of less and less dimension. Finally, we establish the alternation conditions for stationary values of the Rayleigh quotient $\dfrac{(Ax,x)}{(x,x)}$ on these subspaces. The alternation conditions imply that

$$\lambda_{N-k}^{(k)} \geq \lambda_{N-k}, \quad \lambda_1^{(k)} \leq \lambda_{1+k}.$$

Consequently,

$$\max_{\substack{\|x\|\neq 0 \\ (q^{(j)},x)=0 \\ j=1,2,\dots,k}} \frac{(Ax,x)}{(x,x)} \geq \lambda_{N-k}, \qquad \min_{\substack{\|x\|\neq 0 \\ (q^{(j)},x)=0 \\ j=1,2,\dots,k}} \frac{(Ax,x)}{(x,x)} \leq \lambda_{k+1}.$$

The left-hand sides of these inequalities can be regarded as functions of $q^{(j)}$. Without loss of generality, we can assume that the equalities defining the above connections are normalized so that

$$\|q^{(j)}\| = \sum_{i=1}^{N} |q_i^{(j)}|^2 = 1 \quad \text{for all } j.$$

It is obvious that the minimal (maximal) value of the quotient $\dfrac{(Ax,x)}{(x,x)}$ cannot be less than λ_{N-k} (larger than λ_{k+1}). Therefore,

$$\begin{aligned} &\min_{\substack{\|q^{(j)}\|=1 \\ j=1,2,\dots,k}} \max_{\substack{\|x\|\neq 0 \\ j=1,2,\dots,k}} \frac{(Ax,x)}{(x,x)} \geq \lambda_{N-k}, \\ &\min_{\substack{\|q^{(j)}\|=1 \\ j=1,2,\dots,k}} \max_{\substack{\|x\|\neq 0 \\ j=1,2,\dots,k}} \frac{(Ax,x)}{(x,x)} \leq \lambda_{k+1}. \end{aligned} \tag{7.2.1}$$

[1] We suppose that these restrictions are given by linearly independent vectors $q^{(1)}, q^{(2)}$, i.e., they cut out subspaces of H_N of dimension $N-2$ in fact.

7.3. Variational Weber principles

The Rayleigh quotient is constant under a unitary change of variables. The largest and smallest eigenvalues of a matrix as extremal values of the corresponding quadratic form on the unit sphere. Eigenvalues of a matrix as extremal values of the Rayleigh quotient.

A self-adjoint (Hermitian) matrix A is usually associated with the quadratic form (Ax, x) defined by the inner product in a unitary space. Under a unitary change of variables $x = Uy$ $(U^*U = I)$, the values of the quadratic form $(Ax, x) = (AUy, Uy) = (U^*AUy, y)$ and the norm of the vector $\|x\|^2 = (x, x) = (Uy, Uy) = (U^*Uy, y) = (y, y) = \|y\|^2$ do not change. Therefore, under a unitary change of variables, the Rayleigh quotient is also preserved:

$$\frac{(Ax, x)}{(x, x)} = \frac{(U^*AUy, y)}{(y, y)}.$$

Choosing U so that the matrix U^*AU is diagonal, we obtain

$$\frac{(U^*AUy, y)}{(y, y)} = \frac{\sum\limits_{j=1}^{N} \lambda_j |y_j|^2}{\sum\limits_{j=1}^{N} |y_j|^2}.$$

As above, we suppose that the eigenvalues $\lambda_1, \lambda_2, \ldots, \lambda_N$ of A are numbered in ascending order: $\lambda_1 \le \lambda_2 \le \cdots \le \lambda_N$. It is obvious that the largest (λ_N) and the smallest (λ_1) eigenvalues of the matrix A can be found as the extremal values of the quadratic form (Ax, x) on the unit sphere:

$$\lambda_N = \max_{\|x\|=1} (Ax, x), \qquad \lambda_1 = \min_{\|x\|=1} (Ax, x).$$

By variational arguments, for $m < N$ it is possible to find "intermediate" eigenvalues λ_m by the formula

$$\lambda_m = \max_{\substack{\|y\| \neq 0 \\ y_N = \cdots = y_{m+1} = 0}} \frac{\sum\limits_{j=1}^{N} \lambda_j |y_j|^2}{\sum\limits_{j=1}^{N} |y_j|^2}.$$

The equalities

$$y_N = y_{N-1} = \cdots = y_{m+1} = 0$$

can be regarded as the orthogonality conditions $(y, y^{(k)}) = 0$, $k = N, N-1, \ldots, m+1$, for the vector y and the eigenvectors

$$y^{(k)} = (\underbrace{0, \ldots, 0}_{k-1 \text{ zeros}}, 1, \underbrace{0, \ldots, 0}_{N-k \text{ zeros}})^T$$

of the matrix

$$U^*AU = \begin{bmatrix} \lambda_1 & & & 0 \\ & \lambda_2 & & \\ & & \ddots & \\ 0 & & & \lambda_N \end{bmatrix}.$$

The eigenvector $x^{(k)}$ of the original matrix A corresponding to the same eigenvalue λ_k can be found from the equality $x^{(k)} = Uy^{(k)}$. Consequently, the following theorem holds.

THEOREM 7.3.1 (the variational Weber principle).

$$\lambda_m(A) = \max \frac{(Ax, x)}{(x, x)},$$

where the maximum is taken over vectors $x \neq 0$ *such that* $(x, x^{(N)}) = \cdots = (x, x^{(m+1)}) = 0$. *Similarly,*

$$\lambda_m(A) = \min \frac{(Ax, x)}{(x, x)},$$

where the minimum is taken over $x \neq 0$ *such that* $(x, x^1) = \cdots = (x, x^{m+1}) = 0$.

7.4. The variational Courant–Fischer principle

The Courant–Fischer theorem. Comparison of eigenvalues of two Hermitian matrices.

THEOREM 7.4.1 (the variational Courant–Fischer principle). *If* $A = A^*$, *then*

$$\min_{\substack{\|q^{(j)}\|=1 \\ j=1,2,\dots,k}} \max_{\substack{\|x\|\neq 0 \\ (q^{(j)},x)=0}} \frac{(Ax, x)}{(x, x)} = \lambda_{N-k}, \tag{7.4.1}$$

$$\max_{\substack{\|q^{(j)}\|=1 \\ j=1,2,\dots,k}} \min_{\substack{\|x\|\neq 0 \\ (q^{(j)},x)=0}} \frac{(Ax, x)}{(x, x)} = \lambda_{k+1}. \tag{7.4.2}$$

PROOF. Choose a unitary transformation U such that

$$(Ax, x) = (AUy, Uy) = (\Lambda y, y) = \sum_{j=1}^{N} \lambda_j |y_j|^2 \qquad (y = Ux),$$

$$(x, x) = (Uy, Uy) = \sum_{j=1}^{N} |y_j|^2$$

and the linear connections

$$y_N = 0, \quad y_{N-1} = 0, \dots, y_{N-k+1} = 0.$$

These equalities hold provided that

$$y_{N-k-1} = y_{N-k-2} = \cdots = y_1 = 0, \quad y_{N-k} = 1.$$

Furthermore,

$$\frac{(Ax, x)}{(x, x)} = \frac{\lambda_{N-k}}{1} = \lambda_{N-k}.$$

Comparing the last relation with (7.2.1), we obtain (7.4.1). The relation (7.4.2) is established in a similar way. □

The Courant–Fischer theorem implies the following assertion.

COROLLARY 7.4.1. *If* A *and* B *are Hermitian matrices* ($A = A^*$, $B = B^*$) *and* $(Ax, x) \geq (Bx, x)$ *for all* x, *then* $\lambda_j(A) \geq \lambda_j(B)$, $j = 1, 2, \dots, N$.

The inequality $(Ax,x) \geq (Bx,x)$ for Hermitian matrices A and B is usually written in the symbolic form as $A \geq B$.

PROBLEM 7.4.1. Prove that the Hermitian $(N_1+N_2+N_3)\times(N_1+N_2+N_3)$-matrix

$$\begin{bmatrix} A & C & D \\ C^* & -B & E \\ D^* & E^* & 0 \end{bmatrix}$$

has at least N_1 positive and at least N_2 negative eigenvalues if the $N_1 \times N_1$-block $A = A^* > 0$ and the $N_2 \times N_2$-block $B = B^* > 0$ are positive definite.

PROBLEM 7.4.2. Using the inequality $\operatorname{rank}(F^*F - G^*G) \leq \operatorname{rank} F + \operatorname{rank} G$ and Problem 7.4.1, prove that if all columns of an $N \times N_1$-matrix F and all columns of an $N \times N_2$-matrix G are linearly independent and the matrix $[F \vdots G]$ is of full rank, then the matrix $H = F^*F - G^*G$ has exactly N_1 positive, N_2 negative, and N_3 zero eigenvalues, where $N_1 = \operatorname{rank} F$, $N_2 = \operatorname{rank} G$, $N_3 = N - N_1 - N_2$ (we assume that $N \geq N_1 + N_2$).

Problem 7.4.2 is the inertia law for quadratic forms

$$(Hx,x) = \sum_i \Big(\sum_j f_{ij}x_j\Big)^2 - \sum_i \Big(\sum_j g_{ij}x_j\Big)^2$$

in terms of linearly independent linear forms $\sum f_{ij}x_j$ and $\sum g_{ij}x_j$: it asserts that *the number of positive squares as well as the number of negative squares is the same for any representation of the quadratic form in terms of squares of linearly independent linear forms.*

7.5. Inequalities for singular values

Estimates for singular values of the sum $A+B$ of matrices A and B. The case of square matrices. Estimates for singular values of the product AB of matrices A and B. The variational Alakhverdiev principle. The Fan inequality.

We already know (see §§3.3 and 3.4) that an $M \times N$-matrix A admits the singular value decomposition $A = U_M K V_N^*$, where U_M and V_N are unitary square matrices, Σ is the square diagonal matrix with nonnegative diagonal elements $\sigma_1, \sigma_2, \ldots, \sigma_{N_0}$ ($N_0 = \min\{M,N\}$), and the matrix K has the same number of rows and columns as the matrix A and has additional zero columns and rows in comparison with the matrix Σ. In this representation, σ_j, $j = 1,2,\ldots,N_0$, are singular values of A. We enumerate them so that $\sigma_{N_0} \geq \sigma_{N_0-1} \geq \cdots \geq \sigma_1 \geq 0$.

Using the singular value decomposition, it is not hard to verify that $\|A\| = \sigma_{N_0}$, where $N_0 = \min\{M,N\}$. Indeed,

$$\|A\| = \max_{x\neq 0}\frac{\|Ax\|}{\|x\|} = \max_{y\neq 0}\sqrt{\frac{(Ky,Ky)}{(y,y)}} = \max_{y\neq 0}\frac{\sqrt{\sum\limits_{j=1}^{N_0}\sigma_j^2|y_j|^2}}{\sqrt{\sum\limits_{j=1}^{N}|y_j|^2}} = \sigma_{N_0}.$$

Similarly,

$$\min_{x\neq 0}\frac{\|A\|}{\|x\|}=\min_{x\neq 0}\sqrt{\frac{(Ax,Ax)}{(x,x)}}=\begin{cases}\sigma_1\geq 0 & \text{if } M\geq N,\\ 0 & \text{if } M<N.\end{cases}$$

Since $A^{-1}=V_N K^{-1}U_N^*$ for $N=M$, we have $\|A^{-1}\|=[\sigma_1(A)]^{-1}$ for $N=M$ if the matrix A is nonsingular.

Let A and B be $M\times N$-matrices. Construct the $(M+N)\times(M+N)$-matrices

$$\mathfrak{A}=\begin{bmatrix}0 & A\\ A^* & 0\end{bmatrix},\quad \mathfrak{B}=\begin{bmatrix}0 & B\\ B^* & 0\end{bmatrix}.$$

We recall that the eigenvalues of the Hermitian matrix $\mathfrak{A}$ ($\mathfrak{B}$) consist of the singular values of the matrix A (B) up to sign and, perhaps, of several zeros. We have $\|\mathfrak{A}\|=\|A\|$ and $\|\mathfrak{B}\|=\|B\|$. Indeed,

$$\left\|\begin{bmatrix}0 & A\\ A^* & 0\end{bmatrix}\begin{pmatrix}u\\ v\end{pmatrix}\right\|=\sqrt{\|Av\|^2+\|A^*u\|^2}\leq\sqrt{\|A\|^2\|v\|^2+\|A\|^2\|v\|^2}=\|A\|\left\|\begin{pmatrix}u\\ v\end{pmatrix}\right\|.$$

PROBLEM 7.5.1 (easy). Show that $\|\mathfrak{A}\|=\|A\|=\|A^*\|=\|\mathfrak{A}^*\|$.

Thus,

$$-\|B\|\left\|\begin{pmatrix}x\\ y\end{pmatrix}\right\|^2\leq\left(\mathfrak{B}\begin{pmatrix}x\\ y\end{pmatrix},\begin{pmatrix}x\\ y\end{pmatrix}\right)\leq\|B\|\left\|\begin{pmatrix}x\\ y\end{pmatrix}\right\|^2,$$

$$-\|B\|I_{2N}\leq\mathfrak{B}\leq\|B\|I_{2N},$$

$$\mathfrak{A}-\begin{bmatrix}\|B\|I_N & 0\\ 0 & \|B\|I_N\end{bmatrix}\leq\mathfrak{A}+\mathfrak{B}\leq\mathfrak{A}+\begin{bmatrix}\|B\|I_N & 0\\ 0 & \|B\|I_N\end{bmatrix}.$$

Consequently (see Corollary 7.4.1),

$$\lambda_j(\mathfrak{A})-\|B\|\leq\lambda_j(\mathfrak{A}+\mathfrak{B})\leq\lambda_j(\mathfrak{A})+\|B\|.$$

Recalling the rule for the choice of singular values of A and $A+B$ (see §3.5) among eigenvalues of the matrices $\mathfrak{A}$ and $\mathfrak{A}+\mathfrak{B}$, we arrive at the following important assertion.

THEOREM 7.5.1. *The singular values $\sigma_j(A)$ and $\sigma_j(A+B)$ of the matrices A and $A+B$, respectively, are connected by the inequalities*

$$\sigma_j(A)-\|B\|\leq\sigma_j(A+B)\leq\sigma_j(A)+\|B\|,$$
$$|\sigma_j(A+B)-\sigma_j(A)|\leq\|B\|.$$

We establish several important properties of singular values. For simplicity, we restrict ourselves to the case of square $N\times N$-matrices. Let $U^*U=V^*V=I_N$ and $A=U\Sigma V^*$, where

$$\Sigma=\begin{bmatrix}\sigma_N & & & 0\\ & \sigma_{N-1} & & \\ & & \ddots & \\ 0 & & & \sigma_1\end{bmatrix},$$

$$A^*A = V \begin{bmatrix} \sigma_N^2 & & & 0 \\ & \sigma_{N-1}^2 & & \\ & & \ddots & \\ 0 & & & \sigma_1^2 \end{bmatrix} V^*.$$

We recall that

$$\frac{\|Ax\|}{\|x\|} = \sqrt{\frac{(A^*Ax, x)}{(x, x)}} = \sqrt{\frac{\sum_{j=1}^{N} \sigma_j^2 y_j}{\sum_{j=1}^{N} y_j^2}}, \quad \begin{pmatrix} y_1 \\ y_2 \\ \vdots \\ y_N \end{pmatrix} = V \begin{pmatrix} x_1 \\ x_2 \\ \vdots \\ x_N \end{pmatrix}.$$

By Theorem 7.4.1,

$$\sigma_j = \min_{\substack{\|q^{(k)}\|=1 \\ k=1,2,\dots,N-j}} \max_{\substack{\|x\|\neq 0 \\ (q^{(k)},x)=0}} \frac{\|Ax\|}{\|x\|}. \tag{7.5.1}$$

From (7.5.1) we obtain

$$\sigma_j(A)\sigma_1(B) \leq \sigma_j(AB) \leq \sigma_j(A)\|B\| \equiv \sigma_j(A)\sigma_N(B).$$

PROBLEM 7.5.2. Prove that

$$\sigma_j(AB) \leq \|A\|\, \sigma_j(B). \tag{7.5.2}$$

Hint: Use Problem 7.5.1.

In the N-dimensional space, the linear constraints $(q^{(k)}, x) = 0$, $k = 1, 2, \dots,$ $N - j$, determine a linear subspace of dimension at least j. We can assume that the subspace is given by the equation $Kx = 0$, where K is an $N \times N$-matrix of rank at most $N - j$. We write this assumption in the symbolic form as $K \in \mathbb{R}_{N-j}$. Moreover,

$$\sigma_j = \min_{K\in\mathbb{R}_{N-j}} \max_{\substack{Kx=0 \\ \|x\|\neq 0}} \frac{\|Ax\|}{\|x\|} \leq \min_{K\in\mathbb{R}_{N-j}} \max_{\|x\|\neq 0} \frac{\|(A-K)x\|}{\|x\|} = \min_{K\in\mathbb{R}_{N-j}} \|A - K\|,$$

i.e., $\sigma_j(A) \leq \min_{K\in\mathbb{R}_{N-j}} \|A - K\|$. In fact, we have equality $\sigma_j(A) = \min_{K\in\mathbb{R}_{N-j}} \|A - K\|$. Indeed, let $A = U\Sigma V^*$ $(U^*U = V^*V = I)$. Setting

$$K = U \begin{bmatrix} \sigma_N & & & & & & & 0 \\ & \sigma_{N-1} & & & & & & \\ & & \ddots & & & & & \\ & & & \sigma_{j+1} & & & & \\ & & & & 0 & & & \\ & 0 & & & & 0 & & \\ & & & & & & \ddots & \\ & & & & & & & 0 \end{bmatrix} V^*$$

(it is obvious that $K \in \mathbb{R}_{N-j}$), we find that

$$A - K = U \begin{bmatrix} 0 & & & & & & & \\ & 0 & & & & & \mathbf{0} & \\ & & \ddots & & & & & \\ & & & 0 & & & & \\ & & & & \sigma_j & & & \\ & & & & & \sigma_{j-1} & & \\ & \mathbf{0} & & & & & \ddots & \\ & & & & & & & \sigma_1 \end{bmatrix} V^*,$$

$$\|A - K\| = \sigma_j.$$

Thus, we have proved the following theorem.

THEOREM 7.5.2 (the variational Alakhverdiev principle).

$$\sigma_j(A) = \min_{K \in \mathbb{R}_{N-j}} \|A - K\|.$$

Since the rank of the sum of matrices is not greater than the sum of the ranks of the summands, for $T \in \mathbb{R}_s$ we obtain

$$\sigma_r(A+T) = \min_{K \in \mathbb{R}_{N-r}} \|A + T - K\| \geq \min_{K-T \in \mathbb{R}_{N-r+s}} \|A - (K - T)\| = \sigma_{r-s}(A).$$

We change the notation:

$$A + T \longrightarrow A, \quad T \longrightarrow -T, \quad A \longrightarrow A + T.$$

By Theorem 7.5.2, $\sigma_r(A) \geq \sigma_{r-s}(A+T)$ and $\sigma_{r+s}(A) \geq \sigma_r(A+T)$.

COROLLARY 7.5.1. *If a matrix T is of rank s and $1 \leq r - s, r + s \leq N$, then the following inequalities hold:*

$$\sigma_{r-s}(A) \leq \sigma_r(A+T) \leq \sigma_{r+s}(A).$$

Let matrices K_0 and L_0 be such that

$$\sigma_r(A) = \min_{K \in \mathbb{R}_{N-r}} \|A - K\| = \|A - K_0\|, \qquad K_0 \in \mathbb{R}_{N-r},$$
$$\sigma_s(B) = \min_{L \in \mathbb{R}_{N-s}} \|B - L\| = \|B - L_0\|, \qquad L_0 \in \mathbb{R}_{N-s}.$$

Then

$$\|(A+B) - (K_0 + L_0)\| \leq \|A - K_0\| + \|B - L_0\| = \sigma_r(A) + \sigma_s(B),$$
$$K_0 + L_0 \in \mathbb{R}_{(N-r)+(N-s)} = \mathbb{R}_{N-(r+s-N)}.$$

Therefore,

$$\sigma_{r+s-N}(A+B) = \min_{M \in \mathbb{R}_{N-(r+s-N)}} \|A + B - M\|$$
$$\leq \|(A+B) - (K_0 + L_0)\| \leq \sigma_r(A) + \sigma_s(B).$$

Thus, we have established one more inequality for singular values:

$$\sigma_{r+s-N}(A+B) \leq \sigma_r(A) + \sigma_s(B), \quad 2N \geq r + s > N, \ r \leq N, \ s \leq N.$$

Let $K_0 \in \mathbb{R}_{N-r}$ and $L_0 \in \mathbb{R}_{N-s}$. We note that

$$\begin{aligned} AB &= (A - K_0)(B - L_0) + K_0\left(B - \frac{1}{2}L_0\right) + \left(A - \frac{1}{2}K_0\right)L_0 \\ &= (A - K_0)(B - L_0) + K_1 + L_1 = (A - K_0)(B - L_0) + M, \end{aligned}$$

where

$$K_1 = K_0(B - L_0/2) \in \mathbb{R}_{N-r}, \quad L_1 = (A - K_0/2)L_0 \in \mathbb{R}_{N-s},$$
$$M_1 = K_1 + L_1 \in \mathbb{R}_{2N-r-s};$$

indeed, the rank of the product of matrices does not exceed the rank of each factor and the rank of the sum of matrices is not greater than the sum of ranks of the summands. Consequently,

$$\begin{aligned} \sigma_{r+s-N}(AB) = \min_{M \in \mathbb{R}_{2N-r-s}} \|AB - M\| &\le \|AB - M_1\| = \|(A - K_0)(B - L_0)\| \\ &\le \|A - K_0\| \, \|B - L_0\| = \sigma_r(A)\sigma_s(B). \end{aligned}$$

Thus, $\sigma_{r+s-N}(AB) \le \sigma_r(A)\sigma_s(B)$.

Thereby, we have justified the Fan inequalities (obtained in 1951)

$$\sigma_{r+s-N}(A + B) \le \sigma_r(A) + \sigma_s(B), \quad \sigma_{r+s-N}(AB) \le \sigma_r(A)\sigma_s(B),$$

where $r < N$, $s < N$, and $r + s > N$.

7.6. Remark about enumeration of singular values

Different numberings of singular values. Reformulation of the Alakh′-verdiev principle and the Fan inequalities.

Different numberings of singular values are used in different cases. We ordered singular values in ascending order, i.e.,

$$\sigma_N \ge \sigma_{N-1} \ge \cdots \ge \sigma_1 \ge 0. \tag{7.6.1}$$

However, in the literature the descending order is often used:

$$\sigma_1 \ge \sigma_2 \ge \cdots \ge \sigma_N. \tag{7.6.2}$$

The difference in numberings is caused by the following fact. In an infinite-dimensional space, compact operators do not necessarily have a smallest singular value at all. Moreover, such operators can have infinitely many singular values. In a finite-dimensional space, it is convenient to enumerate singular values starting with the smallest one. Later in the volume, we will indicate a numbering of singular values used in the current section.

If singular values are numbered in accordance with (7.6.2), the variational Alakhverdiev principle and the Fan inequalities can be rewritten as follows:

$$\sigma_{r+1}(A) = \min_{K \in \mathbb{R}_r} \|A - K\|,$$
$$\sigma_{r+s-1}(A + B) \le \sigma_r(A) + \sigma_s(B), \quad \sigma_{r+s-1}(AB) \le \sigma_r(A)\sigma_s(B).$$

We continue to use the numbering (7.6.1) of singular values so that greater numbers j correspond to greater singular values σ_j.

7.7. The notion of conditionality for solutions of linear equations

The influence of perturbations of the right-hand side and the coefficients of an equation on a solution. Characteristics of perturbations of a solution. The notion of the condition number. Estimates for perturbations of solutions in terms of the condition number. Remark about round-off errors and accuracy of computation.

Consider a system of N linear equations

$$\sum_{j=1}^{N} a_{ij}x_j = f_j, \quad 1 \le i \le N,$$

with respect to N variables $x_1, x_2, \ldots, x_N$. The system can be written in the vector form

$$Ax = f,$$

where A is a square $N \times N$-matrix with elements a_{ij}. In many cases, information about the conditionality of a solution to the system is very useful. In other words, it is useful to characterize the sensitivity of the solution (the vector x) to a perturbation of the right-hand side (the vector f) and the coefficients (the matrix A).

First we assume the coefficients are fixed. Together with the original system $Ax = f$, we consider the system

$$A(x + \xi) = f + \varphi \tag{7.7.1}$$

with perturbation φ. It is obvious that in (7.7.1) the vector ξ describes the perturbation of the solution x. We try to estimate the relative quantity $\dfrac{\|\xi\|}{\|x\|}$ of the perturbation in terms of the relative quantity $\dfrac{\|\varphi\|}{\|f\|}$ of the perturbation of the right-hand side. We estimate the parameter

$$\mu = \sup_{\varphi, f} \frac{\|\xi\|/\|x\|}{\|\varphi\|/\|f\|} = \sup_{\varphi, f} \frac{\|\xi\|\,\|f\|}{\|\varphi\|\,\|x\|}.$$

The parameter μ depends on A ($\mu = \mu(A)$). We express μ explicitly in terms of singular values $\sigma_j = \sigma_j(A)$. We use the singular value decomposition $A = U\Sigma V^*$ with unitary matrices U, V ($U^*U = V^*V = I_N$) and the square diagonal matrix

$$\Sigma = \begin{bmatrix} \sigma_N & & & 0 \\ & \sigma_{N-1} & & \\ & & \ddots & \\ 0 & & & \sigma_1 \end{bmatrix}, \quad \sigma_j \ge \sigma_{j-1}.$$

Together with the vectors f, φ, x, and ξ, consider the vectors $g = U^*f$, $\psi = U^*\varphi$, $y = V^*x$, and $\eta = V^*\xi$. We note that $\|g\| = \|f\|$, $\|\psi\| = \|\varphi\|$, $\|y\| = \|x\|$, and $\|\eta\| = \|\xi\|$. It is easy to check that the equalities $Ax = f$ and $A(x+\xi) = f+\varphi$ imply the equalities $\Sigma y = g$ and $\Sigma\eta = \psi$. Representing the last equalities componentwise as $\sigma_j y_j = g_j$ and $\sigma_j \eta_j = \psi_j$, we find

$$\|x\| = \|y\| = \sqrt{\sum_{j=1}^{N} |y_j|^2} = \sqrt{\sum_{j=1}^{N} |g_j/\sigma_j|^2} \geq \frac{\sqrt{\sum\limits_{j=1}^{N} g_j^2}}{\max\limits_{1\leq k\leq N} \sigma_k} = \frac{\|g\|}{\sigma_N} = \frac{\|f\|}{\sigma_N},$$

$$\|\xi\| = \|\eta\| = \sqrt{\sum_{j=1}^{N} |\eta_j|^2} = \sqrt{\sum_{j=1}^{N} |\psi_j/\sigma_j|^2} \leq \frac{\sqrt{\sum\limits_{j=1}^{N} |\psi_j|^2}}{\min\limits_{1\leq k\leq n} \sigma_k} = \frac{\|\psi\|}{\sigma_1} = \frac{\|\varphi\|}{\sigma_1}$$

or, in short,

$$\|x\| \geq \frac{\|f\|}{\sigma_N}, \quad \|\xi\| \leq \frac{\|\varphi\|}{\sigma_1}.$$

Consequently,

$$\mu(A) = \sup_{\varphi, f} \frac{\|\xi\|\,\|f\|}{\|\varphi\|\,\|x\|} \leq \frac{\sigma_N(A)}{\sigma_1(A)}.$$

In fact, we have the equality

$$\mu(A) = \frac{\sigma_N(A)}{\sigma_1(A)}.$$

We can verify this assertion if we find vectors f and φ such that

$$\|x\| = \frac{\|f\|}{\sigma_N(A)}, \quad \|\xi\| = \frac{\|\varphi\|}{\sigma_1(A)}.$$

It is obvious that for f and φ we can take the vectors

$$f = U \begin{pmatrix} g_1 \\ 0 \\ \vdots \\ 0 \\ 0 \end{pmatrix}, \quad \varphi = U \begin{pmatrix} 0 \\ 0 \\ \vdots \\ 0 \\ \psi_N \end{pmatrix}, \qquad g_1 \neq 0, \quad \psi_N \neq 0.$$

The above arguments explain why the parameter

$$\mu(A) = \frac{\sigma_N(A)}{\sigma_1(A)} = \|A\|\,\|A^{-1}\| \tag{7.7.2}$$

is referred to as the condition number with respect to the right-hand side of the system of linear equations $Ax = f$.

Using $\mu(A)$, we can characterize the influence of perturbations of the coefficient matrix on the solution x. Therefore, the parameter $\mu(A)$ will be referred to as the *condition number* of the system $Ax = f$.

We note that (7.7.2) implies $\mu(A) = \|A\|\,\|A^{-1}\| = \mu(A^{-1})$. We compare a solution to the perturbed system

$$(A + \Delta)(x + \xi) = f + \varphi \tag{7.7.3}$$

with a solution to the system $Ax = f$. The following relations are obvious:

$$(A + \Delta)\xi = \varphi - \Delta x, \tag{7.7.4}$$

$$\|\varphi\| = \frac{\|\varphi\|}{\|f\|}\,\|f\| = \frac{\|\varphi\|}{\|f\|}\,\|Ax\| \le \frac{\|\varphi\|}{\|f\|}\,\|A\|\,\|x\|, \tag{7.7.5}$$

$$\|\Delta x\| \le \|A\|\,\frac{\|\Delta\|}{\|A\|}\,\|x\|,$$

$$\xi = (A+\Delta)^{-1}(\varphi - \Delta x) = A^{-1}(I_N + \Delta A^{-1})^{-1}(\varphi - \Delta x), \tag{7.7.6}$$

$$\begin{aligned} \frac{\|\xi\|}{\|x\|} &\le \|A^{-1}\|\,\|A\|\,\|(I_N + \Delta A^{-1})^{-1}\|\left(\frac{\|\varphi\|}{\|f\|} + \frac{\|\Delta\|}{\|A\|}\right) \\ &= \mu(A)\,\|(I_N + \Delta A^{-1})^{-1}\|\left(\frac{\|\varphi\|}{\|f\|} + \frac{\|\Delta\|}{\|A\|}\right). \end{aligned} \tag{7.7.7}$$

If $\mu(A)\dfrac{\|\Delta\|}{\|A\|} < 1$, then

$$\|\Delta A^{-1}\| \le \|\Delta\|\,\|A^{-1}\| = \|A\|\,\|A^{-1}\|\,\frac{\|\Delta\|}{\|A\|} = \mu(A)\frac{\|\Delta\|}{\|A\|} < 1.$$

Consequently,

$$\begin{aligned} &(I_N + \Delta A^{-1})^{-1} = I_N + \sum_{k=1}^{\infty}(-1)^k(\Delta A^{-1})^k, \\ &\|(I_N + \Delta A^{-1})^{-1}\| \le 1 + \sum_{k=1}^{\infty}\|\Delta A^{-1}\|^k = \frac{1}{1 - \|\Delta A^{-1}\|} \le \frac{1}{1 - \mu(A)\dfrac{\|\Delta\|}{\|A\|}}. \end{aligned} \tag{7.7.8}$$

For $\mu(A)\dfrac{\|\Delta\|}{\|A\|} < 1$, using (7.7.2)–(7.7.8) and the known estimates for $\dfrac{\|\varphi\|}{\|f\|}$ and $\dfrac{\|\Delta\|}{\|A\|}$ (i.e., the estimates for the relative quantities of perturbations of the right-hand side and the coefficients of the system), we can estimate the perturbation of the solution:

$$\frac{\|\xi\|}{\|x\|} \le \frac{\mu(A)}{1 - \mu(A)\frac{\|\Delta\|}{\|A\|}}\left(\frac{\|\varphi\|}{\|f\|} + \frac{\|\Delta\|}{\|A\|}\right).$$

It is just this estimate which allows us to use $\mu(A)$ in order to estimate the sensitivity of the solution to perturbations of not only the right-hand side, but also of coefficients.

For $\mu(A)\dfrac{\|\Delta\|}{\|A\|} < 1$, the inequalities (see Theorem 7.5.1)

$$\begin{aligned} \sigma_1(A+\Delta) &\ge \sigma_1(A) - \|\Delta\| > \sigma_1(A) - \frac{\|A\|}{\mu(A)} = \sigma_1(A) - \|A\|\frac{\sigma_1(A)}{\sigma_N(A)} \\ &= \sigma_1(A) - \sigma_1(A) = 0, \end{aligned}$$

$$\sigma_N(A+\Delta) \le \sigma_N(A) + \|\Delta\|$$

show that

$$\mu(A+\Delta) \le \frac{\sigma_N(A) + \|\Delta\|}{\sigma_1(A) - \|\Delta\|} = \mu(A)\frac{1 + \dfrac{\|\Delta\|}{\|A\|}}{1 - \mu(A)\dfrac{\|\Delta\|}{\|A\|}}.$$

In other words, if the condition number $\mu(A)$ of the matrix A is not too large, then the condition number $\mu(A+\Delta)$ of the matrix $A+\Delta$ perturbed by a small perturbation Δ cannot be too much greater than $\mu(A)$.

The singular values of matrices A_1 and A_2 such that $A = UA_1V^*$ ($U^*U = V^*V = I$) coincide. Consequently, $\mu(A_1) = \mu(A_2)$. In particular, if we solve the system $Ax = f$ of N linear equations with respect to N variables and, using unitary transformations U and V, reduce the system to the two-diagonal form

$$Dy \equiv \begin{bmatrix} d_1 & b_2 & & & & 0 \\ & d_2 & b_3 & & & \\ & & \ddots & \ddots & & \\ & 0 & & & d_{N-1} & b_N \\ & & & & & d_N \end{bmatrix} \begin{pmatrix} y_1 \\ y_2 \\ y_3 \\ \vdots \\ y_{N-1} \\ y_N \end{pmatrix} = \begin{pmatrix} g_1 \\ g_2 \\ g_3 \\ \vdots \\ g_{N-1} \\ g_N \end{pmatrix} \equiv g \tag{7.7.9}$$

in the same way as in §3.4, then the condition number of the coefficient matrix remains constant: $\mu(A) = \mu(D)$. For the two-diagonal matrix D the singular values and, consequently, the condition number can be computed by the Sturm method (see §§3.4 and 5.7).

It makes sense to solve a system of equations if we hope to obtain an approximate solution sufficiently close to the exact solution. Therefore, it is natural to consider only those systems with condition numbers that are not too large. If $\mu(D)$ is not too large, then the solution y of the transformed system $Dy = g$ is easily found from the recurrence formulas (see §3.2)

$$y_N = \frac{g_N}{d_N}, \quad y_j = \frac{g_j - b_{j+1}y_{j+1}}{d_j}, \quad j = N-1, N-2, \dots, 1. \tag{7.7.10}$$

If information on the accuracy of performance of some arithmetic operations and the condition number $\mu(D)$ is available, we can estimate the accuracy of the computed solution. We do not discuss such an estimate here and refer the reader to the monographs [**15, 26**].

The solution x to the system $Ax = f$ is obtained from y by a unitary transformation $x = V^*y$.

We recommend using unitary (or orthogonal) transformations $D = UAV^*$, $g = Uf$, $x = V^*y$, i.e., transformations with matrices U and V whose conditions numbers have minimal value (equal to 1). This allows us to escape a noncontrollable influence of unavoidable round-off errors (see [**15, 26**]).

PROBLEM 7.7.1. Prove that for a two-diagonal matrix D of the form (7.7.9) the following relations hold:

$$\mathfrak{M}(D) \le \sigma_N(D) = \|D\| \le \sqrt{2}\mathfrak{M}(D),$$

where

$$\mathfrak{M}(D) = \max\{\max_{1\le i\le N-1}(|d_i| + |b_{i+1}|), \max_{2\le i\le N}(|d_i| + |b_i|)\}.$$

PROBLEM 7.7.2. Assume that the influence of round-off errors in computation by formulas (7.7.10) of the two-diagonal system (7.7.9) can be simulated as the perturbation of the coefficients d_j, b_j and the right-hand sides g_j. Instead of d_j,

b_j, and g_j, consider $\overline{d}_j$, $\overline{b}_j$, and $\overline{g}_j$ such that[2] $|\overline{d}_j - d_j| < \varepsilon |d_j|$, $|\overline{b}_j - b_j| < \varepsilon |b_j|$, and $|\overline{g}_j - g_j| < \varepsilon |g_j|$. Deduce an estimate for $\dfrac{\|\overline{y} - y\|}{\|y\|}$, i.e., estimate the relative perturbation of the solution in terms of $\mathfrak{M}(D)$, $\sigma_1(D)$, and ε.

Since

$$|\overline{y}_j - y_j| \le \|\overline{y} - y\|, \quad \|y\| \le \frac{\|f\|}{\sigma_1(D)} \le \left(\frac{\sqrt{N}}{\sigma_1(D)}\right) \max_k |f_k|,$$

from the estimate for $\dfrac{\|\overline{y} - y\|}{\|y\|}$ we can obtain an estimate for the maximal error in terms of $\max\limits_j |\overline{y}_j - y_j|$.

7.8. Approximation by matrices of small rank

Estimate for the change of the determinant under a perturbation of a matrix. Important conclusions in the case of an ill-conditioned matrix. The characteristics of the spectral conditionality and estimates for eigenvalues of a matrix and its submatrices.

Using the singular value decomposition $A = U\Sigma V^*$ $(U^*U = V^*V = I_N)$ of a nonsingular square matrix A, where

$$\Sigma = \begin{bmatrix} \sigma_N & & & 0 \\ & \sigma_{N-1} & & \\ & & \ddots & \\ 0 & & & \sigma_1 \end{bmatrix}, \quad \sigma_j \ge \sigma_{j-1},$$

and the equalities

$$\det A = \det U \det \Sigma \det \det V^*, \quad |\det U| = |\det V^*| = 1,$$

we establish the relation $|\det A| = \sigma_N(A)\sigma_{N-1} \dots \sigma_1(A) > 0$. If

$$\|\Delta\| < \sigma_1(A) = \frac{\|A\|}{\mu(A)},$$

then $\sigma_1(A + \Delta) \ge \sigma_1(A) - \|\Delta\| > 0$. Consequently, the perturbed matrix $A + \Delta$ is nonsingular.

We estimate the quantity

$$\frac{\det(A + \Delta) - \det A}{\det A},$$

which presents the relative change of $\det A$ under the perturbation of A by a small additional term Δ:

$$\left|\frac{\det(A + \Delta) - \det A}{\det A}\right| = |\det[A^{-1}(A + \Delta)] - 1| = |\det(I + A^{-1}\Delta) - 1|$$

$$= \left|\prod_{j=1}^{N}[1 + \lambda_j(A^{-1}\Delta)] - 1\right| \le \prod_{j=1}^{N}[1 + |\lambda_j(A^{-1}\Delta)|] - 1.$$

[2] For personal computers the standard accuracy is characterized by the values $\varepsilon \sim 10^{-15} \div 10^{-16}$.

We used the fact that

$$\prod_{j=1}^{N}(1+\lambda_j)-1$$

is a polynomial (the sum of monomials) in $\lambda_1, \lambda_2, \ldots, \lambda_N$. If λ_j is replaced by the module $|\lambda_j|$ in this polynomial, then the module of each monomial remains constant, whereas the sum of the monomials can only increase. For $x_i \geq 0$ we have

$$1+x_j \leq 1+x_j+\frac{1}{2!}x_j^2+\frac{1}{3!}x_3^3+\cdots=e^{x_j},$$
$$\prod_{j=1}^{N}(1+x_j) \leq \exp\left(\sum_{j=1}^{N} x_j\right).$$

Therefore,

$$\left|\frac{\det(A+\Delta)-\det A}{\det A}\right| \leq \prod_{j=1}^{N}[1+|\lambda_j(A^{-1}\Delta)|]-1 \leq \exp\left(\sum_{j=1}^{N}|\lambda_j(A^{-1}\Delta)|\right)-1.$$

In §6.3, we proved a number of important consequences of the Weyl inequalities. In particular, we proved that for an $N \times N$-matrix B

$$\sum_{j=1}^{N}|\lambda_j(B)| \leq \sum_{j=1}^{N}\sigma_j(B).$$

Therefore,

$$\begin{aligned}\sum_{j=1}^{N}|\lambda_j(A^{-1}\Delta)| &\leq \sum_{j=1}^{N}\sigma_j(A^{-1}\Delta) \leq \|\Delta\|\sum_{j=1}^{N}\sigma_j(A^{-1}) \\ &= \frac{\|\Delta\|}{\|A\|}\|A\|\,\|A^{-1}\|\sum_{j=1}^{N}\frac{\sigma_j(A^{-1})}{\sigma_N(A^{-1})} = \frac{\|\Delta\|}{\|A\|}\mu(A)\sum_{j=1}^{N}\frac{\sigma_j(A^{-1})}{\sigma_N(A^{-1})},\end{aligned}$$

where $\mu(A)$ is the condition number (see §7.7) of the linear system $Ax = f$ with the coefficient matrix A. We used the inequality $\sigma_j(A^{-1}\Delta) \leq \|\Delta\|\sigma_j(A^{-1})$ (see (7.5.2)) and the relations $\|A\| = \sigma_N(A)$, $\|A^{-1}\| = \sigma_1^{-1}(A) = \sigma_N(A^{-1})$. Thus,

$$\left|\frac{\det(A+\Delta)-\det A}{\det A}\right| \leq \exp\left(\frac{\Delta}{\|A\|}\mu(A)\sum_{j=1}^{N}\frac{\sigma_j(A^{-1})}{\sigma_N(A^{-1})}\right)-1. \tag{7.8.1}$$

Introduce the notation

$$\nu^{(1)}(C) = \sum_{j=1}^{N}\frac{\sigma_j(C)}{\sigma_N(C)}.$$

It is obvious that $\nu^{(1)}(C) \leq N$ and $\nu^{(1)}(A^{-1}) \leq N$. Using the inequalities

$$e^x - 1 < xe^x < \frac{x}{1-x}, \quad x > 0,$$

we rewrite (7.8.1) and simplify the result:

$$\left|\frac{\det(A+B)-\det A}{\det A}\right| \le \begin{cases} \exp\left(\frac{\|\Delta\|}{\|A\|}\mu(A)\nu^{(1)}(A^{-1})\right)-1, \\ \frac{\|\Delta\|}{\|A\|}\mu(A)\nu^{(1)}(A^{-1})\exp\left(\frac{\|\Delta\|}{\|A\|}\mu(A)\nu(A^{-1})\right), \\ \dfrac{\mu(A)\nu^{(1)}(A^{-1})\frac{\|\Delta\|}{\|A\|}}{1-\mu(A)\nu^{(1)}(A^{-1})\frac{\|\Delta\|}{\|A\|}} \le \dfrac{N\mu(A)\frac{\|\Delta\|}{\|A\|}}{1-N\mu(A)\frac{\|\Delta\|}{\|A\|}}. \end{cases} \tag{7.8.2}$$

All the estimates (7.8.2) are somewhat rough. However, they adequately express the fact that for an ill-conditioned matrix (the condition number $\mu(A)$ is too large) an insignificant perturbation of elements can produce a change of the determinant which is too large.

To illustrate this, we give an example. Consider the following matrix of 30th order:

$$A = \begin{bmatrix} 1 & 2 & & & & 0 \\ & 1 & 2 & & & \\ & & 1 & 2 & & \\ & & & \ddots & \ddots & \\ & & & & 1 & 2 \\ 0 & & & & & 1 \end{bmatrix}.$$

Only the elements on the main diagonal and the diagonal above it differ from zero. The determinant of A is equal to 1 and the norm of A is equal to 3 with large accuracy ($\sigma_{30}(A) = \|A\| \approx 3$). The singular value $\sigma_1(A)$ turns out to be very small: $\sigma_1(A) \approx 0.138 \cdot 10^{-8}$ and $\mu(A) = 2.18 \cdot 10^9$. The matrix A is very ill conditioned. Therefore, it is natural to expect that a small perturbation $A + \Delta$ of A with small norm $\|\Delta\|$ leads to significant perturbation of the determinant.

For Δ we take the matrix with the single nonzero element -2^{-29} in the lower left corner. We have

$$\|\Delta\| = 2^{-29} \approx 0.18 \cdot 10^{-8},$$

$$\det[A+\Delta] = \underbrace{1 \cdot 1 \cdot \dots \cdot 1}_{30 \text{ times}} - 2^{-29}\underbrace{2 \cdot 2 \cdot \dots \cdot 2}_{29 \text{ times}} = 0.$$

Consequently,

$$\left|\frac{\det(A+\Delta)-\det A}{\det A}\right| = 1.$$

Later in the volume, together with the condition number $\mu(A)$ and the parameter $\nu^{(1)}(A)$, we consider the parameters

$$\nu^{(k)}(C) = \sqrt[k]{\sum_{j=1}^{N}\left(\frac{\sigma_j(C)}{\sigma_N(C)}\right)^k}, \quad k > 0.$$

The inequalities $\sigma_N(C) \geq \sigma_j(C) \geq \sigma_1(C)$ imply the estimate $\nu^{(k)}(C) \leq \sqrt[k]{N}$. On the other hand, for $k \geq 1$ we have

$$\begin{aligned}\nu^{(1)}(C) &= \sum_{j=1}^{N} \frac{\sigma_j(C)}{\sigma_N(C)} \\ &\leq \sqrt[k]{\sum_{j=1}^{N}\left(\frac{\sigma_j(C)}{\sigma_N(C)}\right)^k} \Big(\underbrace{1^{k/(k-1)} + 1^{k/(k-1)} + \cdots + 1^{k/(k-1)}}_{N \text{ terms}}\Big)^{1-1/k} \\ &= N^{1-1/k} \sqrt[k]{\sum_{j=1}^{N}\left(\frac{\sigma_j(C)}{\sigma_N(C)}\right)^k} = N^{1-1/k}\nu^{(k)}(C).\end{aligned}$$

Recalling the inequalities (see §6.3)

$$\sum_{j=N-n+1}^{N} |\lambda_j^p(C)| \leq \sum_{j=N-n+1}^{N} \sigma_j^p(C), \quad N \geq n \geq 1, \tag{7.8.3}$$

and using $\nu^{(p)}(C)$, we can estimate the eigenvalues of the matrix C as follows:

$$\sqrt[p]{\sum_{j=N-n+1}^{N} |\lambda_j^p|} \leq \nu^{(p)}(C)\,\|C\|, \quad 1 \leq n \leq N, \tag{7.8.4}$$

where the eigenvalues are ordered so that $|\lambda_{j+1}| \geq |\lambda_j|$. In particular, (7.8.4) means that the number k of the eigenvalues in the annulus $\|C\| \geq |\lambda| \geq \rho\|C\|$ ($\rho < 1$) can be estimated in terms of $\nu^{(p)}(C)$:

$$k \leq \left[\frac{\nu^{(p)}(C)}{\rho}\right]^p. \tag{7.8.5}$$

We emphasize that the estimate is independent of the dimension N of the space in which the operator corresponding to the matrix C acts. It is interesting to note that we can also use (7.8.3)–(7.8.5) in order to estimate the principal minors of C, not only C itself. For definiteness, let C_n be the upper principal $n \times n$-minor of C_N. The corresponding inequality in (5.4.4) can be written as follows:

$$\sum_{j=1}^{n} \sigma_j^p(C_n) \leq \sum_{j=N-n+1}^{N} \sigma_j^p(C_N). \tag{7.8.6}$$

Consequently,

$$\nu^{(p)}(C_n) \leq \frac{\|C_N\|}{\|C_n\|}\nu^{(p)}(C_N).$$

In view of (7.8.5) and (7.8.6), the parameters $\nu^{(p)}(A)$, $0 < p < \infty$, can be referred to as the *spectral condition number*. In accordance with the variational Alakhverdiev principle (see Theorem 7.5.2) an operator (matrix) A can be approximated by the matrix $A^{(n)}$ of rank n with error $\sigma_{N-n}(A)$ ($\|A - A^{(n)}\| \leq \sigma_{N-n}(A)$). Using the equality

$$\sum_{j=1}^{N} \sigma_j^p(A) = [\nu^{(p)}(A)]^p \|A\|^p,$$

we can estimate the number n of the singular values that are greater than $\delta\|A\|$. It is obvious that

$$\left[\nu^{(p)}(A)\right]^p \|A\|^p = \sum_{j=1}^{N} \sigma_j^p(A) \geq \sum_{\sigma_j > \delta\,\|A\|} \sigma_j^p(A) \geq n\delta^p \,\|A\|^p.$$

In other words,

$$n \leq \frac{\sqrt[p]{\nu^{(p)}(A)}}{\delta}, \qquad n_{\text{eff}} = \frac{\sqrt[p]{\nu^{(p)}(A)}}{\delta},$$

where n_{eff} is the effective rank of an operator that approximates A with relative accuracy δ. The question about the possibility of approximating operators in N-dimensional spaces by operators of small rank (of small dimension) often arises in applications.

CHAPTER 8

Resolvent and Dichotomy of Spectrum

8.1. Projections onto invariant subspaces

Invariant subspaces of a matrix A. Bases for invariant subspaces. Projections onto invariant subspaces. Properties of projections. Connection between projections P and $I - P$.

Let a matrix A be represented in the form

$$A = T \begin{bmatrix} A_{11} & 0 \\ 0 & A_{22} \end{bmatrix} T^{-1}, \tag{8.1.1}$$

where A_{11} and A_{22} have no common eigenvalues. Then the columns of the rectangular blocks T_1 and T_2 in the representation

$$T = [T_1 \vdots T_2], \tag{8.1.2}$$

form bases for the invariant subspaces of A associated with the eigenvalues of A_{11} and A_{22}, respectively.

The choice of a basis for a subspace is not unique. For example, if the columns of T_1 form a basis for a subspace, then the columns of $\widetilde{T}_1 = T_1 K_1$, where K_1 is a nonsingular square matrix, also form a basis for the same subspace. Therefore, in order to determine subspaces, projections are often used because they are independent of the choice of the basis.

A matrix P is called a *projection* if $P^2 = P$. It is obvious that the projection P transforms (projects) a vector x into a vector $y = Px$ and projects the vector $y = Px$ into itself, i.e., $Py = P^2x = y$. A vector x can be represented as $x = y + z$, where $Py = y$ and $Pz = 0$. It is obvious that $y = Px$ and $z = x - Px = (I - P)x$. The matrix $I - P$ is also a projection: $(I-P)^2 = I - 2P + P^2 = I - 2P + P = I - P$. We say that *$P$ projects a vector x into the vector $y = Px$ parallel to the vector $z = (I - P)x$*. The entire N-dimensional space can be expressed as the direct sum of the subspace of vectors $y = Px$ and the subspace of vectors $x - Px$. Let these subspaces have dimensions N_1 and N_2 ($N_1 + N_2 = N$) and let the columns of T_1 and T_2 form bases for these subspaces. It is easy to see that

$$P = \left[T_1 \vdots T_2\right] \begin{bmatrix} I_{N_1} & 0 \\ 0 & 0 \end{bmatrix} [T_1 \vdots T_2]^{-1}, \quad I - P = [T_1 \vdots T_2] \begin{bmatrix} 0 & 0 \\ 0 & I_{N_2} \end{bmatrix} [T_1 \vdots T_2]^{-1}.$$

The projections do not change if we take the columns of $\widetilde{T}_1 = T_1 K_1$ and $\widetilde{T}_2 = T_2 K_2$ instead of the columns of T_1 and T_2,

Since the columns of the matrix

$$T = [\underset{N_1 \text{ columns}}{T_1} \ \vdots \ \underset{N_2 \text{ columns}}{T_2}] \quad (N_1 + N_2 = N)$$

are linearly independent, we can represent T in the form

$$T = U \begin{bmatrix} I_{N_1} & M \\ 0 & I_{N_2} \end{bmatrix} \begin{bmatrix} K_{11} & 0 \\ 0 & K_{22} \end{bmatrix}, \tag{8.1.3}$$

where U is a unitary matrix ($U^*U = I_N$) and K_{11} and K_{22} are nonsingular square $N_1 \times N_1$- and $N_2 \times N_2$-blocks, respectively. Indeed, consider the singular value decomposition

$$T_1 = U \begin{bmatrix} D_1 \\ 0 \end{bmatrix} V_1^* = U \begin{bmatrix} D_1 V_1^* \\ 0 \end{bmatrix} = U \begin{bmatrix} K_{11} \\ 0 \end{bmatrix},$$

where U, V_1 are unitary matrices, D_1 is a diagonal matrix, and $K_{11} = D_1 V_1^*$ is a square $N_1 \times N_1$-matrix. It is obvious that

$$U^*T = \begin{bmatrix} K_{11} & K_{12} \\ 0 & K_{22} \end{bmatrix} = \begin{bmatrix} I_{N_1} & [K_{12}K_{22}^{-1}] \\ 0 & I_{N_2} \end{bmatrix} \begin{bmatrix} K_{11} & 0 \\ 0 & K_{22} \end{bmatrix} = \begin{bmatrix} I_{N_1} & M \\ 0 & I_{N_2} \end{bmatrix} \begin{bmatrix} K_{11} & 0 \\ 0 & K_{22} \end{bmatrix}.$$

The representation (8.1.3) is established.

Using the representations

$$T = U \begin{bmatrix} I_{N_1} & M \\ 0 & I_{N_2} \end{bmatrix} \begin{bmatrix} K_{11} & 0 \\ 0 & K_{22} \end{bmatrix},$$

$$T^{-1} = \begin{bmatrix} K_{11}^{-1} & 0 \\ 0 & K_{22}^{-1} \end{bmatrix} \begin{bmatrix} I_{N_1} & -M \\ 0 & I_{N_2} \end{bmatrix} U^*,$$

we can write any projection

$$P = T \begin{bmatrix} I_{N_1} & 0 \\ 0 & 0 \end{bmatrix} T^{-1}$$

onto an N_1-dimensional subspace in the following canonical form:

$$P = U \begin{bmatrix} I_{N_1} & M \\ 0 & I_{N-N_1} \end{bmatrix} \begin{bmatrix} I_{N_1} & 0 \\ 0 & 0 \end{bmatrix} \begin{bmatrix} I_{N_1} & -M \\ 0 & I_{N-N_1} \end{bmatrix} U^*. \tag{8.1.4}$$

It is easy to deduce the following simple but important formulas:

$$N_1 = \operatorname{tr} P, \quad N_2 = \operatorname{tr}(I_N - P).$$

A projection P is a projection onto an invariant subspace of a matrix A if $APx = PAPx$ for any x, i.e., if $AP = PAP$. If both $I - P$ and P are projections onto invariant subspaces of A, then $A(I-P) = (I-P)A(I-P)$. We simplify the last equality: $A - AP = A - PA - AP + PAP$ and $0 = PAP - PA$. Thus, it is equivalent to the equality $PA = PAP$. If both P and $I - P$ are the projections onto invariant subspaces of A, then $PA = AP$, $P^2 = P$. Multiplying the equality $PA = AP$ from the right or from the left by P, we obtain $PAP = AP^2 = AP$ and $PAP = P^2A = PA$. In view of the representation (8.1.1) of the matrix A and the corresponding decomposition (8.1.2) of the matrix T, it is easy to verify that

$$\begin{aligned} P &= [T_1 \vdots T_2] \begin{bmatrix} I_{N_1} & 0 \\ 0 & 0 \end{bmatrix} [T_1 \vdots T_2]^{-1}, \\ I - P &= \begin{bmatrix} T_1 \vdots T_2 \end{bmatrix} \begin{bmatrix} 0 & 0 \\ 0 & I_{N_2} \end{bmatrix} [T_1 \vdots T_2]^{-1} \end{aligned} \tag{8.1.5}$$

are projections onto invariant subspaces of A.

The following theorem presents one more important property of projections.

THEOREM 8.1.1. *If P_1 and P_2 are projections such that $\|P_1 - P_2\| < 1$, then $\operatorname{tr} P_1 = \operatorname{tr} P_2$ (i.e., P_1 and P_2 are projections onto subspaces of the same dimension).*

PROOF. The singular values of the matrices $I - P_1 + P_2$ and $I - P_2 + P_1$ differ from zero. Indeed,

$$\begin{aligned}\sigma_N(I) = \sigma_{N-1}(I) = \cdots = \sigma_1(I) = 1,\\ \sigma_N(P_2 - P_1) = \|P_2 - P_1\| = \rho < 1,\\ \sigma_j(P_2 - P_1) \le \sigma_N(P_2 - P_1) = \rho,\\ \sigma_j[I \pm (P_2 - P_1)] \ge \sigma_j(I) - \|P_2 - P_1\| = 1 - \rho > 0.\end{aligned}$$

In other words, the matrices $I + P_2 - P_1$ and $I + P_1 - P_2$ are nonsingular. It is obvious that

$$\operatorname{rank}[P_1(I - P_1 + P_2)] = \operatorname{rank} P_1, \quad \operatorname{rank}[P_2(I - P_2 + P_1)] = \operatorname{rank} P_2.$$

Since P_1 and P_2 are projections, we have

$$\begin{aligned}P_1(I - P_1 + P_2) = P_1 - P_1^2 + P_1P_2 = P_1P_2,\\ P_2(I - P_2 + P_1) = P_2 - P_2^2 + P_2P_1 = P_2P_1,\\ \operatorname{rank} P_1 = \operatorname{rank}[P_1(I - P_1 + P_2)] = \operatorname{rank} P_1P_2,\\ \operatorname{rank} P_2 = \operatorname{rank}[P_2(I - P_2 + P_1)] = \operatorname{rank} P_2P_1.\end{aligned}$$

As is known,

$$\begin{aligned}\operatorname{rank} P_1P_2 \le \min\{\operatorname{rank} P_1, \operatorname{rank} P_2\},\\ \operatorname{rank} P_2P_1 \le \min\{\operatorname{rank} P_1, \operatorname{rank} P_2\}.\end{aligned}$$

Thus, we arrive at the relations

$$\begin{aligned}\operatorname{rank} P_1 = \operatorname{rank} P_1P_2 \le \min\{\operatorname{rank} P_1, \operatorname{rank} P_2\},\\ \operatorname{rank} P_2 = \operatorname{rank} P_2P_1 \le \min\{\operatorname{rank} P_1, \operatorname{rank} P_2\},\end{aligned}$$

which are consistent only if P_1 and P_2 have the same rank. □

Suppose that a projection P is represented in the form

$$P = U \begin{bmatrix} I & M \\ 0 & I \end{bmatrix} \begin{bmatrix} I & 0 \\ 0 & 0 \end{bmatrix} \begin{bmatrix} I & -M \\ 0 & I \end{bmatrix} U^* = U \begin{bmatrix} I & -M \\ 0 & 0 \end{bmatrix} U^*,$$

where $U^*U = I$ and U is a unitary matrix. It is obvious that the norm of P coincides with the norm of the matrix $[I \vdots -M]$. Let

$$M = W\Sigma V^*, \quad WW^* = I_k, \quad VV^* = I_l,$$

where Σ is the "diagonal" matrix with k rows and l columns:

$$\Sigma = \begin{bmatrix} \sigma_k & & & 0 \\ & \sigma_{k-1} & & \\ & & \ddots & \\ 0 & & & \sigma_1 \end{bmatrix} \quad \text{if } k = l,$$

$$\Sigma = \left[\begin{array}{c|c} \begin{matrix} \sigma_k & & & 0 \\ & \sigma_{k-1} & & \\ & & \ddots & \\ 0 & & & \sigma_1 \end{matrix} & 0 \end{array}\right] \quad \text{if } k < l,$$

$$\Sigma = \left[\begin{array}{c} \begin{matrix} \sigma_l & & & 0 \\ & \sigma_{l-1} & & \\ & & \ddots & \\ 0 & & & \sigma_1 \end{matrix} \\ \hline 0 \end{array}\right] \quad \text{if } k > l.$$

In this case, for

$$\widetilde{V} = \begin{bmatrix} W & 0 \\ 0 & V \end{bmatrix}$$

we have

$$\|[I \vdots - M]\| = \|W[I \vdots - \Sigma]\widetilde{V}^*\| = \|[I \vdots - \Sigma]\| = \max_j \sqrt{1 + \sigma_j^2}$$
$$= \sqrt{1 + (\max_j \sigma_j)^2} = \sqrt{1 + \|M\|^2}.$$

Therefore, $\|P\| = \sqrt{1 + \|M\|^2} \geq 1$. We note that

$$I - P = U \begin{bmatrix} 0 & M \\ 0 & I \end{bmatrix} U^*.$$

Consequently,

$$\|I - P\| = \left\| \begin{bmatrix} M \\ I \end{bmatrix} \right\| = \sqrt{1 + \|M\|^2}.$$

Thus,

$$\|P\| = \|I - P\| = \sqrt{1 + \|M\|^2} \geq 1.$$

The equality $\|P\| = 1$ is possible only if $\|M\| = 0$. Furthermore, P and $I - P$ are projections onto mutually orthogonal subspaces or, which is equivalent, $(Px, [I - P]y) = ([P - P^*P]x, y) = 0$ for any x and y. In particular, for $y =$

$[P - P^*P]x$ we have

$$\|[P - P^*P]x\|^2 = ([P - P^*P]x, [P - P^*P]x) = 0,$$
$$P = P^*P = [P^*P]^* = P^*.$$

From $P = P^*$ and $P^2 = P$ it follows that $(Px, [I - P]y) = (x, P^*[I - P]y) = (x, [P - P^2]y) = 0$, i.e., the projections Px and $[I - P]y$ are orthogonal.

THEOREM 8.1.2. *If P is a projection such that $P \neq 0$ and $P \neq I$, then $\|P\| = \|I - P\| \geq 1$. The equality $\|P\| = 1$ holds only if $P = P^*$, i.e., if the projection P is orthogonal.*

PROBLEM 8.1.1. Show that if P is a projection ($P^2 = P$), then

$$\Pi = P[P^*P + (I - P^*)(I - P)]^{-1}P^*$$

is also a projection ($\Pi^2 = \Pi$); moreover, it is an orthogonal projection onto the same subspace as P: $\Pi P = P$, $P\Pi = \Pi$, $\Pi = \Pi^*$.

8.2. Integral representation of projections

Projections onto mutually complementary invariant subspaces of a block matrix A. Representation of a projection as an integral along a contour surrounded the eigenvalues of a matrix. Justification of integral representation of projections.

Suppose that a matrix A is represented in the form (8.1.1) and the eigenvalues of A_{11} and A_{22} are distinct. Then

$$P = T \begin{bmatrix} I_k & 0 \\ 0 & 0 \end{bmatrix} T^{-1}, \quad I_N - P = T \begin{bmatrix} 0 & 0 \\ 0 & I_{N-k} \end{bmatrix} T^{-1}$$

commute with A. Moreover, P and $I_N - P$ are projections ($P^2 = P$, $(I_N - P)^2 = I_N - P$, and $PA = AP$) onto mutually complementary invariant subspaces of A. Let us divide the eigenvalues $\lambda_j(A)$ of the matrix A into two groups so that $\lambda_1, \lambda_2, \ldots, \lambda_k$ are the eigenvalues of A_{11} and $\lambda_{k+1}, \lambda_{k+2}, \ldots, \lambda_N$ are the eigenvalues of A_{22}. Take a closed contour γ that separates these groups in the complex plane. Then the projections P and $I - P$ can be found using a complex contour integral along γ.

Let the domain containing the eigenvalues of the first group be to the left if we move in the counterclockwise direction. Then

$$P = \frac{1}{2\pi i} \oint_{\gamma} (\lambda I_N - A)^{-1} \, d\lambda. \tag{8.2.1}$$

In view of the equality

$$\frac{1}{2\pi i} \oint_{\gamma} (\lambda I_N - A)^{-1} \, d\lambda$$
$$= T \begin{bmatrix} \frac{1}{2\pi i} \oint_{\gamma} (\lambda I_k - A_{11})^{-1} \, d\lambda & 0 \\ 0 & \frac{1}{2\pi i} \oint_{\gamma} (\lambda I_{N-k} - A_{22})^{-1} \, d\lambda \end{bmatrix} T^{-1},$$

to prove (8.2.1) it suffices to prove the relations

$$\frac{1}{2\pi i}\oint_\gamma (\lambda I_k - A_{11})^{-1}\, d\lambda = I_k, \quad \frac{1}{2\pi i}\oint_\gamma (\lambda I_{N-k} - A_{22})^{-1}\, d\lambda = 0.$$

The last equality is almost obvious. Indeed, the domain bounded by the contour γ contains no eigenvalues of A_{22}, i.e., the matrix $\lambda I_{N-k} - A_{22}$ is nonsingular everywhere in this domain. Therefore, the inverse matrix $(\lambda I_{N-k} - A_{22})^{-1}$ is a bounded analytic (even rational) function of λ in this domain. (More exactly, each element of the matrix $(\lambda I_{N-k} - A_{22})^{-1}$ has these properties.) By the Cauchy theorem, the integral under consideration vanishes.

To compute the first integral, we must take into account that the contour surrounds all the eigenvalues of A_{11}. Therefore, the value of the integral remains unchanged if we deform the contour in such a way that the domain bounded by this contour becomes larger. This is true because of the Cauchy theorem about integrals of analytic functions. Taking a suitable deformation, we can transform the contour to a circle of sufficiently large radius α. If $|\lambda| = \alpha > \|A_{11}\|$, then

$$\begin{aligned}(\lambda I_k - A_{11})^{-1} &= \frac{1}{\lambda}\left[\left(I_k - \frac{1}{\lambda}A_{11}\right)^{-1}\right] = \frac{1}{\lambda}I_k + \frac{1}{\lambda^2}A_{11}\left(\sum_{m=0}^{\infty}\frac{1}{\lambda^m}A_{11}^m\right)\\ &= \frac{1}{\lambda}I_k + \Delta_1(\lambda),\end{aligned}$$

$$\|\Delta_1(\lambda)\| = \|(\lambda I_k - A_{11})^{-1} - \frac{1}{\lambda}I_k\| \le \frac{\|A_{11}\|}{\alpha^2}\,\frac{1}{1 - \|A_{11}\|/\alpha}$$

on this circle. Compute the integral along the circle:

$$\begin{aligned}\frac{1}{2\pi i}\oint_{|\lambda|=\alpha} (\lambda I_k - A_{11})^{-1}\, d\lambda &= \frac{1}{2\pi i}\int_0^{2\pi}\frac{1}{\alpha e^{i\varphi}}\, d[\alpha e^{i\varphi}] I_k + \frac{1}{2\pi i}\int_0^{2\pi}\Delta_1(\alpha e^{i\varphi})\, d\alpha e^{i\varphi}\\ &= I_k + \frac{1}{2\pi i}\int_0^{2\pi}\Delta_1(\alpha e^{i\varphi})\, d\alpha e^{i\varphi}.\end{aligned}$$

Since

$$\begin{aligned}\left\|\frac{1}{2\pi i}\int_0^{2\pi}\Delta_1(\alpha e^{i\varphi})\, d\alpha e^{i\varphi}\right\| &\le \frac{1}{2\pi}\,\frac{\|A_{11}\|}{\alpha^2}\,\frac{1}{1 - \|A_{11}\|/\alpha}\alpha \cdot 2\pi\\ &= \frac{\|A_{11}\|}{\alpha}\,\frac{1}{1 - \|A_{11}\|/\alpha},\end{aligned}$$

we see that

$$\left\|\frac{1}{2\pi i}\oint_{|\lambda|=\alpha} (\lambda I_k - A_{11})^{-1}\, d\lambda - I_k\right\| \le \frac{\|A_{11}\|}{\alpha}\,\frac{1}{1 - \|A\|/\alpha}$$

for $|\alpha| > \|A_{11}\|$. The integral is independent of α (for $\alpha > \|A_{11}\|$). Therefore,

$$\frac{1}{2\pi i}\oint_{|\lambda|=\alpha} (\lambda I_k - A_{11})^{-1}\, d\lambda = I_k.$$

As was mentioned,

$$\frac{1}{2\pi i}\oint_{\gamma}(\lambda I_k - A_{11})^{-1}\,d\lambda = \frac{1}{2\pi i}\oint_{|\lambda|=\alpha}(\lambda I_k - A_{11})^{-1}\,d\lambda.$$

Thus, (8.2.1) is proved.

We consider a modification of (8.2.1) under the condition that the eigenvalues of A_{11} lie in the left half-plane ($\operatorname{Re}\lambda < 0$) and the eigenvalues of A_{22} lie in the right half-plane ($\operatorname{Re}\lambda > 0$). Take $\alpha > \|A\|$. For a contour surrounding the eigenvalues in the left half-plane we can take the boundary of a half-disk, i.e., the segment $\lambda = it$, $-\alpha \le t \le \alpha$, and the half-circle $\lambda = \alpha e^{i\varphi}$, $\pi/2 \le \varphi \le 3\pi/2$. Indeed, for $|\lambda| = \alpha > \|A\|$ the matrix $(\lambda I - A)^{-1}$ is invertible because of the following estimate for $\Delta(\lambda) = (\lambda I_N - A)^{-1} - (1/\lambda)I_N$:

$$\|\Delta(\lambda)\| \le \frac{\|A\|}{\alpha^2}\,\frac{1}{1-\|A\|/\alpha}. \tag{8.2.2}$$

This estimate is similar to that for $\Delta_1(\lambda)$. In particular, (8.2.2) implies that all λ's lie in the disk $\lambda \le \|A\|$, i.e., in the disk $\lambda < \alpha$. Acting as above, we obtain the estimates

$$\left\|\frac{1}{2\pi i}\int_{\pi/2}^{(3\pi)/2}\Delta(\alpha e^{i\varphi})\,d(\alpha e^{i\varphi})\right\| \le \frac{1}{2\pi}\,\frac{\|A\|}{\alpha^2}\,\frac{1}{1-\|A\|/\alpha}\,\pi a = \frac{1}{2}\,\frac{\|A\|}{\alpha}\,\frac{1}{1-\|A\|/\alpha},$$

$$\left\|\oint_{\substack{|\lambda|=a\\ \operatorname{Re}\lambda<0}}(\lambda\, I_N - A)^{-1}\,d\lambda - \frac{1}{2}\,I_N\right\| \le \frac{1}{2}\,\frac{\|A\|}{\alpha}\,\frac{1}{1-\|A\|/\alpha}.$$

Therefore,

$$\begin{aligned} P &= \frac{1}{2}I_N + \frac{1}{2\pi i}\int_{-ia}^{+ia}(\lambda I_N - A)^{-1}\,d\lambda + O(\|A\|/\alpha) \\ &= \frac{1}{2}I_N + \frac{1}{2\pi}\int_{-a}^{+a}(itI - A)^{-1}\,dt + O(\|A\|/\alpha) \end{aligned}$$

or

$$P = \frac{1}{2}I_N + \frac{1}{2\pi}\lim_{a\to\infty}\int_{-a}^{+a}(itI_N - A)^{-1}\,dt.$$

It is obvious that

$$I - P = \frac{1}{2}I_N - \frac{1}{2\pi}\lim_{a\to\infty}\int_{-a}^{+a}(itI_N - A)^{-1}\,dt.$$

We recall that in mathematical analysis the expression

$$\lim_{a\to\infty}\int_{-a}^{+a}f(t)\,dt$$

is called the *principal value of improper integral*

$$\int_{-\infty}^{+\infty} f(t)\,dt.$$

Thus, we have shown that projections onto invariant subspaces of A associated with the eigenvalues of A lying in the left half-plane and in the right half-plane, respectively, can be computed using the principal value of the contour integral of $(\lambda I - A)^{-1}$ along the imaginary axis.

8.3. Dichotomy of spectrum

The resolvent of a matrix. An estimate for the difference of resolvents in the case of close matrices. An estimate for the difference of projections in the case of close matrices. Dichotomy of the ε-spectrum (spectrum) of a matrix. The norm of the resolvent as a qualitative characteristic of dichotomy.

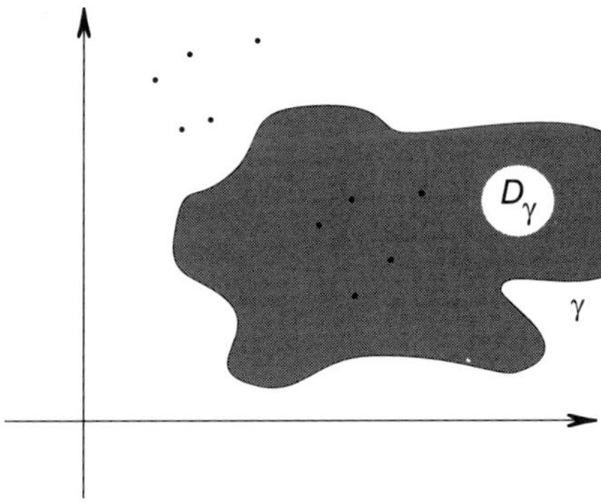

FIGURE 8.3.1

Let D_γ be the domain bounded by the contour γ (Figure 8.3.1). Consider a projection $P_\gamma(A)$ that commutes with a matrix A, i.e., $P_\gamma(A)A = AP_\gamma(A)$, $P_\gamma(A)^2 = P_\gamma(A)$, onto the invariant subspace of A associated with all $\lambda_j(A) \in D_\gamma$:

$$P_\gamma(A) = \frac{1}{2\pi i}\oint_\gamma (\lambda I - A)^{-1}\,d\lambda.$$

Consider another matrix B and a similar projection

$$P_\gamma(B) = \frac{1}{2\pi i}\oint_\gamma (\lambda I - B)^{-1}\,d\lambda,$$

where the integral is taken along the same contour γ. Suppose that the matrices B and A are close in the following sense:

$$\|B - A\| \le \varepsilon\|A\|.$$

Assume that the norm of the matrix $(\lambda I - A)^{-1}$ on the contour γ is estimated as follows:

$$\max_{\lambda\in\gamma} \|(\lambda I - A)^{-1}\| = \frac{m_\gamma(A)}{\|A\|}.$$

The expression $(\lambda I - A)^{-1}$ is called the *resolvent* of the matrix A. We estimate the norm $\|(\lambda I - A)^{-1} - (\lambda I - B)^{-1}\|$ in terms of ε and $m_\gamma(A)$. The chain of the equalities

$$\begin{aligned}(\lambda I - A)^{-1} - (\lambda I - B)^{-1} &= (\lambda I - A)^{-1}[I - (\lambda I - A)(\lambda I - B)^{-1}]\\ &= (\lambda I - A)^{-1}[I - (\lambda I - B)(\lambda I - B)^{-1} + (A - B)(\lambda I - B)^{-1}]\\ &= (\lambda I - A)^{-1}(A - B)(\lambda I - B)^{-1}\\ &= (\lambda I - A)^{-1}(A - B)(\lambda I - A)^{-1}\\ &\quad + (\lambda I - A)^{-1}(A - B)[(\lambda I - B)^{-1} - (\lambda I - A)^{-1}]\end{aligned}$$

leads to the inequality

$$\begin{aligned}\|(\lambda I - A)^{-1} - (\lambda I - B)^{-1}\| &\le \|(\lambda I - A)^{-1}\|^2\, \|A - B\|\\ &\quad + \|(\lambda I - A)^{-1}\|\, \|A - B\|\, \|(\lambda I - A)^{-1} - (\lambda I - B)^{-1}\|.\end{aligned}$$

If

$$\|A - B\|\, \|(\lambda I - A)^{-1}\| \le \varepsilon\|A\|\frac{m_\gamma(A)}{\|A\|} = \varepsilon m_\gamma(A) < 1,$$

then

$$\|(\lambda I - A)^{-1} - (\lambda I - B)^{-1}\| \le \frac{\|(\lambda I - A)^{-1}\|^2\|A - B\|}{1 - \|(\lambda I - A)^{-1}\|\, \|A - B\|} \le \frac{m_\gamma^2(A)\varepsilon}{\|A\|(1 - \varepsilon m_\gamma(A))}.$$

Therefore, for $\varepsilon m_\gamma(A) < 1$ the integral representations of projections yield the estimate

$$\|P_\gamma(A) - P_\gamma(B)\| = \left\|\frac{1}{2\pi i}\oint_\gamma [(\lambda I - A)^{-1} - (\lambda I - B)^{-1}]\, d\lambda\right\| \le \frac{l_\gamma}{2\pi\|A\|}\,\frac{m_\gamma^2(A)\varepsilon}{1 - \varepsilon m_\gamma(A)},$$

where

$$l_\gamma = \oint_\gamma |d\lambda|.$$

This estimate guarantees that the projections $P_\gamma(A)$ and $P_\gamma(B)$ practically coincide provided that

$$\varepsilon m_\gamma^2(A)\frac{l_\gamma}{\|A\|} \ll 1.$$

Assume that at all points λ of the closed contour γ we have

$$\|(\lambda I - B)^{-1}\| = \|(\lambda I - A - \Delta)\| < \infty, \tag{8.3.1}$$

where $\|\Delta\| \le \varepsilon\|A\|$. The inequality (8.3.1) means that the intersection of the contour γ and the ε-spectrum Λ_ε of A is empty. This assumption can be expressed as the inequality $\varepsilon m_\gamma(A) < 1$. The contour γ divides $\Lambda_\varepsilon(A)$ into two parts: the first part contains the eigenvalues lying inside the domain bounded by γ, whereas the second part consists of the eigenvalues lying outside this domain. In this case, we say that the *contour* γ *realizes the dichotomy of the* ε*-spectrum and that of the spectrum of the matrix* A.

To find projections $P_\gamma(A)$ and $I - P_\gamma(A)$ for such a dichotomy, it is necessary that $\varepsilon m_\gamma^2(A)l_\gamma/2\pi\|A\|$ be considerably less than 1. For a given contour and a parameter ε, it makes sense to characterize the dichotomy quality by the parameter

$$m_\gamma(A) = \max_{\lambda\in\gamma} \|A\| \, \|(\lambda I - A)^{-1}\|,$$

which is an estimate for the norm of the resolvent.

8.4. Matrix functions and integral representations

Matrix functions as contour integrals. Matrix polynomials. Projections as matrix functions. Some relations for matrix polynomials.

Contour integrals of the resolvent of a matrix are used for representing not only projections onto invariant subspaces, but also for general functions of matrix. For example, consider the integral

$$I(A) = \frac{1}{2\pi i} \oint_\gamma f(\lambda)(\lambda I - A)^{-1}\, d\lambda,$$

where the closed contour γ surrounds the spectrum of the matrix A and the function $f(\lambda)$ is analytic in a connected domain containing γ. The integral coincides with a matrix function $f(A)$ that can be defined algebraically by a special procedure from matrix theory (see, e.g., [**9**]).

We briefly present ideas that can serve as the basis of the definition of a matrix function in terms of a contour integral.

A polynomial represents the simplest analytic function:

$$P(\lambda) = a_0 + a_1\lambda + a_2\lambda^2 + \cdots + a_R\lambda^R.$$

It is natural to give an analogous definition of a matrix polynomial, i.e.,

$$P(A) = a_0 I + a_1 A + a_2 A^2 + \cdots + a_R A^R.$$

By the Cayley–Hamilton theorem asserting that every matrix satisfies the characteristic equation

$$0 = \det(\lambda I - A) = \lambda^N + p_{N-1}\lambda^{N-1} + p_{N-2}\lambda^{N-2} + \cdots + p_N,$$

we conclude that the matrix polynomials $P(A)$ and $Q(A)$ coincide with ($P(A) = Q(A)$) provided that $P(\lambda) - Q(\lambda)$ is divisible by $\det(\lambda I - A)$. In other words, P and Q coincide if every zero of $\det(\lambda I - A)$ is zero (of the same or larger multiplicity) of $P(\lambda) - Q(\lambda)$. This assertion can be formulated in another way: if

$$P(\lambda_j) = Q(\lambda_j), P'(\lambda_j) = Q'(\lambda_j), \ldots, P^{(\nu_j-1)}(\lambda_j) = Q^{(\nu_j-1)}(\lambda_j)$$

for every eigenvalue $\lambda = \lambda_j(A)$ of multiplicity ν_j, then $P(A) = Q(A)$. A definition of a matrix function must preserve this property of matrix polynomials. It is natural to set

$$f(A) = g(A) \tag{8.4.1}$$

if

$$f(\lambda_j) = g(\lambda_j),\ f'(\lambda_j) = g'(\lambda_j), \ldots, f^{(\nu_j-1)}(\lambda_j) = g^{(\nu_j-1)}(\lambda_j)$$

for all $\lambda_j = \lambda_j(A)$. We note that (8.4.1) holds if $f(\lambda)$ and $g(\lambda)$ are analytic functions in a connected domain that contains a closed nonselfintersecting contour γ surrounding the eigenvalues $\lambda_j = \lambda_j(A)$ and the matrix functions $f(A)$ and $g(A)$ are defined as the integrals

$$f(A) = \frac{1}{2\pi i}\oint_{\gamma} f(\lambda)(\lambda I - A)^{-1}\,d\lambda, \quad g(A) = \frac{1}{2\pi i}\oint_{\gamma} g(\lambda)(\lambda I - A)^{-1}\,d\lambda.$$

Indeed,

$$(\lambda I - A)^{-1} = \frac{1}{\det(\lambda I - A)}(\lambda^{N-1}B_{N-1} + \lambda^{N-2}B_{N-2} + \cdots + \lambda B_1 + B_0),$$

and for every eigenvalue λ_j the multiplicity of the zero of $f(\lambda) - g(\lambda)$ is not less than that of the polynomial $\det(\lambda I - A)$. Hence the function $\dfrac{f(\lambda) - g(\lambda)}{\det(\lambda I - A)}$ has removable singularities at λ_j. Therefore,

$$f(A) - g(A) = \frac{1}{2\pi i}\oint_{\gamma} \frac{f(\lambda) - g(\lambda)}{\det(\lambda I - A)}[\lambda^{N-1}B_{N-1} + \lambda^{N-2}B_{N-2} + \dots$$
$$+ \lambda B_1 + B_0]\,d\lambda = 0.$$

We can assume that the function $f(\lambda)$ is piecewise analytic. More exactly, there are several closed nonselfintersecting contours γ_j, $j = 1, 2, \dots, k$, such that each of them lies in one of the connected domains G_j where $f(\lambda)$ is analytic and surrounds the eigenvalues of A lying in G_j. Suppose that each eigenvalue of A lies inside a domain bounded by one of the contours γ_j, $j = 1, \dots, k$. Set

$$f(A) = \sum_{j=1}^{k} \frac{1}{2\pi i}\oint_{\gamma_j} f(\lambda)(\lambda I - A)^{-1}\,d\lambda.$$

It is clear that the matrix $f(A)$ does not change under those deformations of γ_j that leave γ_j in the analyticity domain G_j and have the eigenvalues lying in G_j inside the deformed contour. To find $f(A)$, it suffices to know the values of $f(\lambda)$ only in small neighborhoods of the eigenvalues of A. More exactly, $f(A)$ can be determined by the values of the function $f(\lambda)$ and its derivatives at points of the spectrum of A. Moreover, it suffices to know a finite number of derivatives.

The algebraic definition (see [**9**]) of a matrix function is based on the replacement of $f(\lambda)$ by a polynomial such that it takes the value $f(\lambda_j)$ at points λ_j of the spectrum of A and its derivatives coincide with $f'(\lambda_j), \dots, f^{(\nu_j - 1)}(\lambda_j)$. As an example, consider disjoint connected domains G_1 and G_2 in the complex plane. Let γ_1 and γ_2 be closed contours lying in G_1 and G_2 and surrounding the eigenvalues lying in G_1 and G_2, respectively. Suppose that every eigenvalue of A lies in one of these domains.

In $G_1 \cup G_2$ consider the piecewise constant function

$$f(\lambda) = \begin{cases} 1 & \text{if } \lambda \in G_1, \\ 0 & \text{if } \lambda \in G_2. \end{cases}$$

By definition, the matrix function $f(A)$ can be introduced as follows:

$$f(A) = \frac{1}{2\pi i} \oint_{\gamma_1} (\lambda I - A)^{-1} \, d\lambda = P_{\gamma_1}(A).$$

To the function $1 - f(\lambda)$ we associate the matrix function

$$\begin{aligned} I - f(A) &= \frac{1}{2\pi i} \oint_{\gamma_2} [1 - f(\lambda)](\lambda I - A)^{-1} \, d\lambda = \frac{1}{2\pi i} \oint_{\gamma_2} (\lambda I - A)^{-1} \, d\lambda \\ &= I - P_{\gamma_1}(A) = P_{\gamma_2}(A). \end{aligned}$$

We see that the projections $P_{\gamma_1}(A)$ and $P_{\gamma_2}(A)$ are functions of matrix A in accordance with the above definition.

Consider the matrices $I_{\gamma_j}^{(k)}(A)$, $j = 1, 2$, $k = 0, 1, 2, \ldots$, defined by the integrals

$$I_{\gamma_j}^{(k)}(A) = \frac{1}{2\pi i} \oint_{\gamma_j} \lambda^k (\lambda I - A)^{-1} \, d\lambda.$$

Note that

$$I_{\gamma_j}^{(0)}(A) = \frac{1}{2\pi i} \oint_{\gamma_j} (\lambda I - A)^{-1} \, d\lambda = P_{\gamma_j}(A).$$

It is easy to see that $I_{\gamma_j}^{(k+1)}(A) = A I_{\gamma_j}^{(k)}(A)$. Indeed,

$$\begin{aligned} I_{\gamma_j}^{(k+1)}(A) - A I_{\gamma_j}^{(k)}(A) &= \frac{1}{2\pi i} \oint_{\gamma_j} \lambda^{k+1} (\lambda I - A)^{-1} \, d\lambda - A \oint_{\gamma_j} \lambda^k (\lambda I - A)^{-1} \, d\lambda \\ &= \frac{1}{2\pi i} \oint_{\gamma_j} \lambda^k (\lambda I - A)(\lambda I - A)^{-1} \, d\lambda = \frac{1}{2\pi i} \oint_{\gamma_j} \lambda^k \, d\lambda = 0. \end{aligned}$$

We used the Cauchy theorem asserting that the integral vanishes if an integrand is an analytic function in the domain bounded by the integration contour. Thus,

$$I_{\gamma_j}^{(0)}(A) = P_{\gamma_j}(A), \quad I_{\gamma_j}^{(k+1)}(A) = A I_{\gamma_j}^{(k)}(A).$$

Consequently, $I_{\gamma_j}^{(k)}(A) = A^k P_{\gamma_j}(A)$. If the contour γ surrounds the eigenvalues of A, then $P_\gamma(A) = I$ and $I_\gamma^{(k)}(A) = A^k$. We have proved the equality

$$A^k = \frac{1}{2\pi i} \oint_{\gamma} \lambda^k (\lambda I - A)^{-1} \, d\lambda,$$

which implies that

$$\frac{1}{2\pi i} \oint_{\gamma} P(\lambda)(\lambda I - A)^{-1} \, d\lambda = P(A) = a_0 I + a_1 A + \cdots + a_R A^R,$$

where $P(\lambda) = a_0 + a_1 \lambda + \cdots + a_R \lambda^R$ is an arbitrary polynomial. Thus, the above definition of a matrix function leads to the natural correspondence between the polynomials $P(\lambda)$ and $P(A)$.

8.5. Matrix exponential and matrix powers

A matrix exponential as an integral along a contour surrounding the spectrum of a matrix. Properties of the matrix exponential. Derivatives and Taylor series for the matrix exponential. An estimate of the growth of the matrix exponential at infinity. Estimates for the norms of powers of a matrix.

An important example is presented by the matrix function e^{tA} depending on the real parameter t:

$$e^{tA} = \frac{1}{2\pi i} \oint_{\gamma} e^{t\lambda}(\lambda I - A)^{-1}\, d\lambda, \tag{8.5.1}$$

where the contour γ surrounds the spectrum of A. Using the equalities

$$\frac{d}{dt}[e^{t\lambda}(\lambda I - A)^{-1}] = \lambda e^{t\lambda}(\lambda I - A)^{-1} = e^{t\lambda} I + A e^{t\lambda}(\lambda I - A)^{-1},$$

we find that

$$\frac{d}{dt} e^{tA} = \frac{1}{2\pi i} \oint_{\gamma} e^{t\lambda} d\lambda I + \frac{1}{2\pi i} A \oint_{\gamma} e^{t\lambda}(\lambda I - A)^{-1}\, d\lambda = A e^{tA},$$

i.e., the matrix $Y(t) = e^{tA}$ satisfies the differential equation

$$\frac{d}{dt} Y(t) = AY(t).$$

We note that $Y(0) = I$. Indeed,

$$Y(0) = \frac{1}{2\pi i} \oint_{\gamma} (\lambda I - A)^{-1}\, d\lambda = I$$

since the contour γ surrounds the spectrum of A. The matrix function $Y(t) = e^{tA}$ is called the *matrix exponential.*

By (8.5.1), the matrices e^{tA} and A commute, i.e., $Ae^{tA} = e^{tA}A$. Differentiating the product $e^{tA}e^{-tA}$:

$$\frac{d}{dt}(e^{tA}e^{-tA}) = Ae^{tA}e^{-tA} + e^{tA}(-A)e^{tA}e^{-tA} = (A - A)e^{tA}e^{-tA} = 0,$$

we see that it is independent of t. Consequently,

$$e^{tA}e^{-tA} = e^{0A}e^{-0A} = I \cdot I = I.$$

In other words, the matrix e^{tA} is nonsingular and the equality $[e^{tA}]^{-1} = e^{-tA}$ holds. It is easy to derive formulas for the derivatives of an arbitrary order n:

$$\frac{d^n}{dt^n} e^{tA} = A^n e^{tA}$$

and prove that e^{tA} is represented as the Taylor series

$$e^{tA} = I + tA + \frac{t^2}{2!}A^2 + \frac{t^3}{3!}A^3 + \dots, \tag{8.5.2}$$

which implies the estimate

$$\|e^{tA}\| \le 1 + |t|\, \|A\| + \frac{|t|^2}{2!} \|A\|^2 + \dots = e^{|t|\, \|A\|}.$$

Since $1 = \|I\| = \|e^{tA}e^{-tA}\| \le \|e^{tA}\|\,\|e^{-tA}\|$, the following estimates hold:

$$e^{-|t|\,\|A\|} \le \|e^{tA}\| \le e^{|t|\,\|A\|}.$$

From (8.5.2) it follows that $(e^{tA})^* = e^{tA^*}$.

PROBLEM 8.5.1. Compute the matrix exponentials e^{tB}, e^{tC}, and $e^{t(B+C)}$ for the matrices

$$B = \begin{bmatrix} 1 & 0 \\ 0 & 2 \end{bmatrix}, \quad C = \begin{bmatrix} 0 & 1 \\ 0 & 0 \end{bmatrix}.$$

Verify the relations $e^{tB}e^{tC} \ne e^{t(B+C)}$ and $e^{t(B+C)} \ne e^{tC}e^{tB}$.

PROBLEM 8.5.2. Prove that $A_1A_2 = A_2A_1$ implies $e^{t(A_1+A_2)} = e^{tA_1}e^{tA_2} = e^{tA_2}e^{tA_1}$.

Suppose that the eigenvalues $\lambda_j(A)$ of an $N \times N$-matrix A lie in the left half-plane ($\operatorname{Re}\lambda_j(A) \le -\sigma < 0$, $j = 1, 2, \dots, N$). We can assume that in (8.5.1) the contour γ is to the left of the line $\operatorname{Re}\lambda = -\sigma/2 < 0$ and that for $t > 0$

$$|e^{t\lambda}| < e^{-t\sigma/2}, \quad \|(\lambda I - A)^{-1}\| < K' = \text{const}$$

everywhere on γ. Hence for $t > 0$ we have

$$\|e^{tA}\| \le K' l(\gamma) e^{-t\sigma/2} = K e^{-t\sigma/2}.$$

Thus, under the above assumption about the spectrum of the matrix A, the norm of the matrix exponential e^{tA} decreases exponentially as t increases.

In the sequel, we will use similar estimates for the norms of powers A^k of a matrix A provided that the spectrum of A lies in the unit disk ($|\lambda_j(A)| \le \rho < 1$). If for γ we take the circle $\lambda = \sqrt{\rho}e^{i\varphi}$ so that the eigenvalues of A lie in the disk, then we find

$$\text{(8.5.3)}\quad A^k = \frac{1}{2\pi i}\oint_\gamma \lambda^k(\lambda I - A)^{-1}d\lambda = \frac{1}{2\pi i}\int_0^{2\pi} (\sqrt{\rho}e^{i\varphi})^k(\sqrt{\rho}e^{i\varphi}I - A)^{-1}\sqrt{\rho}de^{i\varphi}.$$

Since the resolvent of A is bounded on γ, from (8.5.3) it follows that

$$\|(\sqrt{\rho}e^{i\varphi}I - A)^{-1}\| \le K.$$

Consequently, $\|A^k\| \le K\rho^{(k+1)/2}$, i.e., the norms $\|A^k\|$ decrease not slower than the geometric series with ratio $\sqrt{\rho}$ as k increases.

PROBLEM 8.5.3. Using the Schur theorem (Theorem 4.2.1) and the matrix Taylor series (8.5.2), prove that the eigenvalues $\lambda_j(e^{tA})$ of the matrix e^{tA} are connected with the eigenvalues $\lambda_j(A)$ of the matrix A by the formula $\lambda_j(e^{tA}) = e^{t\lambda_j(A)}$.

8.6. Estimate for the resolvent of a matrix

Estimate of the norm of the resolvent of a square matrix in terms of singular values and eigenvalues. The proof of the estimate.

In this section, we prove[1] the following important estimate for the norm of the resolvent of a square $N \times N$-matrix A in terms of the singular values $\sigma_j = \sigma_j(A)$

[1] We follow **[19]** where analogous estimates obtained originally by Lidskii and Matsaev were improved. Unlike **[19]**, we consider only the finite-dimensional case.

$(\sigma_j \geq \sigma_{j-1} \geq 0)$ and the eigenvalues $\lambda_j(A)$:

$$\|(\lambda I - A)^{-1}\| = \frac{1}{|\lambda|}\left\|\left(I - \frac{1}{\lambda}A\right)^{-1}\right\|.$$

THEOREM 8.6.1. *The following estimate holds:*

$$\left\|\left(I - \frac{1}{\lambda}A\right)^{-1}\right\| \leq \frac{\prod_{j=2}^{N}\left[1 + \frac{\sigma_j(A)}{|\lambda|}\right]}{\left|\prod_{j=1}^{N}\left(1 - \frac{\lambda_j(A)}{\lambda}\right)\right|}. \tag{8.6.1}$$

Sometimes, it is convenient to use the following equivalent form of (8.6.1):

$$\left\|\left(I - \frac{1}{\lambda}A\right)^{-1}\right\| \leq \frac{\prod_{j=2}^{N}\left[1 + \frac{\sigma_j(A)}{|\lambda|}\right]}{\left|\det\left(I - \frac{1}{\lambda}A\right)\right|}. \tag{8.6.2}$$

We first establish two preliminary results on which the proof of the theorem will be based.

As was mentioned in §7.5, for a matrix T of rank r we have $\sigma_j(A+T) \leq \sigma_{j+r}(A)$, $1 \leq j \leq N-r$. In particular, the matrix $T = xy^*$, where x and y are N-dimensional vectors, has rank 1. Thus, $\sigma_j(A + xy^*) \leq \sigma_{j+1}(A)$, $j = 1, 2, \ldots, N-1$.

The second result asserts that the determinant of the matrix $I + zu^*$, which is the sum of the matrix zu^* of rank 1 and the unitary matrix I, can be computed by the formula $\det(I + zu^*) = 1 + (z, u)$. Indeed, choosing an orthogonal matrix P such that $Pz = (\xi_1, 0, 0, \ldots, 0)^T$, $Pu = (\nu_1, \nu_2, \ldots, \nu_N)^T$, and $PP^* = I_N$, we obtain

$$\begin{aligned}\det(I + zu^*) &= \det(P(I + zu^*)P^*) = \det(I + (Pz)(Pu)^*)\\ &= \det\begin{bmatrix} 1+\xi_1\overline{\nu}_1 & \xi_1\overline{\nu}_2 & \xi_1\overline{\nu}_3 & \ldots & \xi_1\overline{\nu}_N \\ 0 & 1 & 0 & \ldots & 0 \\ 0 & 0 & 1 & \ldots & 0 \\ \ldots & \ldots & \ldots & \ldots & \ldots \\ 0 & 0 & 0 & \ldots & 1 \end{bmatrix}\\ &= 1 + \xi_1\overline{\nu}_1 = 1 + (Pz, Pu) = 1 + (z, u).\end{aligned}$$

PROOF OF THEOREM 8.6.1. Consider the matrix $A_1 = A + \rho xy^*$, where x, y are vectors and ρ is a parameter. Using the inequality (6.3.3)

$$\prod_{j=1}^{N}[1 + |\mu|\,|\lambda_j(A_1)|] \leq \prod_{j=1}^{N}[1 + |\mu|\,|\sigma_j(A_1)|],$$

the obvious estimate

$$|\det(I - \mu A_1)| = \left|\prod_{j=1}^{N}[1 - \mu\lambda_j(A_1)]\right| \leq \prod_{j=1}^{N}[1 + |\mu||\lambda_j(A_1)|]$$

and the relations

$$\sigma_N(A_1) \leq \sigma_N(A) + |\rho|\,\|x\|\,\|y\|, \quad \sigma_{j-1}(A_1) \leq \sigma_j(A), \quad 2 \leq j \leq N,$$

we obtain the estimate

$$|\det\left((I-\mu A)-\mu\rho x y^*\right)| \leq [1+|\mu|\sigma_N(A)+|\mu|\,|\rho|\,\|x\|\,\|y\|]\prod_{j=2}^{N}[1+|\mu|\sigma_j(A)].$$

It is convenient to rewrite the last estimate in the form

$$\begin{aligned} &\left|\det\left[I-\rho\mu(I-\mu A)^{-1}xy^*\right]\right| \\ &\leq \frac{[1+|\mu|\sigma_N(A)+|\mu|\,|\rho|\,\|x\|\,\|y\|]\prod_{j=2}^{N}[1+|\mu|\sigma_j(A)]}{|\det(I-\mu A)|}. \end{aligned} \tag{8.6.3}$$

We compute the determinant on the left-hand side of (8.6.3) as follows:

$$\det\left(I-\rho\mu(I-\mu A)^{-1}xy^*\right)=1-\rho\mu((I-\mu A)^{-1}x,y).$$

Dividing the result by $\rho\mu$, we obtain the estimate

$$\left|\frac{1}{\rho\mu}-((I-\mu A)^{-1}x,y)\right| \leq \frac{\left[\dfrac{1+|\mu|\sigma_N(A)}{\rho|\mu|}+\|x\|\,\|y\|\right]\prod_{j=2}^{N}[1+|\mu|\sigma_j(A)]}{|\det(I-\mu A)|}.$$

Passing to a limit as $\rho\to\infty$, we find

$$\left|((I-\mu A)^{-1}x,y)\right| \leq \frac{\prod_{j=2}^{N}[1+|\mu|\sigma_j(A)]}{|\det(I-\mu A)|}\,\|x\|\,\|y\|.$$

For x and y take the right and left unit singular vectors of the matrix $(I-\mu A)^{-1}$ corresponding to the largest singular value $\sigma_N((I-\mu A)^{-1})=\|(I-\mu A)^{-1}\|$. Since $((I-\mu A))^{-1}x,y)=\|(I-\mu A)^{-1}\|\|x\|\|y\|=\|(I-\mu A)^{-1}\|$, we can set $\mu=1/\lambda$ and arrive at the estimate (8.6.2) which is equivalent to the required estimate (8.6.1). □

CHAPTER 9

Quadratic Forms in the Spectrum Dichotomy Problem

9.1. Integral criteria for the dichotomy quality

The use of the norms of Hermitian matrices in estimating the dichotomy quality. The parameter $\kappa_\gamma(A)$ as a criterion for the dichotomy quality. Two-sided estimates for $\kappa_\gamma(A)$ in terms of the parameter $\mu_\gamma(A)$. Special contours and criteria for the dichotomy quality. Contours of finite length. A special case of an infinite length contour.

As a criterion for the dichotomy quality by some contour γ, instead of the parameter

$$m_\gamma(A) = \|A\| \max_{\lambda\in\gamma} \|(\lambda I - A)^{-1}\|$$

(see §8.3), it is convenient to take some (integral) characteristics. For example, we can take the number $\kappa_\gamma(A) = \|H_\gamma(A)\|$ or $\operatorname{tr} H_\gamma(A)$, where

$$H_\gamma = H_\gamma^* = \frac{\|A\|^2}{l_\gamma} \oint_\gamma (\overline{\lambda} I - A^*)^{-1}(\lambda I - A)^{-1}\, |d\lambda|.$$

Since $H_\gamma^* = H_\gamma \geq 0$, we have

$$H_\gamma = UDU^*,$$

where D is a diagonal matrix with nonnegative diagonal elements d_i. On the other hand, from the representation of U it follows that

$$\operatorname{tr} H_\gamma = \operatorname{tr} D = \sum_{i=1}^{N} d_i \leq N \max_i d_i = N\|D\| = N\|H_\gamma\|,$$

$$\|H_\gamma(A)\| \leq \operatorname{tr} H_\gamma(A) \leq N\|H_\gamma(A)\|.$$

The following two-sided estimates hold:

(9.1.1)

$$\sqrt{\|H_\gamma(A)\|} \leq m_\gamma(A) \leq \sqrt{\|H_\gamma(A)\| + \left(\frac{l_\gamma}{4\|A\|}\|H_\gamma(A)\|\right)^2} + \frac{l_\gamma}{4\|A\|}\|H_\gamma(A)\|,$$

(9.1.2)
$$\frac{m_\gamma^2(A)}{1 + \dfrac{l_\gamma m_\gamma(A)}{2\|A\|}} \leq \|H_\gamma(A)\| \leq m_\gamma^2(A).$$

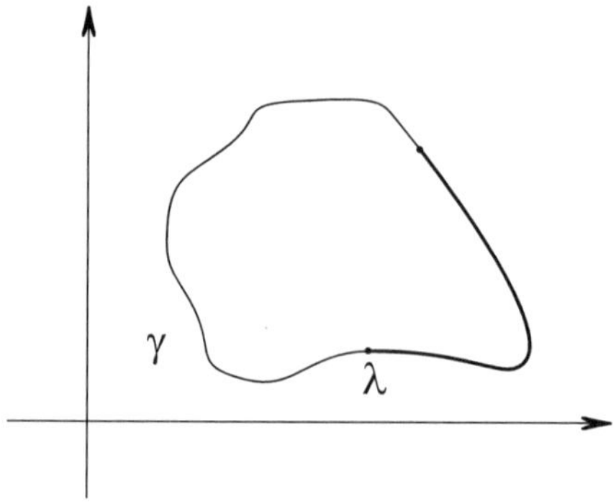

FIGURE 9.1.1

Consider a contour γ of finite length l_γ. Below we will indicate necessary modifications for a special contour of infinite length (namely, the case of the imaginary axis separating eigenvalues in the right and the left half-planes).

We establish estimates connecting H_γ and $m_\gamma(A)$. Choose a point $\lambda_0 \in \gamma$ such that

$$\|(\lambda_0 I - A)^{-1}\| = \max_{\lambda\in\gamma} \|(\lambda I - A)^{-1}\|.$$

After that we choose a vector $x_0 \neq 0$ such that

$$\|(\lambda_0 I - A)^{-1} x_0\| = \|(\lambda_0 I - A)^{-1}\| \, \|x_0\|.$$

It is easy to check the equality

$$(\lambda I - A)^{-1} - (\lambda_0 I - A)^{-1} = (\lambda_0 - \lambda)(\lambda_0 I - A)^{-1}(\lambda I - A)^{-1}$$

and deduce the estimates

$$\|(\lambda I - A)^{-1} x_0\| \geq \|(\lambda_0 I - A)^{-1}\| \, \|x_0\| - |\lambda - \lambda_0| \, \|(\lambda_0 I - A)^{-1}\| \, \|(\lambda I - A)^{-1} x_0\|,$$

$$\|(\lambda I - A)^{-1} x_0\| \geq \frac{\|(\lambda_0 I - A)^{-1}\| \, \|x_0\|}{1 + |\lambda - \lambda_0| \, \|(\lambda_0 I - A)^{-1}\|}.$$

Suppose that the point $\lambda \in \gamma$ can be joined with the point λ_0 by an arc of γ of length $s(\lambda)$ (Figure 9.1.1).

It is obvious that $s(\lambda) \geq |\lambda - \lambda_0|$. Therefore,

$$\|(\lambda I - A)^{-1} x_0\| \geq \frac{\|(\lambda_0 I - A)^{-1}\| \, \|x_0\|}{1 + s(\lambda)\|(\lambda_0 I - A)^{-1}\|}.$$

If

$$J = \oint_\gamma (\bar{\lambda} I - A^*)^{-1}(\lambda I - A)^{-1} \, |d\lambda|,$$

then

$$(Jx_0, x_0) = \oint_\gamma \|(\lambda I - A)^{-1} x_0\|^2 |d\lambda| \geq 2 \int_0^{l_\gamma/2} \frac{\|(\lambda_0 I - A)^{-1}\|^2 \|x_0\|^2 \, ds(\lambda)}{[1 + s(\lambda)\|(\lambda_0 I - A)^{-1}\|]^2}.$$

We integrate in both directions along the contour from the point λ_0 to the points located at the distance $l_\gamma/2$ from it. Computing the above integral, we obtain the estimates

$$(9.1.3)\qquad \begin{aligned} (Jx_0, x_0) &\geq l_\gamma \frac{\|(\lambda_0 I - A)^{-1}\|^2}{1 + \frac{l_\gamma}{2}\|(\lambda_0 I - A)^{-1}\|}\|x_0\|^2, \\ \|J\| &\geq l_\gamma \frac{\max\limits_{\lambda\in\gamma}\|(\lambda I - A)^{-1}\|^2}{1 + \frac{l_\gamma}{2}\max\limits_{\lambda\in\gamma}\|(\lambda I - A)^{-1}\|}. \end{aligned}$$

Recall that

$$\max_{\lambda\in\gamma}\|(\lambda I - A)^{-1}\| = \|(\lambda_0 I - A)^{-1}\|,$$

$$J = \oint_\gamma (\bar{\lambda} I - A^*)^{-1}(\lambda I - A)^{-1}\,|d\lambda|.$$

It is obvious that

$$(9.1.4)\qquad l_\gamma \max_{\lambda\in\gamma}\|(\lambda I - A)^{-1}\|^2 \geq \|J\|.$$

Changing the notation, we put

$$H_\gamma(A) = \frac{\|A\|^2}{l_\gamma} J = \frac{\|A\|^2}{l_\gamma}\oint_\gamma (\bar{\lambda} I - A^*)^{-1}(\lambda I - A)^{-1}\,|d\lambda|,$$

$$m_\gamma(A) = \|A\|\max_{\lambda\in\gamma}\|(\lambda I - A)^{-1}\|$$

and rewrite (9.1.3) and (9.1.4) in the form

$$(9.1.5)\qquad m_\gamma^2(A) \geq \|H_\gamma(A)\| \geq \frac{m_\gamma^2(A)}{1 + \frac{l_\gamma m_\gamma(A)}{2\|A\|}},$$

$$(9.1.6)\qquad m_\gamma(A) \geq \sqrt{\|H_\gamma(A)\|}.$$

From (9.1.5) and (9.1.6) we obtain the inequality

$$(9.1.7)\qquad m_\gamma^2(A) \leq \|H_\gamma(A)\| + \frac{l_\gamma m_\gamma(A)}{2\|A\|}\|H_\gamma(A)\|.$$

Rewrite (9.1.7) in the form

$$\left[m_\gamma(A) - \frac{l_\gamma\|H_\gamma(A)\|}{4\|A\|}\right]^2 \leq \|H_\gamma(A)\| + \left(\frac{l_\gamma\|H_\gamma(A)\|}{4\|A\|}\right)^2.$$

We arrive at the required estimates (9.1.1) and (9.1.2).

Together with the Hermitian matrices

$$H_\gamma(A) = \frac{\|A\|^2}{l_\gamma}\oint_\gamma (\bar{\lambda} I - A^*)^{-1}(\lambda I - A)^{-1}\,|d\lambda|,$$

we consider more general matrices

$$H = \frac{\|A\|^2}{l_\gamma}\oint_\gamma (\bar{\lambda} I - A^*)^{-1} F (\lambda I - A)^{-1}\,|d\lambda|,$$

where F is a Hermitian positive definite matrix ($F = F^* > 0$).

Now we consider integrals along a special contour γ. Let a matrix A have no pure imaginary eigenvalues, i.e., the norm $\|(itI - A)^{-1}\|$ is bounded for all real t. Since $\sigma_1(itI - A) \geq \sigma_1(itI) - \|A\| = |t| - \|A\|$, for $t > \|A\|$ we have

$$\|(itI - A)^{-1}\| = \frac{1}{\sigma_1(itI) - \|A\|} \leq \frac{1}{|t| - \|A\|},$$
$$\|(-itI - A^*)^{-1} F(itI - A)^{-1}\| \leq \frac{\|F\|}{(|t| - \|A\|)^2},$$

$$\left\| \int_{-\infty}^{-2\|A\|} (-itI - A^*)^{-1} F(itI - A)^{-1}\, dt + \oint_{2\|A\|}^{+\infty} (-itI - A^*)^{-1} F(itI - A)^{-1}\, dt \right\|$$
$$\leq 2\|F\| \int_{2\|A\|}^{\infty} \frac{dt}{(t - \|A\|)^2} = \frac{2\|F\|}{\|A\|}.$$

It is clear that the integral

$$\int_{-\infty}^{+\infty} (-itI - A^*)^{-1} F(itI - A)^{-1}\, dt = \int_{-\infty}^{+\infty} (itI + A^*)^{-1} F(-itI + A)^{-1}\, dt$$

converges. Since the integration interval is infinite, the normalizing factor must be changed. As a criterion for the dichotomy quality for the spectrum of the matrix A we take the norm $\|H_F\|$ of the matrix

$$H_F(A) = \frac{\|A\|}{\pi} \int_{-\infty}^{+\infty} (itI + A^*)^{-1} F(-itI + A)^{-1}\, dt.$$

It is easy to see that $H_F(A) = H_F(\rho A)$ for $\rho \neq 0$, i.e., $\|H_F(\rho A)\|$ is the same for all nonzero matrices ρA. We note that the integration along the imaginary axis was used to compute the projection

$$P(A) = \frac{1}{2} I + \lim_{a \to \infty} \frac{1}{2\pi i} \int_{-a}^{+a} (itI - A)^{-1}\, dt$$

onto the invariant subspace of A corresponding to the eigenvalues of A lying in the left half-plane.

Circles present another special type of contours often used in practice. The integral

$$H_F = \frac{1}{2\pi} \int_{0}^{2\pi} (A^* - re^{-i\theta} I)^{-1} F(A - re^{i\theta} I)^{-1}\, d\theta$$

converges if the circle $|\lambda| = r$ ($\lambda = re^{i\theta}$) contains no points of the spectrum of A. The norm $\|H_F\|$ is taken as a criterion for the dichotomy quality for the spectrum of A by the circle $|\lambda| = r$. Let $P(A)$ be the projection onto the invariant subspace of A corresponding to the eigenvalues in the disk $|\lambda| < r$. Then $P(A)$ admits the following representation:

$$P(A) = \frac{1}{2\pi} \int_0^{2\pi} \left(I - \frac{e^{-i\theta}}{r} A \right)^{-1} d\theta.$$

Indeed,

$$P(A) = \frac{1}{2\pi} \oint_{|\lambda|=r} (\lambda I - A)^{-1} d\lambda = \frac{1}{2\pi i} \int_0^{2\pi} (re^{i\theta} I - A)^{-1} re^{i\theta} i \, d\theta$$
$$= \frac{1}{2\pi} \int_0^{2\pi} \left(I - \frac{e^{-i\theta}}{r} A \right)^{-1} d\theta.$$

9.2. Historical remarks

Polynomials and quadratic forms. Connection between the number of negative and positive eigenvalues of a matrix and the number of pairs of conjugate complex roots of the polynomial. Nonsingularity of a matrix and eigenvalues. Representation of a quadratic form in terms of linearly independent forms.

Quadratic forms were first studied by Hermite (in 1856) who investigated the location of roots of algebraic equations. At about the same time, similar questions were studied by Borkhardt.

Suppose that the polynomial

$$P(\lambda) = \lambda^N + p_1 \lambda^{N-1} + p_2 \lambda^{N-2} + \cdots + p_{N-1} \lambda + p_N$$

has no multiple roots. Then the Vandermonde determinant

$$\det \begin{bmatrix} 1 & \lambda_1 & \lambda_1^2 & \dots & \lambda_1^{N-1} \\ 1 & \lambda_2 & \lambda_2^2 & \dots & \lambda_2^{N-1} \\ \vdots & \vdots & \vdots & \ddots & \vdots \\ 1 & \lambda_N & \lambda_N^2 & \dots & \lambda_N^{N-1} \end{bmatrix}$$

does not vanish. Consequently, the linear forms

$$\Lambda_j = \sum_{k=1}^{N} \lambda_j^{k-1} x_k, \quad j = 1, 2, \dots, N,$$

in N real variables $x_1, x_2, \dots, x_N$ are linearly independent. Suppose that among the roots λ_j, $j = 1, 2, \dots, N$, there are exactly $N - 2p$ real numbers and p pairs of conjugate complex numbers. Choose a numbering so that $\lambda_1, \lambda_2, \dots, \lambda_{N-2p}$ and the corresponding linear forms Λ_1, $\Lambda_2, \dots, \Lambda_{N-2p}$ are real. Let $\lambda_{N-2p+k} = \overline{\lambda}_{N-k} = \alpha_{N-k} - i\beta_{N-k}$, $k = 1, 2, \dots, p$. Then $\Lambda_{N-2p+k} = \overline{\Lambda}_{N-k}$. Denote the real and imaginary parts of the linear form $\Lambda_{N-2p+k} = \overline{\Lambda}_{N-k}$ by $\frac{1}{\sqrt{2}} L_k$ and $\frac{1}{\sqrt{2}} M_k$, respectively, i.e.,

$$\Lambda_{N-2p+k} = \frac{1}{\sqrt{2}} (L_k + iM_k), \quad \Lambda_{N-k} = \frac{1}{\sqrt{2}} (L_k - iM_k).$$

Note that the real forms $\Lambda_1, \Lambda_2, \dots, \Lambda_{N-2p}, L_1, L_2, \dots, L_k, M_1, M_2, \dots, M_k$ are linearly independent. Construct the quadratic forms

$$(Cx, x) = \sum_{j=1}^{N} [\Lambda_j(x)]^2, \tag{9.2.1}$$

$$(Dx, x) = \sum_{j=1}^{N} \lambda_j [\Lambda_j(x)]^2. \tag{9.2.2}$$

In (9.2.1) and (9.2.2), the first $N-2p$ terms are real, whereas the remaining terms can be expressed as the sums of conjugate complex pairs:

$$[\Lambda_{N-2p+k}]^2 + [\Lambda_{N-k}]^2, \quad \lambda_{N-2p+k}[\Lambda_{N-2p+k}]^2 + \lambda_{N-k}[\Lambda_{N-k}]^2.$$

Therefore, the sums (9.2.1) and (9.2.2) are real, i.e., the elements c_{ij} and d_{ij} of the matrices C and D corresponding to the quadratic forms (Cx, x) and (Dx, x) are real. It is easy to see that

$$\begin{aligned}\sum_{k,l=1}^{N} c_{kl}x_k x_l &= \sum_{j=1}^{N}\left(\sum_{k=1}^{N}\lambda_j^{k-1}x_k\right)\left(\sum_{l=1}^{N}\lambda_j^{l-1}x_l\right) = \sum_{j,k,l=1}^{N}\lambda_j^{k+l-2}x_k x_l \\ &= \sum_{k,l=1}^{N}\left(\sum_{j=1}^{N}\lambda_j^{k+l-2}\right)x_k x_l = \sum_{k,l=1}^{N} s_{k+l-2}x_k x_l.\end{aligned}$$

In other words, $c_{kl} = s_{k+l-2}$, where $s_m = \sum_{j=1}^{N}\lambda_j^m$ are power sums of roots. Similarly, $d_{kl} = s_{k+l-1}$. We note that

$$[\Lambda_{N-2p+k}]^2 + [\Lambda_{N-k}]^2 = \frac{1}{2}(L_k + iM_k)^2 + \frac{1}{2}(L_k - iM_k)^2 = L_k^2 - M_k^2.$$

Therefore,

$$(Cx, x) = \sum_{j=1}^{N-2p}[\Lambda_j(x)]^2 + \sum_{k=1}^{p}[L_k(x)]^2 - \sum_{k=1}^{p}[M_k(x)]^2.$$

From the inertia law for quadratic forms, we conclude that the matrix C has exactly p negative eigenvalues.

Thus, the following proposition holds (see Problem 7.4.2).

PROPOSITION 9.2.1. *The number of negative eigenvalues of a symmetric nonsingular matrix*

$$C = \begin{bmatrix} s_0 & s_1 & s_2 & \dots & s_{N-1} \\ s_1 & s_2 & s_3 & \dots & s_N \\ s_2 & s_3 & s_4 & \dots & s_{N+1} \\ \vdots & \vdots & \vdots & \ddots & \vdots \\ s_{N-1} & s_N & s_{N+1} & \dots & s_{2N-2} \end{bmatrix}$$

coincides with the number of pairs of conjugate complex roots of the polynomial $P(\lambda)$.

The matrix C is nonsingular because the polynomial $P(\lambda)$ has no multiple roots. The matrix D is nonsingular, i.e., the quadratic form

$$(Dx,x) = \sum_{j=1}^{N-2p} \lambda_j \Lambda_j^2 + \sum_{k=1}^{p} \frac{1}{2}(\alpha_k + i\beta_k)(L_k + iM_k)^2 + \sum_{k=1}^{p} \frac{1}{2}(\alpha_k - i\beta_k)(L_k - iM_k)^2$$
$$= \sum_{j=1}^{N-2p} \lambda_j \Lambda_j^2 + \sum_{k=1}^{p} [\alpha_k(L_k^2 - M_k^2) - 2\beta_k L_k M_k]$$

is nongenerate if the eigenvalues λ_j, $1 \le j \le N-2p$, do not vanish and all second order matrices of the form

$$\begin{bmatrix} \alpha_k & -\beta_k \\ -\beta_k & -\alpha_k \end{bmatrix}, \quad 1 \le k \le p,$$

are nonsingular. Since the determinant of such a matrix is equal to $-\alpha_k^2 - \beta_k^2 = -|\lambda_{N-2p+k}|^2 = -|\lambda_{N-k}|^2$, the above assertion can be formulated as follows.

PROPOSITION 9.2.2. *A matrix D is nonsingular if all its eigenvalues λ_j, $1 \le j \le N$, of D are distinct and nonzero.*

It is easy to see that for $\alpha_k^2 + \beta_k^2 \ne 0$ the expression $\alpha_k(L_k - M_k)^2 - 2\beta_k L_k M_k$ is the difference of two squares. Indeed, we have

$$\alpha_k(L_k - M_k)^2 - 2\beta_k L_k M_k = \alpha_k \left(L_k - \frac{\beta_k}{\alpha_k} M_k \right)^2 - \frac{\alpha_k^2 + \beta_k^2}{\alpha_k} M_k^2$$

for $\alpha_k \ne 0$ and

$$\alpha_k(L_k - M_k)^2 - 2\beta_k L_k M_k = -2\beta_k L_k M_k = \frac{\beta_k}{2}(L_k - M_k)^2 - \frac{\beta_k}{2}(L_k + M_k)^2$$

for $\alpha_k = 0$.

Finally, we can make the following conclusion.

The quadratic form (Dx, x) is represented as the sum of positive and negative squares of linearly independent forms. Moreover, the number of positive squares coincides with the sum of the number of positive real roots and the number of pairs of conjugate complex roots of the polynomial $P(\lambda)$.

Thus, given the number of positive and negative eigenvalues of matrices C and D, it is possible to find the number of positive and negative real roots of the polynomial $P(\lambda)$ and the number of pairs of its conjugate complex roots.

PROBLEM 9.2.1. Consider the quadratic form

$$(Ex, x) = \sum_{j=1}^{N} (1 - \lambda_j^2) \left(\sum_{k=1}^{N} \lambda_j^{k-1} x_k \right)^2,$$

where λ_j are the roots of the polynomial $P(\lambda)$. How are the roots λ_j located if the matrix E is nonsingular and the number of negative and positive eigenvalues is known? How to compute the coefficients of the quadratic form (Ex, x) if the power sums of roots are known?

We recall the well-known recurrence formulas (see, for example, [**23**]) for computing power sums $s_1, s_2, \dots$ from given coefficients $p_1, p_2, \dots, p_N$ of the polynomial $P(\lambda) = \lambda_N + p_1\lambda^{N-1} + p_2\lambda^{N-2} + \dots + p_{N-1}\lambda + p_N$ ($p_k = 0$ for $k > N$):

$$\begin{aligned} -s_1 &= p_1, \\ -s_2 &= 2p_2 + s_1p_1, \\ -s_3 &= 3p_3 + s_2p_1 + s_1p_2, \\ &\dots\dots\dots \\ -s_j &= jp_j + s_{j-1}p_1 + s_{j-1}p_2 + \dots + s_1p_{j-1}. \end{aligned}$$

Having computed s_j, we can use the Sylvester rule to find the number of positive and negative eigenvalues of C and D. Moreover, we can use rational operations without actually computing the roots of the equation $P(\lambda) = 0$. As a result, we obtain the precise information about the location of these roots. We emphasize that no computation of roots is necessary in this approah.

9.3. Lyapunov theorems

Solution of the Lyapunov equation. The representation of a solution in the integral form. The unique solvability of the Lyapunov equation depending on the location of eigenvalues of a matrix. The discrete Lyapunov equation as an analogue of the Lyapunov equation.

An important advantage in further development of the idea of using quadratic forms in the study of the location of eigenvalues of a matrix A was due to Lyapunov. He was interested in a criterion that all roots of the characteristic polynomial lie in the left half-plane. In this case (see §4.4), the matrix equation

$$XA + A^*X + C = 0 \tag{9.3.1}$$

suggested by Lyapunov is uniquely solvable for any matrix C; moreover, the solution X is a Hermitian matrix ($X = X^*$) if $C = C^*$. As was shown by Lyapunov, the solution X is represented by the formula

$$X = \int_0^\infty e^{tA^*} C e^{tA}\, dt.$$

We prove this formula. First note that

$$\|e^{tA^*} C e^{tA}\| \le \|e^{tA^*}\|\, \|C\|\, \|e^{tA}\| \le \|C\|\, \|e^{tA}\|^2.$$

If $C = C^*$, then

$$(e^{tA^*} C A^{tA})^* = e^{tA^*} C e^{tA}.$$

Recall (see §8.5) that in the case $\operatorname{Re}\lambda_j(A) \le -\sigma < 0$, $j = 1, 2, \dots, N$, there exists a constant K such that $\|e^{tA}\| \le K e^{-\sigma t/2}$. Therefore, $\|e^{tA^*} C e^{tA}\| \le K^2 \|C\| e^{-\sigma t}$. Consequently, the integral converges and the matrix X is self-adjoint ($X = X^*$) if $C = C^*$.

Since for each t the matrix exponentials e^{tA^*} and e^{tA} are nonsingular and the matrix $C = C^*$ is positive definite, we see that the integrand and, consequently, the result of integration is a positive definite matrix, i.e., $X = X^* > 0$.

Set

$$Z(s) = \int_s^\infty e^{tA^*} C e^{tA}\, dt = \int_0^\infty e^{(t+s)A^*} C e^{(t+s)A}\, dt.$$

It is obvious that $Z(0) = X$ and

$$\begin{aligned}\frac{d}{ds} Z(s) &= -e^{sA^*} C e^{sA} = A^* \left[\int_0^\infty e^{(t+s)A^*} C e^{(t+s)A}\, dt\right] + \left[\int_0^\infty e^{(t+s)A^*} C e^{(t+s)A}\, dt\right] A \\ &= A^* Z(s) + Z(s) A,\end{aligned}$$

i.e., $Z(s)A + A^*Z(s) + e^{sA^*}Ce^{sA} = 0$. Therefore, for $s = 0$ the matrix X satisfies equation (9.3.1). In view of the unique solvability, equation (9.3.1) has no other solutions except for the solution presented by the integral

$$X = \int_0^\infty e^{tA^*} C e^{tA}\, dt.$$

In fact, we have also established that the positive definiteness of $C = C^*$ implies the positive definiteness of the solution X ($X = X^* > 0$) (under the condition $\operatorname{Re} \lambda_j(A) < 0$, $j = 1, 2, \ldots, N$). Conversely, if we find Hermitian positive definite matrices X and C ($X = X^* > 0$ and $C = C^* > 0$) satisfying (9.3.1), then we can conclude that the eigenvalues of A lie in the left half-plane. Indeed, let $x(t)$ be an arbitrary solution to the vector differential equation $\dfrac{dx}{dt} = Ax$. Compute the t-derivative of the quadratic form $(Xx(t), x(t))$:

$$\begin{aligned}\frac{d}{dt}(Xx(t), x(t)) &= (X\dot{x}(t), x(t)) + (Xx(t), \dot{x}(t)) \\ &= (XAx(t), x(t)) + (Xx(t), Ax(t)) \\ &= ((XA + A^*X)x(t), x(t)) = -(Cx(t), x(t)).\end{aligned}$$

Since the quadratic forms (Cx, x) and (Xx, x) are positive definite, we have

$$\begin{gathered}(Cx, x) \geq \mu(Xx, x), \quad \mu > 0, \\ \xi_1\|x\|^2 \geq (Xx, x) \geq \xi_0\|x\|^2, \quad \xi_0 > 0.\end{gathered}$$

Therefore,

$$\begin{gathered}e^{\mu t}\left[\frac{d}{dt}(Xx(t), x(t)) + \mu(Xx(t), x(t))\right] \leq 0, \\ \frac{d}{dt}\left[e^{\mu t}(Xx(t), x(t))\right] \leq 0, \\ e^{\mu t}(Xx(t), x(t)) \leq (Xx(0), x(0)), \\ (Xx(t), x(t)) \leq e^{-\mu t}(Xx(0), x(0)), \\ \|x(t)\|^2 \leq \frac{\xi_1}{\xi_0} e^{-\mu t}\|x(0)\|^2.\end{gathered}$$

The last inequality holds only if the eigenvalues of the matrix A lie in the left half-plane.

THEOREM 9.3.1 (Lyapunov). *If the eigenvalues of a matrix A lie in the left half-plane, then for any self-adjoint positive definite matrix C ($C = C^* > 0$) the matrix equation $XA + A^*X + C = 0$ is uniquely solvable. Furthermore, the solution X is a self-adjoint positive definite matrix ($X = X^* > 0$).*

Conversely, if there exist positive definite matrices X and C ($X = X^* > 0$ and $C = C^* > 0$) satisfying the matrix equation $XA + A^*X + C = 0$, then the eigenvalues of A lie in the left half-plane.

Consider the equation

$$X - A^*XA = C, \tag{9.3.2}$$

which looks like the Lyapunov equation (see (9.3.1)). As is known (see §4.4), equation (9.3.2) is uniquely solvable for any matrix C provided that the eigenvalues of the matrix A lie in the unit disk ($|\lambda_j(A)| \le \rho < 1$). The infinite sum

$$S = C + A^*CA + (A^*)^2CA^2 + \dots$$

converges because of the estimate

$$\|(A^*)^kCA^k\| \le \|C\|\,\|A^k\|^2 \le \frac{K\|C\|}{2\pi}\rho^{k+1},$$

which is a consequence of the inequality (see §8.5) $\|A^k\| = \|[A^*]^k\| \le K\,\rho^{(k+1)/2}$. It is easy to verify that $S - A^*SA = C$, i.e., S coincides with the unique solution X to equation (9.3.2). If $C = C^* > 0$ is a positive definite matrix, then the solution X is also a positive definite matrix. On the other hand, if there exist two positive definite matrices $C = C^* > 0$ and $X = X^* > 0$ satisfying the matrix equation (9.3.2), then the eigenvalues of A lie in the unit disk.

Let us prove this important assertion. Consider an eigenvalue λ_0 of A, the corresponding eigenvector $x_0 \neq 0$ ($Ax_0 = \lambda_0 x_0$), and the quadratic form $(Xx_0, x_0) > 0$. We note that

$$\begin{aligned}|\lambda_0|^2(Xx_0,x_0) &= (X\lambda_0x_0,\lambda_0x_0) = (XAx_0,Ax_0) = (A^*XAx_0,x_0)\\ &= (Xx_0,x_0) - (Cx_0,x_0).\end{aligned}$$

Therefore,

$$\begin{aligned}(1-|\lambda_0|^2)(Xx_0,x_0) &= (Cx_0,x_0) > 0,\\ 1-|\lambda_0|^2 &= \frac{(Cx_0,x_0)}{(Xx_0,x_0)} > 0.\end{aligned}$$

The proof is complete.

Note that the Lyapunov equation (9.3.1) is used in the study of the decrease of solutions to the differential equation $\dot{x}(t) = Ax(t)$ as t increases. In a similar way, the matrix equation (9.3.2) is used in the analysis of the iterative (discrete) processes of the form $x_{n+1} = Ax_n$. If X and C are positive definite matrices connected by equation (9.3.2), then

$$\begin{aligned}(Xx_k,x_k) &= (XAx_{k-1},Ax_{k-1}) = (A^*XAx_{k-1},x_{k-1})\\ &= (Xx_{k-1},x_{k-1}) - (Cx_{k-1},x_{k-1})\\ &\le (Xx_{k-1},x_{k-1}) - \frac{1}{\|C^{-1}\|\,\|X\|}(Xx_{k-1},x_{k-1}),\end{aligned}$$

$$(Xx_k, x_k) \le \left(1 - \frac{1}{\|C^{-1}\|\,\|X\|}\right)(Xx_{k-1}, x_{k-1}).$$

Since X is positive definite and x_0 is arbitrary, we have

$$\|C^{-1}\|\,\|X\| > 1,$$

$$(Xx_n, x_n) \le \left(1 - \frac{1}{\|C^{-1}\|\,\|X\|}\right)^n (Xx_0, x_0).$$

It is clear that the decrease of $\|x_n\|$ can be characterized by the estimates

$$\frac{\|x_n\|^2}{\|X^{-1}\|^{-1}} \le (Xx_n, x_n),$$

$$\|x_n\| \le \left(1 - \frac{1}{\|C^{-1}\|\,\|X\|}\right)^{n/2} \sqrt{\|X\|\,\|X^{-1}\|}\,\|x_0\|.$$

The last estimate is similar to the above estimate

$$\|x(t)\| \le \sqrt{\xi_1/\xi_0 e^{-\mu t/2}}\|x_0\|$$

for a solution x to the differential equation $\dot{x} = Ax$. Due to this analogy, equation (9.3.2) is sometimes referred to as the *Lyapunov equation in the discrete case* or the *discrete Lyapunov equation*

PROBLEM 9.3.1. For a sequence of vectors $\{x_n\}$ connected by the relations $x_n = Ax_{n-1}$ prove the more exact estimate

$$\|x_n\| \le \left(1 - \frac{1}{\omega}\right)^{n/2} \sqrt{\|X\|\,\|X^{-1}\|}\,\|x_0\|$$

if $X - A^*XA = C$, where X and C are positive definite matrices and ω is the largest root of the equation $\det(X - \omega C) = 0$.

CHAPTER 10

Matrix Equations and Projections

10.1. Solutions to the Lyapunov equations

A formula for a solution to the matrix Lyapunov equation. A similar representation of a solution to the discrete Lyapunov equation.

In §4.4, we studied the matrix Lyapunov equation $XA + A^*X + C = 0$. Recall that the equation is uniquely solvable for any matrix C provided that the eigenvalues of the matrix A lie in the left-half plane. In §9.3, we represented a solution X to the matrix Lyapunov equation in the following integral form:

$$X = \int_0^\infty e^{tA^*} C e^{tA} dt.$$

In this section, under the same assumption on the location of the eigenvalues of A, we derive another integral representation:

$$X = \frac{1}{2\pi} \int_{-\infty}^{+\infty} (-itI - A^*)^{-1} C (itI - A)^{-1} dt. \tag{10.1.1}$$

We recall (see §8.2) that if a matrix A has no purely imaginary eigenvalues, then

$$\frac{1}{2} I + \lim_{a \to \infty} \frac{1}{2\pi} \int_{-a}^{+a} (itI - A)^{-1}\, dt = P,$$

where P is the projection onto the invariant subspace of the matrix A corresponding to the eigenvalues $\lambda_j(A)$ lying in the left half-plane. If the spectrum of A lies in the left half-plane, then $P = I$. Consequently,

$$\lim_{a \to \infty} \frac{1}{2\pi} \int_{-a}^{+a} (itI - A)^{-1}\, dt = \frac{1}{2} I,$$

$$\lim_{a \to \infty} \frac{1}{2\pi} \int_{-a}^{+a} (-itI - A^*)^{-1}\, dt = \left(\frac{1}{2} I\right)^* = \frac{1}{2} I.$$

The second equality is obtained from the first one by replacing A with A^*. Therefore,

$$\lim_{a \to \infty} \frac{1}{2\pi} \int_{-a}^{+a} [C(itI - A)^{-1} + (-itI - A^*)^{-1} C]\, dt = \frac{1}{2} C + \frac{1}{2} C = C.$$

On the other hand,

$$\begin{aligned}
&A^*(-itI - A^*)^{-1}C(itI - A)^{-1} + (-itI - A^*)^{-1}C(itI - A)^{-1}A \\
&\quad = (A^* + itI)(-itI - A^*)^{-1}C(itI - A)^{-1} \\
&\qquad - it(-itI - A^*)^{-1}C(itI - A)^{-1} + it(-itI - A^*)^{-1}C(itI - A)^{-1} \\
&\qquad + (-itI - A^*)^{-1}C(itI - A)^{-1}(A - itI) \\
&\quad = -C(itI - A)^{-1} - (-itI - A^*)^{-1}C.
\end{aligned} \tag{10.1.2}$$

Hence,

$$A^*\left[\lim_{a\to\infty} \frac{1}{2\pi} \int_{-a}^{+a} (-itI - A^*)^{-1}C(itI - A)^{-1}\, dt\right] + \left[\lim_{a\to\infty} \frac{1}{2\pi} \int_{-a}^{+a} (-itI - A^*)^{-1}C(itI - A)^{-1}\, dt\right] A = -C.$$

Thus, the solution X to the matrix Lyapunov equation $A^*X + XA + C = 0$ admits the integral representation

$$X = \lim_{a\to\infty} \frac{1}{2\pi} \int_{-a}^{+a} (-itI - A^*)^{-1}C(itI - A)^{-1}\, dt,$$

which coincides with (10.1.1). The integral on the right-hand side of (10.1.1) converges. Hence, it is not necessary to compute the principal value of this integral. We emphasize that a solution to the Lyapunov equation is unique.

A similar representation holds for a solution X to the discrete Lyapunov equation

$$X - A^*XA = C$$

under the natural assumption that the eigenvalues $\lambda_j(A)$ of A lie in the unit disk ($|\lambda_j(A)| \le \rho < 1$). To obtain this representation, we consider the expression

$$\begin{aligned}
(e^{-i\varphi}I - A^*)^{-1}C(e^{i\varphi}I - A)^{-1} &= (I - e^{i\varphi}A^*)^{-1}C(I - e^{-i\varphi}A)^{-1} \\
&= \left[\sum_{k=0}^{\infty} e^{ik\varphi}(A^*)^k\right] C \left[\sum_{l=0}^{\infty} e^{-il\varphi}A^l\right].
\end{aligned}$$

In view of the estimate $\|A^k\| = \|(A^*)^k\| \le \text{const}\, \rho^{(k+1)/2}$ (see §8.5), we can integrate term-by-term:

$$\begin{aligned}
\frac{1}{2\pi}\int_0^{2\pi} (e^{-i\varphi}I - A^*)^{-1}C(e^{i\varphi}I - A)^{-1} d\varphi &= \frac{1}{2\pi} \sum_{k,l=0}^{\infty} \int_0^{2\pi} e^{i(k-l)\varphi}(A^*)^k C A^l \\
&= \sum_{k=0}^{\infty} (A^*)^k C A^k = I + A^*CA + (A^*)^2CA^2 + \cdots.
\end{aligned}$$

This sum is a solution X to the discrete Lyapunov equation (see §9.3). Thus, we have justified the representation

$$X = \frac{1}{2\pi}\int_0^{2\pi} (e^{-i\varphi}I - A^*)^{-1}C(e^{i\varphi}I - A)d\varphi.$$

Sometimes, it is convenient to interchange A and A^* in the discrete Lyapunov equation. Under the same assumption that the eigenvalues of A lie in the unit disk, a solution J to the equation $J - AJA^* = C$ can be represented as follows:

$$J = \frac{1}{2\pi}\int_0^{2\pi} (e^{i\varphi}I - A)^{-1}C(e^{i\varphi} - A)^{-*}\,d\varphi.$$

Hereinafter we use the abbreviated notation $(e^{i\varphi}I - A)^{-*} = ((e^{i\varphi}I - A)^*)^{-1}$.

10.2. A generalization of the Lyapunov equation

The solvability of a matrix equation in the case where the coefficient matrix has no purely imaginary eigenvalues. An integral representation of a solution. Uniqueness of a solution. The solvability of a matrix equation and location of the eigenvalues of the coefficient matrix.

Suppose that every eigenvalue of a matrix A is located in the right half-plane or in the left half-plane, but not on the imaginary axis. Then (see §8.2)

$$\lim_{a\to\infty}\int_{-a}^{+a} (itI - A)^{-1}dt = P - \frac{1}{2}I,$$

$$\lim_{a\to\infty}\int_{-a}^{+a} (itI - A^*)^{-1}dt = P^* - \frac{1}{2}I,$$

where P is the projection onto the invariant subspace of A corresponding to the eigenvalues in the left half-plane. By (10.1.2) and the assumption on eigenvalues of A, we have

$$\begin{aligned} A^*\frac{1}{2\pi}&\int_{-\infty}^{+\infty} (-itI - A^*)^{-1}C(itI - A)^{-1}\,dt \\ &+ \frac{1}{2\pi}\int_{-\infty}^{\infty} (-itI - A^*)^{-1}C(itI - A)^{-1}\,dtA \\ = \lim_{a\to\infty}&\left[\frac{1}{2\pi}\int_{-a}^{+a} A(-itI - A^*)^{-1}C(itI - A)^{-1}\,dt\right. \\ &\left.+ \frac{1}{2\pi}\int_{-a}^{a} (-itI - A^*)^{-1}C(itI - A)^{-1}A\,dt\right] \end{aligned}$$

$$= -C \lim_{a\to\infty} \frac{1}{2\pi} \int_{-a}^{+a} (itI - A)^{-1}\, dt - \left[\lim_{a\to\infty} \int_{-a}^{+a} (-itI - A^*)^{-1}\, dt \right] C$$

$$= -C(P - \frac{1}{2}I) - (P^* - \frac{1}{2}I)C = C - CP - P^*C.$$

Thus, the matrix

$$X = \frac{1}{2\pi} \int_{-\infty}^{+\infty} (-itI - A^*)^{-1} C (itI - A)^{-1}\, dt$$

satisfies the equation $A^*X + XA = C - CP - P^*C$. We note that

$$C - CP - P^*C = (I - P^*)C(I - P) - P^*CP.$$

Indeed,

$$(I - P^*)C(I - P) - P^*CP = C - P^*C - CP + P^*CP - P^*CP = C - P^*C - CP.$$

Therefore, the equation for X can be rewritten as follows:

$$A^*X + XA = (I - P^*)C(I - P) - P^*CP. \tag{10.2.1}$$

Equation (10.2.1) generalizes the matrix Lyapunov equation, but, in addition to X, it contains the unknown projection P ($P^2 = P$ and $PA = AP$). It is obvious that $X = X^* > 0$ if $C = C^* > 0$. Thus, we have proved the following proposition.

PROPOSITION 10.2.1. *If a matrix A has no purely imaginary eigenvalues, then the system of matrix equations*

$$A^*X + XA = (I - P^*)C(I - P) - P^*CP, \tag{10.2.2}$$

$$PA = AP, \quad P^2 = P, \tag{10.2.3}$$

is solvable with respect to X and P; moreover, the following integral representations hold:

$$P = \frac{1}{2}I + \lim_{a\to\infty} \int_{-a}^{+a} (itI - A)^{-1}\, dt,$$

$$X = -\frac{1}{2\pi} \int_{\infty}^{+\infty} (itI + A^*)^{-1} C (itI - A)^{-1}\, dt.$$

If $C = C^*$, then $X = X^*$ and if $C = C^* > 0$, then $X = X^* > 0$.

The above system of matrix equations looks like the Lyapunov equation and can be regarded as its generalization. However, this system has an essential deficiency, namely, the solution X is not always unique even if $X = X^* > 0$.

REMARK 10.2.1. Assume that a matrix A has no eigenvalues on the imaginary axis, but $\lambda_1(A) = -\overline{\lambda_2(A)}$. In this case, $\overline{\lambda_2(A)}$ is an eigenvalue (denote it by $\lambda_1(A^*)$) of the matrix A^* ($\lambda_1(A) = -\lambda_1(A^*)$). As we know, if the matrices A and B have a common eigenvalue, then for certain right-hand sides F the matrix Sylvester equation $XA - BX = F$ is not solvable. Indeed, the Sylvester equation is a system of linear equations with respect to elements of X. The number of equations coincides with the number of variables. Since for some right-hand sides these equations can have no solutions, the homogeneous equations have nonzero solutions, i.e., there exists a matrix $Y \neq 0$ of the same size as the matrix X such that $YA - BY = 0$. For $B = -A^*$ the Sylvester equation becomes the Lyapunov equation $XA + A^*X = F$. Therefore, in the case $\lambda_1(A) = -\lambda_1(A^*)$, a solution (even if it exists) is not unique. If the matrix A has no purely imaginary eigenvalues, then for $C = C^* > 0$ there exists a solution X $(X = X^* > 0)$ to the system (10.2.2), (10.2.3). However, in the case $\lambda_1(A) = -\lambda_1(A^*)$, the equation $A^*Y + YA = 0$ also has a nonzero solution. Simultaneously, $A^*Y^* + Y^*A = 0$. It is obvious that either $Y_1 = Y^* + Y \neq 0$ or $Y_2 = i(Y^* - Y) \neq 0$ (or both Y_1 and Y_2 are nonzero), $Y_1^* = Y_1$, $Y_2^* = Y_2$. In other words, we can assume that the nonzero solution Y to the equation $A^*Y + YA = 0$ is self-adjoint $(Y = Y^*)$. For any ε,

$$A^*(X + \varepsilon Y) + (X + \varepsilon Y)A = (I - P^*)C(I - P) - P^*CP,$$

and for a sufficiently small ε the matrix $X + \varepsilon Y$ is positive definite because X is positive definite.

If $X = X^* > 0$, then $X_1 = P^*XP + (I - P^*)X(I - P) = X_1^* > 0$. Indeed, since $x = Px + (I - P)x$, for $x \neq 0$ at least one of the vectors Px and $(I - P)x$ differs from zero. Consequently, at least one of the terms of the sum

$$(XPx, Px) + (X(I-P)x, (I-P)x) = ([P^*XP + (I-P^*)X(I-P)]x, x) = (X_1x, x)$$

is positive. We note that $PA = AP$ and $A^*P^* = P^*A^*$. Multiplying the equation

$$A^*X + XA = (I - P^*)C(I - P) - P^*CP$$

from the left by P^* and from the right by P and then from the left by $(I - P^*)$ and from the right by $(I - P)$, we arrive at the equalities

$$\begin{gathered} A^*P^*XP + P^*XPA = -P^*CP, \\ A^*(I - P^*)X(I - P^*) + (I - P^*)X(I - P)A = (I - P^*)C(I - P), \\ A^*X_1 + X_1A^* = -P^*CP + (I - P^*)C(I - P). \end{gathered}$$

Thus, we obtain the following result.

PROPOSITION 10.2.2. *If a pair P, X $(X = X^* > 0)$ is a solution to equations* (10.2.2), (10.2.3), *then the pair P, X_1, where $X_1 = P^*XP + (I-P^*)X(I-P) = X_1^*$, is also a solution to equations* (10.2.2), (10.2.3).

We note that $X_1P = P^*XP = (P^*XP)^* = P^*X_1$. If we add the equation $XP = P^*X$ to the system (10.2.2), (10.2.3), then the solution $X = X^* > 0$ to the extended system is unique. Let us write this system $(C = C^* > 0)$:

$$\begin{gathered} A^*X + XA + P^*CP - (I - P^*)C(I - P) = 0, \\ PA = AP, \quad P^2 = P, \quad XP = P^*X, \quad X = X^* > 0. \end{gathered} \tag{10.2.4}$$

Later we will show that the matrix

$$X = \frac{1}{2\pi} \int_{-\infty}^{+\infty} (itI + A^*)^{-1} C (itI - A)^{-1} dt \tag{10.2.5}$$

automatically satisfies the equation $XP = P^*X$ and, consequently, $X_1 = P^*XP + (I - P^*)X(I - P) = X$. Thereby, the integral (10.2.5) is a solution to the extended system (10.2.4).

LEMMA 10.2.1. *If the eigenvalues of an $N_1 \times N_1$-matrix A_1 and of an $N_2 \times N_2$-matrix A_2 lie in the left half-plane, then for any $N_1 \times N_2$-matrix B we have*

$$\int_{-\infty}^{\infty} (itI_{N_1} - A_1)^{-1} B (itI_{N_2} - A_2)^{-1}\, dt = 0. \tag{10.2.6}$$

PROOF. Equation (10.2.6) is equivalent to the condition $I_1(R) \to 0$ as $R \to \infty$, where

$$I_1(R) = \int_{-iR}^{iR} (zI_{N_1} - A_1)^{-1} B (zI_{N_2} - A_2)^{-1}\, dz.$$

Denote by $I_2(R)$ the integral along the half-circle Γ_R of radius R and center $z = 0$ that lies in the right half-plane and joins the points $z = \mp iR$, i.e.,

$$\begin{aligned} I_2(R) &= \int_{\Gamma_R} (zI_{N_1} - A_1)^{-1} B (zI_{N_2} - A_2)^{-1}\, dz \\ &= \frac{1}{R} \int_{-\pi/2}^{+\pi/2} \left(e^{i\varphi} I_{N_1} - \frac{1}{R} A_1 \right)^{-1} B \left(e^{i\varphi} I_{N_2} - \frac{1}{R} A_2 \right)^{-1} ie^{i\varphi}\, d\varphi. \end{aligned}$$

It is obvious that $I_2(R) = O(1/R)$ as $R \to \infty$. Since $I_2(R) - I_1(R)$ is an integral of a matrix whose elements are analytic functions without poles inside the closed integration contour, we have $I_2(R) - I_1(R) = 0$. Therefore, $I_1(R) = O(1/R)$. □

We give a variant of Lemma 10.2.1 without proof because it can be proved in the same way as earlier.

LEMMA 10.2.2. *If the eigenvalues of an $N_1 \times N_1$-matrix A_1 and of an $N_2 \times N_2$-matrix A_2 lie in the right half-plane, then*

$$\int_{-\infty}^{+\infty} (itI_{N_1} - A_1)^{-1} B (itI_{N_2} - A_2)^{-1} dt = 0.$$

Let A be an arbitrary square matrix without eigenvalues on the imaginary axis. The matrix A can be represented in the form

$$A = T \begin{bmatrix} A_+ & 0 \\ 0 & A_- \end{bmatrix} T^{-1},$$

where T is a nonsingular matrix, A_+ is a square $N_1 \times N_1$-matrix with eigenvalues in the right half-plane, and A_- is a square $N_2 \times N_2$-matrix with eigenvalues in the left half-plane. The projection P onto the invariant subspace of A corresponding to the eigenvalues in the left half-plane has the form

$$T \begin{bmatrix} 0 & 0 \\ 0 & I_{N_2} \end{bmatrix} T^{-1};$$

moreover, if $\lambda = it$, then

(10.2.7)

$$\begin{aligned}
&(I-P)^*(\overline{\lambda} I - A^*)^{-1}C(\lambda I - A)^{-1}P \\
&\quad = T^{-*}\begin{bmatrix}(-itI_{N_1}-A_+^*)^{-1} & 0\\ 0 & 0\end{bmatrix}\begin{bmatrix}B_{11} & B_{12}\\ B_{21} & B_{22}\end{bmatrix}\begin{bmatrix}0 & 0\\ 0 & (itI_{N_2}-A_-^*)^{-1}\end{bmatrix}T^{-1}\\
&\quad = -T^{-*}\begin{bmatrix}0 & (itI_{N_1}+A_+^*)^{-1}B_{12}(itI_{N_2}-A_-^*)^{-1}\\ 0 & 0\end{bmatrix}T^{-1},
\end{aligned}$$

where $B = T^*CT$. Using Lemma 10.2.1, from (10.2.7) we deduce the equality

$$(I-P)^*\Bigl[\int_{-\infty}^{\infty}(-itI - A^*)^{-1}C(itI-A)^{-1}\,dt\Bigr]P = 0.$$

Consequently,

$$X = \frac{1}{2\pi}\int_{-\infty}^{\infty}(-itI - A^*)^{-1}C(itI-A)^{-1}\,dt$$

satisfies the equation $(I-P)^*XP = 0$. The equality $P^*X(I-P) = 0$ or equivalently $P^*X = P^*XP$, is proved in a similar way. These equalities imply the required relations $P^*X = XP$ and $X = P^*XP + (I-P)^*X(I-P)$. Indeed,

$$\begin{aligned}
X &= [P^* + (I-P)^*]X[P + (I-P)]\\
&= P^*XP + (I-P^*)X(I-P) + (I-P)^*XP + P^*X(I-P)\\
&= P^*XP + (I-P^*)X(I-P),\\
P^*X &= (P^*)^2XP = P^*XP = (P^*XP)^* = XP,
\end{aligned}$$

where the relations $X = X^*$, $P^2 = P$, and $(P^*P)^2 = P^*$ are used.

Thus, we have proved the following theorem.

THEOREM 10.2.1. *Let a matrix A have no purely imaginary eigenvalues. Then for any matrix C the matrices*

$$X = \frac{1}{2\pi}\int_{-\infty}^{\infty}(-itI - A^*)^{-1}C(itI-A)^{-1}\,dt, \tag{10.2.8}$$

$$P = \frac{1}{2}I + \lim_{\alpha\to\infty}\int_{-a}^{+a}(itI-A)^{-1}\,dt \tag{10.2.9}$$

form a solution to the matrix equations (10.2.4).

The positive definiteness of the matrix X under the condition $C = C^* > 0$ follows from the integral representation. The existence of such a solution allows us to conclude that the matrix A has no purely imaginary eigenvalues, although no preliminary information about A is known. Indeed, the above equations imply that

P and $I-P$ are projections onto complementary invariant subspaces of the matrix A. Consequently,

$$A = T\begin{bmatrix} A_1 & 0 \\ 0 & A_2 \end{bmatrix} T^{-1},$$
$$P = T\begin{bmatrix} I_{N_1} & 0 \\ 0 & 0 \end{bmatrix} T^{-1}.$$

Introduce the notation

$$Y = T^*XT = \begin{bmatrix} Y_{11} & Y_{12} \\ Y_{21} & Y_{22} \end{bmatrix},$$
$$B = T^*CT = \begin{bmatrix} B_{11} & B_{12} \\ B_{21} & B_{22} \end{bmatrix}.$$

Let us verify that the equations $A^*X + XA + P^*CP - (I - P^*)C(I - P) = 0$ and $XP = P^*X$ in (10.2.4) written in the above notation are equivalent to the following equations:

$$\begin{bmatrix} A_1^* & 0 \\ 0 & A_2^* \end{bmatrix}\begin{bmatrix} Y_{11} & Y_{12} \\ Y_{21} & Y_{22} \end{bmatrix} + \begin{bmatrix} Y_{11} & Y_{12} \\ Y_{21} & Y_{22} \end{bmatrix}\begin{bmatrix} A_1 & 0 \\ 0 & A_2 \end{bmatrix} + \begin{bmatrix} B_{11} & 0 \\ 0 & -B_{22} \end{bmatrix} = \begin{bmatrix} 0 & 0 \\ 0 & 0 \end{bmatrix},$$
$$\begin{bmatrix} Y_{11} & Y_{12} \\ 0 & 0 \end{bmatrix} = \begin{bmatrix} Y_{11} & 0 \\ Y_{21} & 0 \end{bmatrix},$$

which implies

$$Y_{12} = 0, \quad Y_{21} = 0,$$
$$A_1^*Y_{11} + Y_{11}A_1 + B_{11} = 0 \qquad (B_{11} = B_{11}^* > 0),$$
$$A_2^*Y_{22} + Y_{22}A_2 - B_{22} = 0 \quad (B_{22} = B_{22}^* > 0).$$

In view of the Lyapunov theorem and the positive definiteness of Y_{11} and Y_{22}, the eigenvalues of A_1 and A_2 lie in the left and right half-planes, respectively. By the Lyapunov theorem, Y_{11} and Y_{22} are uniquely determined. The matrix A has no purely imaginary eigenvalues and the number of the eigenvalues in the left half-plane is equal to $\operatorname{tr} P$. For X and P the integral representations (10.2.8) and (10.2.9) hold. Thus, we have proved the following theorem.

THEOREM 10.2.2. *If for a positive definite matrix $C = C^* > 0$ there exist matrices P and X $(X = X^*)$ satisfying* (10.2.4) *and, in addition, the matrix X is positive definite, then the matrix A has no purely imaginary eigenvalues and the matrices X and P are represented in the form* (10.2.8) *and* (10.2.9), *respectively.*

Equations (10.2.4) were suggested by Bulgakov who formulated and proved Theorem 10.2.1 as well as a modification of Theorem 10.2.2.

PROBLEM 10.2.1. Let a matrix A have no purely imaginary eigenvalues. Let P be the projection $(P = P^2)$ onto the invariant subspace corresponding to the eigenvalues in the left half-plane. Under the assumption $PA = AP$, prove that for any matrix C (not necessarily symmetric) the system

$$A^*X + XA + P^*CP - (I - P)^*C(I - P) = 0, \quad XP = P^*X$$

has a unique solution X represented in the integral form (10.2.8).

10.3. Matrix pencil regular on the unit circle

The definition of a matrix pencil regular on the unit circle. Examples of matrix pencils. Representations of matrices forming a matrix pencil. The definition of the infinite eigenvalue and of the corresponding eigenvector. Reducing subspaces and representations of projections.

We say that a pair of square $N \times N$-matrices A and B form a *pencil regular on the unit circle* if for any real number φ the matrix $e^{i\varphi}B - A$ is nonsingular.

By a *pencil* one usually means the expression $\lambda B - A$ with formal parameter λ. A pencil is said to be *regular* if $\det(\lambda B - A)$ is not equal to zero identically, i.e., there exists λ such that the matrix $\lambda B - A$ is nonsingular. The regularity on the unit circle is equivalent to the condition $\det(\lambda B - A) \neq 0$ for λ such that $|\lambda| = 1$. The same condition can be formulated as the condition that the matrix

$$A - \frac{1 - it}{1 + it}B$$

is nonsingular for all real numbers t. Regarding $t = \infty$ as a real number, we set

$$A - \frac{1 - i\infty}{1 + i\infty}B = A + B.$$

EXAMPLE 10.3.1. The following pencil is not regular:

$$\lambda \begin{bmatrix} 1 & 0 \\ 0 & 0 \end{bmatrix} - \begin{bmatrix} 0 & 0 \\ 1 & 0 \end{bmatrix}.$$

EXAMPLE 10.3.2. The following pencils are regular:

$$\lambda \begin{bmatrix} 12 & 0 \\ 2 & 3 \end{bmatrix} - \begin{bmatrix} 0 & 18 \\ 0 & 0 \end{bmatrix}, \qquad \lambda \begin{bmatrix} 1 & -3 \\ 2 & 1 \end{bmatrix} - \begin{bmatrix} 5 & 0 \\ 0 & -1 \end{bmatrix}.$$

The determinant of the first pencil vanishes for $\lambda = -1 = e^{\pm i\pi}$. Therefore, it is not regular on the unit circle.

If a pencil $\lambda B - A$ is regular on the unit circle, then the matrix $A - [(1 - it)/(1 + it)]B$ is nonsingular and for every real finite number t we have

$$\det(itI + (A + B)^{-1}(A - B)) = (1 + it)^N [\det(A + B)]^{-1} \det\left(A - \frac{1 - it}{1 + it}B\right) \neq 0.$$

Consequently, the imaginary axis gives a dichotomy of the spectrum of the matrix $(A + B)^{-1}(A - B)$. Moreover, there exists a nonsingular matrix Q such that

$$(A + B)^{-1}(A - B) = Q^{-1} \begin{bmatrix} K_- & 0 \\ 0 & K_+ \end{bmatrix} Q \tag{10.3.1}$$

and the eigenvalues of the $N_1 \times N_1$-matrix K_- and $N_2 \times N_2$-matrix K_+ ($N_1 + N_2 = N$) lie in the left and right half-planes, respectively. The projection

$$P = Q^{-1} \begin{bmatrix} I_{N_1} & 0 \\ 0 & 0 \end{bmatrix} Q$$

onto the invariant subspace of the matrix $(A + B)^{-1}(A - B)$ corresponding to the eigenvalues in the left half-plane can be computed by the formula (see §8.2)

$$P = \frac{1}{2}I + \lim_{a \to \infty} \frac{1}{2\pi} \int_{-a}^{+a} [itI - (A + B)^{-1}(A - B)]^{-1} dt. \tag{10.3.2}$$

Using the equalities

$$A - B = (A+B)Q^{-1}\begin{bmatrix} K_- & 0 \\ 0 & K_+ \end{bmatrix} Q,$$

$$A + B = (A+B)Q^{-1}\begin{bmatrix} I_{N_1} & 0 \\ 0 & I_{N_2} \end{bmatrix} Q,$$

the first of which is equivalent to (10.3.1) and the second is trivial, we establish the following representations:

$$\begin{aligned} A &= \frac{1}{2}(A+B)Q^{-1}\begin{bmatrix} I_{N_1}+K_- & 0 \\ 0 & I_{N_2}+K_+ \end{bmatrix} Q \\ &= \left\{\frac{1}{2}(A+B)Q^{-1}\begin{bmatrix} I_{N_1}-K_- & 0 \\ 0 & I_{N_2}+K_+ \end{bmatrix}\right\} \\ &\quad\times \begin{bmatrix} (I_{N_1}-K_-)^{-1}(I_{N_1}+K_-) & 0 \\ 0 & I_{N_2} \end{bmatrix} Q, \end{aligned}$$

$$\begin{aligned} B = &\left\{\frac{1}{2}(A+B)Q^{-1}\begin{bmatrix} I_{N_1}-K_- & 0 \\ 0 & I_{N_2}+K_+ \end{bmatrix}\right\} \\ &\times \begin{bmatrix} I_{N_1} & 0 \\ 0 & (I_{N_2}+K_+)^{-1}(I_{N_2}-K_+) \end{bmatrix} Q. \end{aligned}$$

We used the fact that since the spectra of the matrices K_- and K_+ lie in the left and right half-planes, respectively, the blocks $(I_{N_1} - K_-)$ and $(I_{N_2} + K_+)$ are nonsingular and, consequently, invertible. Introduce the notation

$$T = \frac{1}{2}(A+B)Q^{-1}\begin{bmatrix} I_{N_1}-K_- & 0 \\ 0 & I_{N_2}+K_+ \end{bmatrix} \qquad (\det T \neq 0),$$

$$A_1 = (I_{N_1}-K_-)^{-1}(I_{N_1}+K_-), \quad B_2 = (I_{N_2}+K_+)^{-1}(I_{N_2}-K_+).$$

We find

$$A = T\begin{bmatrix} A_1 & 0 \\ 0 & I_{N_2} \end{bmatrix} Q, \quad B = T\begin{bmatrix} I_{N_1} & 0 \\ 0 & B_2 \end{bmatrix} Q, \tag{10.3.3}$$

where the matrices T and Q, as well as the square blocks A_1 and B_2, are nonsingular and the eigenvalues of A_1 and B_2 are of module less than 1, i.e., all eigenvalues lie in the unit disk. Indeed, to each eigenvalue λ_j of A_1 we associate the eigenvector x $(x \neq 0)$ such that

$$\begin{gathered} (I_{N_1}-K_-)^{-1}(I_{N_1}+K_-)x \equiv A_1 x = \lambda_j x, \\ (I_{N_1}+K_-)x = \lambda_j (I_{N_1}-K_-)x, \\ (1-\lambda_j)x = -(1+\lambda_j)K_- x. \end{gathered}$$

Therefore,

$$K_- x = \frac{\lambda_j - 1}{\lambda_j + 1}x = (\alpha + i\beta)x;$$

moreover, $\alpha < 0$. Consequently,

$$|\lambda_j| = \left|\frac{1+\alpha+i\beta}{1-\alpha-i\beta}\right| = \sqrt{\frac{(1+\alpha)^2+\beta^2}{(1-\alpha)^2+\beta^2}} < 1.$$

Similarly, we verify that the spectrum of the matrix B_2 lies in the unit disk.

If a pair of matrices A and B form a pencil regular on the unit circle, then it can be represented in the form (10.3.3) with nonsingular matrices T and Q and blocks A_1 and B_1 whose spectra lie in the unit disk.

The above representation is connected with the decomposition of the N-dimensional complex Euclidean space into two complementary subspaces. The operators

$$P = Q^{-1}\begin{bmatrix} I_{N_1} & 0 \\ 0 & 0 \end{bmatrix} Q, \quad I - P = Q^{-1}\begin{bmatrix} 0 & 0 \\ 0 & I_{N_2} \end{bmatrix} Q \tag{10.3.4}$$

are projections onto these subspaces. By saying that *subspaces reduce the pencil $\lambda B - A$ from the right*, we mean that the following property holds. If two vectors Px and Py lying in one of these subspaces satisfy the equality $APx = BPy$, then

$$\begin{bmatrix} A_1 & 0 \\ 0 & 0 \end{bmatrix} QPx = \begin{bmatrix} I_{N_1} & 0 \\ 0 & 0 \end{bmatrix} QPy.$$

Similarly, for the vectors $(I - P)u$ and $(I - P)v$, connected by the relation $A(I - P)u = B(I - P)v$, we have

$$\begin{bmatrix} 0 & 0 \\ 0 & I_{N_2} \end{bmatrix} Q(I-P)u = \begin{bmatrix} 0 & 0 \\ 0 & B_2 \end{bmatrix} Q(I-P)v.$$

A complex number λ is called a *finite eigenvalue of the pencil $\lambda B - A$* if $\det(\lambda B - A) = 0$. In this case, there exists a nonzero vector x, called an *eigenvector of the pencil*, such that $\lambda Bx = Ax$. Moreover, Ax and Bx cannot vanish simultaneously; otherwise, $e^{i\varphi}Bx = Ax$ for all real φ, which leads to the equality $\det(e^{i\varphi}B - A) = 0$. But such an equality is impossible because the pencil is regular on the unit circle.

We say that the pencil $\lambda B - A$ has the *infinite eigenvalue* $\lambda = \infty$ if there exists a vector $x \neq 0$ such that $Bx = 0$. Moreover, $Ax \neq 0$ since the pencil is regular. The vector x is called an *eigenvector corresponding to the eigenvalue* $\lambda = \infty$. For an eigenvalue λ different from zero and infinity the corresponding eigenvector x satisfies both conditions $Ax \neq 0$ and $Bx \neq 0$.

At least one of the projections Px and $(I - P)x$ of a nonzero vector x must be nonzero. It turns out that if x is an eigenvector of a pencil regular on the unit circle, then either $Px = 0$ or $(I - P)x = 0$. Indeed,

$$\det(\lambda B - A) = \det T \det Q \det(\lambda I_{N_1} - A_1) \det(\lambda B_2 - I_{N_2}).$$

Therefore, if $\det(\lambda B - A) = 0$ and $|\lambda| < 1$, then $\det(\lambda B_2 - I_{N_2}) = 0$. For $\lambda = \infty$ we have $\det B_2 = 0$. It is obvious that $Px \neq 0$, $(I - P)x = 0$ for $|\lambda| < 1$ and $Px = 0$, $(I - P)x \neq 0$ for $|\lambda| > 1$ (and $\lambda = \infty$).

If a pencil is regular on the unit circle, then its eigenvectors lie in one of the subspaces that reduce the pencil and are determined by the projections P and $I - P$. The projection P projects onto the reducing subspace corresponding to the eigenvalues in the unit disk, whereas the projection $I - P$ projects onto the subspace corresponding to the eigenvalues outside the unit disk.

We indicate another integral representation of P which is more convenient than the representation (10.3.2):

$$P = \frac{1}{2\pi}\int_0^{2\pi} (B - e^{-i\varphi}A)^{-1}B\,d\varphi. \tag{10.3.5}$$

Indeed, in view of (10.3.3), we have

(10.3.6)

$$\begin{aligned}(B - e^{-i\varphi}A)^{-1}Bd\varphi &= \frac{1}{i}(e^{i\varphi}B - A)^{-1}Bde^{i\varphi}\\ &= \frac{1}{i}Q^{-1}\begin{bmatrix}(e^{i\varphi}I_{N_1} - A_1)^{-1}de^{i\varphi} & 0\\ 0 & (e^{i\varphi}B_2 - I_{N_2})^{-1}B_2de^{i\varphi}\end{bmatrix}Q,\end{aligned}$$

$$\begin{aligned}&\frac{1}{2\pi}\int_0^{2\pi}(B - e^{-i\varphi}A)^{-1}Bd\varphi\\ &\quad = Q^{-1}\begin{bmatrix}\frac{1}{2\pi i}\oint_\gamma(zI_{N_1} - A_1)^{-1}dz & 0\\ 0 & \frac{1}{2\pi i}\oint_\gamma(zB_2 - I_{N_2})^{-1}B_2dz\end{bmatrix}Q;\end{aligned}$$

here γ is the unit circle. Since the eigenvalues of B_2 lie in the unit disk, the elements of the matrix $(zB_2-I_{N_2})^{-1}B_2$ are bounded analytic (rational) functions of z defined in the disk and on the boundary. Therefore,

$$\frac{1}{2\pi i}\oint_\gamma(zB_2 - I_{N_2})^{-1}B_2dz = 0.$$

On the other hand, as was mentioned in §8.2, the fact that the eigenvalues of the matrix A_1 are surrounded by the contour γ implies that

$$\frac{1}{2\pi i}\oint_\gamma(zI_{N_1} - A_1)^{-1}dz = I_{N_1}.$$

Therefore,

$$\frac{1}{2\pi}\int_0^{2\pi}(B - e^{-i\varphi}A)^{-1}Bd\varphi = Q^{-1}\begin{bmatrix}I_{N_1} & 0\\ 0 & 0\end{bmatrix}Q = P.$$

The representation (10.3.5) is proved.

If P is known, then it is not hard to find the matrices T and Q in (10.3.3) and (10.3.4). We note that T and Q are not uniquely determined. From the representations of P and $I-P$ (see (10.3.4)) we see that for the first N_1 columns of Q^{-1} we can take the basis vectors for a reducing subspace (i.e., the image under the projection P) and for the last N_2 columns of Q^{-1} we can take the basis vectors for the complementary reducing subspace (i.e., the image under the projection $I-P$). The determination of Q is not unique because the choice of basis vectors is not unique.

We describe the procedure of the determination of the matrix Q. It begins with the computation of the singular value decompositions of the matrices corresponding to the projections P and $I-P$:

$$P = [\underbrace{\widetilde{W}_1}_{N_1 \text{ columns}} \vdots \underbrace{\widetilde{\widetilde{W}}_1}_{N_2 \text{ columns}}] \begin{bmatrix} \sigma_N^{(1)} & & & & & & \\ & \sigma_{N-1}^{(1)} & & & & \mathbf{0} & \\ & & \ddots & & & & \\ & & & \sigma_{N-N_1+1}^{(1)} & & & \\ & & & & 0 & & \\ & \mathbf{0} & & & & \ddots & \\ & & & & & & 0 \end{bmatrix} U_1,$$

$$I - P = [\underbrace{\widetilde{W}_2}_{N_2 \text{ columns}} \vdots \underbrace{\widetilde{\widetilde{W}}_2}_{N_1 \text{ columns}}] \begin{bmatrix} \sigma_N^{(2)} & & & & & \mathbf{0} \\ & \ddots & & & & \\ & & \sigma_{N-N_2+1}^{(2)} & & & \\ & & & 0 & & \\ & \mathbf{0} & & & \ddots & \\ & & & & & 0 \end{bmatrix} U_2.$$

Here U_1, U_2, $W_1 = [\widetilde{W}_1 \vdots \widetilde{\widetilde{W}}_1]$, and $W_2 = [\widetilde{W}_2 \vdots \widetilde{\widetilde{W}}_2]$ are unitary matrices and $\sigma_j^{(1)}$, $N \geq j \geq N - N_1 + 1$, and $\sigma_j^{(2)}$, $N \geq j \geq N - N_2 + 1$, are nonzero singular values of the matrices P and $I - P$. The columns of the blocks $\widetilde{W}_1$ and $\widetilde{W}_2$ of the matrices W_1 and W_2 form a basis for reducing subspaces. Therefore, for Q we can take $Q = [\widetilde{W}_1 \vdots \widetilde{W}_2]^{-1}$. If we find the matrices Q and P, then, using (10.3.3) and (10.3.4), we obtain the equalities

$$BP = T\begin{bmatrix} I_{N_1} & 0 \\ 0 & 0 \end{bmatrix} Q, \quad A(I-P) = T\begin{bmatrix} 0 & 0 \\ 0 & I_{N_2} \end{bmatrix} Q, \quad A(I-P) + BP = TQ.$$

Thus,

$$T = [A(I-P) + BP]Q^{-1} = [A(I-P) + BP][\widetilde{W}_1 \vdots \widetilde{W}_2].$$

10.4. Generalization of the discrete Lyapunov equation

Representations of solutions to the discrete Lyapunov equation. Matrix equations for the circle dichotomy problem. A theorem about a pencil regular on the unit circle. Criteria for the circle dichotomy of the spectrum of a matrix.

Let $N \times N$-matrices A and B form a pencil regular on the unit circle. We will use the representations from §10.3:

$$A = T\begin{bmatrix} A_1 & 0 \\ 0 & I_{N_2} \end{bmatrix} Q, \quad B = T\begin{bmatrix} I_{N_1} & 0 \\ 0 & B_2 \end{bmatrix} Q, \tag{10.4.1}$$

where $|\lambda_j(A_1)| < 1$, $|\lambda_j(B_2)| < 1$, $\det T \neq 0$, $\det Q \neq 0$. We recall that the projection P onto the space that reduces the pencil and corresponds to the eigenvalues λ

(roots of the characteristic equation $\det(A - \lambda B) = 0$) lie in the unit disk and can be written as follows:

$$P = Q^{-1}\begin{bmatrix} I_{N_1} & 0 \\ 0 & 0 \end{bmatrix} Q. \tag{10.4.2}$$

We begin with the study of the integral

$$\begin{aligned} Z &= \frac{1}{2\pi}\int_0^{2\pi}(A - e^{i\varphi}B)^{-1}F(A - e^{i\varphi}B)^{-*}d\varphi \\ &= Q^{-1}\Big\{\frac{1}{2\pi}\int_0^{2\pi}\begin{bmatrix}(A_1 - e^{i\varphi}I_{N_1})^{-1} & 0 \\ 0 & (I_{N_2} - e^{i\varphi}B_2)^{-1}\end{bmatrix} \\ &\qquad \times T^{-1}FT^{-*}\begin{bmatrix}(A_1 - e^{i\varphi}I_{N_1})^{-*} & 0 \\ 0 & (I_{N_2} - e^{i\varphi}B_2)^{-*}\end{bmatrix}d\varphi\Big\}Q^{-*} \\ &= Q^{-1}\begin{bmatrix} J_{11} & J_{12} \\ J_{21} & J_{22}\end{bmatrix}Q^{-*}, \end{aligned} \tag{10.4.3}$$

where

$$\begin{aligned} J_{11} &= \frac{1}{2\pi}\int_0^{2\pi}(A_1 - e^{i\varphi}I_{N_1})^{-1}D_{11}(A_1 - e^{i\varphi}I_{N_1})^{-*}d\varphi, \\ \begin{bmatrix} D_{11} & D_{12} \\ D_{21} & D_{22}\end{bmatrix} &= T^{-1}FT^{-*}, \\ J_{12} &= \frac{1}{2\pi}\int_0^{2\pi}(A_1 - e^{i\varphi}I_{N_1})^{-1}D_{12}(I_{N_2} - e^{i\varphi}B_2)^{-*}d\varphi, \\ J_{21} &= \frac{1}{2\pi}\int_0^{2\pi}(I_{N_2} - e^{i\varphi}B_2)^{-1}D_{21}(A_1 - e^{i\varphi}I_{N_1})^{-*}d\varphi, \\ J_{22} &= \frac{1}{2\pi}\int_0^{2\pi}(I_{N_2} - e^{i\varphi}B_2)^{-1}D_{22}(I_{N_2} - e^{i\varphi}B_2)^{-*}d\varphi. \end{aligned} \tag{10.4.4}$$

We show that $J_{12} = 0$. The equality $J_{21} = 0$ can be proved in a similar way. The integral J_{12} can be written as a contour integral along the unit circle, i.e.,

$$\begin{aligned} J_{12} &= \frac{1}{2\pi i}\int_0^{2\pi}(A_1 - e^{i\varphi}I_{N_1})^{-1}D_{12}(e^{i\varphi}I_{N_2} - B_2^*)^{-1}de^{i\varphi} \\ &= \frac{1}{2\pi i}\oint_{|z|=1}(A_1 - zI_{N_1})^{-1}D_{12}(zI_{N_2} - B_2^*)^{-1}dz \\ &= -\frac{1}{2\pi i}\oint_{|z|=1}\left(\frac{1}{z}A_1 - I_{N_1}\right)^{-1}D_{12}\left(I_{N_2} - \frac{1}{z}B_2^*\right)^{-1}d\frac{1}{z}. \end{aligned}$$

For $|z| \geq 1$ the integrand is a matrix whose elements are analytic functions because the eigenvalues of the matrices A_1 and B_2 lie in the unit disk. Therefore, without any change of the value of the integral, we can replace the contour $|z| = 1$ by the circle $|z| = R$, where the radius $R > 1$ is arbitrarily large as desired, i.e.,

$$J_{12} = \frac{1}{2\pi Ri}\int_0^{2\pi}\left(\frac{e^{-i\varphi}}{R}A_1 - I_{N_1}\right)^{-1} D_{12}\left(I_{N_2} - \frac{e^{-i\varphi}}{R}B_2^*\right)^{-1} de^{-i\varphi} = O\left(\frac{1}{R}\right).$$

Since R is arbitrary, we obtain $J_{12} = 0$. Thus,

$$Z = Q^{-1}\begin{bmatrix} J_{11} & 0 \\ 0 & J_{22}\end{bmatrix} Q^{-*}.$$

Using the representation (10.4.2) of the projection P, we verify that

$$PZ = Q^{-1}\begin{bmatrix} J_{11} & 0 \\ 0 & 0\end{bmatrix} Q^{-*} = ZP^* = PZP^*,$$

$$(I-P)Z = Q^{-1}\begin{bmatrix} 0 & 0 \\ 0 & J_{22}\end{bmatrix} Q^{-*} = Z(I-P)^* = (I-P)Z(I-P^*).$$

Consequently,

(10.4.5)

$$Z = \frac{1}{2\pi}\int_0^{2\pi}(A - e^{i\varphi}B)^{-1}F(A - e^{i\varphi}B)^{-*}d\varphi = PZP^* + (I-P)Z(I-P)^*.$$

From (10.4.5) and the elementary equalities

$$BPZP^*B^* = T\begin{bmatrix} J_{11} & 0 \\ 0 & 0\end{bmatrix} T^*,$$

$$B(I-P)Z(I-P)^*B^* = T\begin{bmatrix} 0 & 0 \\ 0 & B_2J_{22}B_2^*\end{bmatrix} T^*,$$

$$APZP^*A^* = T\begin{bmatrix} A_1J_{11}A_1^* & 0 \\ 0 & 0\end{bmatrix} T^*,$$

$$A(I-P)Z(I-P)^*A^* = T\begin{bmatrix} 0 & 0 \\ 0 & J_{22}\end{bmatrix} T^*$$

it follows that

$$BZB^* - AZA^* = T\begin{bmatrix} J_{11} - A_1J_{11}A_1^* & 0 \\ 0 & B_2J_{22}B_2^* - J_{22}\end{bmatrix} T^*.$$

Since

$$J_{11} = \frac{1}{2\pi}\int_0^{2\pi}(e^{i\varphi}I_{N_1} - A_1)^{-1}D_{11}(e^{i\varphi}I_{N_1} - A_1)^{-*}d\varphi,$$

$$J_{22} = \frac{1}{2\pi}\int_0^{2\pi}(e^{i\varphi}I_{N_2} - B_2)^{-1}D_{22}(e^{i\varphi}I_{N_2} - B_2)^{-*}d\varphi$$

and the spectra of the matrices A_1 and B_2 lie in the unit disk, we conclude that J_{11} and J_{22} are solutions to the discrete Lyapunov equations (see §§4.4, 9.3, and 10.1)

$$J_{11} - A_1J_{11}A_1^* = D_{11}, \quad J_{22} - B_2J_{22}B_2^* = D_{22}.$$

Consequently,

$$BZB^* - AZA^* = T\begin{bmatrix} D_{11} & 0 \\ 0 & -D_{22} \end{bmatrix} T^*. \tag{10.4.6}$$

We emphasize the following fact that is important later in this volume. It turns out that for $F = ACA^*$ or $F = BCB^*$ the matrix Z computed by (10.4.5) satisfies the following equalities:

$$\begin{aligned} PZ = PZP^* &= \frac{1}{2\pi}\int_0^{2\pi}(A - e^{i\varphi}B)^{-1}APCP^*A^*(A - e^{i\varphi}B)^{-*}d\varphi, \\ PZ = PZP^* &= \frac{1}{2\pi}\int_0^{2\pi}(A - e^{i\varphi}B)^{-1}BPCP^*B^*(A - e^{i\varphi}B)^{-*}d\varphi. \end{aligned} \tag{10.4.7}$$

These equalities are obvious because the projection P commutes with $(A-e^{i\varphi}B)^{-1}A$ and $(A-e^{i\varphi}B)^{-1}B$. The equalities (10.4.7) remain valid with P replaced by $I-P$. The integral under consideration has the form

$$H = \frac{1}{2\pi}\int_0^{2\pi}(A - e^{i\varphi}B)^{-1}(ACA^* + BCB^*)(A - e^{i\varphi}B)^{-*}d\varphi, \tag{10.4.8}$$

i.e., we choose $F = ACA^* + BCB^*$, where C is an arbitrary matrix. By the above arguments, we can, leaving H unchanged, replace $F = ACA^* + BCB^*$ by $F = APCP^*A^* + A(I-P)C(I-P)^*A^* + BPCP^*B^* + B(I-P)C(I-P)^*B^*$; moreover

$$\begin{aligned} \begin{bmatrix} D_{11} & D_{12} \\ D_{21} & D_{22} \end{bmatrix} &= T^{-1}(ACA^* + BCB^*)T^{-*} \\ &= \begin{bmatrix} A_1 & 0 \\ 0 & 0 \end{bmatrix} QCQ^* \begin{bmatrix} A_1^* & 0 \\ 0 & 0 \end{bmatrix} + \begin{bmatrix} 0 & 0 \\ 0 & I_{N_2} \end{bmatrix} QCQ^* \begin{bmatrix} 0 & 0 \\ 0 & I_{N_2} \end{bmatrix} \\ &\quad + \begin{bmatrix} I_{N_1} & 0 \\ 0 & 0 \end{bmatrix} QCQ^* \begin{bmatrix} I_{N_1} & 0 \\ 0 & 0 \end{bmatrix} + \begin{bmatrix} 0 & 0 \\ 0 & B_2 \end{bmatrix} QCQ^* \begin{bmatrix} 0 & 0 \\ 0 & B_2^* \end{bmatrix}. \end{aligned}$$

Consequently,

$$\begin{aligned} &D_{12} = 0, \quad D_{21} = 0, \quad T\begin{bmatrix} D_{11} & 0 \\ 0 & 0 \end{bmatrix} T^* = APCP^*A^* + BPCP^*B^*, \\ &T\begin{bmatrix} 0 & 0 \\ 0 & D_{22} \end{bmatrix} T^* = A(I-P)C(I-P)^*A^* + B(I-P)C(I-P)^*B^*. \end{aligned}$$

The matrix H of the form (10.4.8) is an integral of type Z (see (10.4.5)). Using (10.4.6), we conclude that the matrix H and the projection

$$P = \frac{1}{2\pi}\int_0^{2\pi}(B - e^{-i\varphi}A)^{-1}Bd\varphi$$

are connected by the following matrix equations:

$$\begin{aligned} BHB^* - AHA^* = A[PCP^* &- (I-P)C(I-P)^*]A^* \\ &+ B[PCP^* - (I-P)C(I-P)^*]B^*, \end{aligned} \tag{10.4.9}$$

$$PH = HP^*, \quad P^2 = P, \quad P(A+B)^{-1}B = (A+B)^{-1}BP. \tag{10.4.10}$$

These equations generalize the discrete Lyapunov equation, and their properties are similar to the properties of the equations which were studied in the problem about the spectrum dichotomy by the imaginary axis (see §10.2). Equations (10.4.9) and (10.4.10) are referred to as *matrix equations for the circle dichotomy problem.*

THEOREM 10.4.1. *If a matrix $A+B$ is nonsingular and there exists a matrix P and Hermitian positive definite matrices H and C ($H = H^* > 0$, $C = C^* > 0$) satisfying equations* (10.4.9) *and* (10.4.10), *then the pencil $\lambda B - A$ is regular on the unit circle and the operator defined by the matrix P is the projection onto the right reducing subspace corresponding to the eigenvalues in the unit disk.*

PROOF. Consider the case $A+B = I$. Equations (10.4.9) and (10.4.10) imply that P and $I-P$ are projections onto the complementary invariant subspaces of A. Consequently, there exists a basis (we will write matrix relations in this basis) such that the following representations hold:

$$A = \begin{bmatrix} A_{11} & 0 \\ 0 & A_{22} \end{bmatrix}, \quad P = \begin{bmatrix} I_{N_1} & 0 \\ 0 & 0 \end{bmatrix}, \quad I-P = \begin{bmatrix} 0 & 0 \\ 0 & I_{N_2} \end{bmatrix},$$

$$B = (I-A) = \begin{bmatrix} I_{N_1} - A_{11} & 0 \\ 0 & I_{N_2} - A_{22} \end{bmatrix} = \begin{bmatrix} B_{11} & 0 \\ 0 & B_{22} \end{bmatrix};$$

moreover,

$$PA = AP = \begin{bmatrix} A_{11} & 0 \\ 0 & 0 \end{bmatrix}, \quad (I-P)B = B(I-P) = \begin{bmatrix} 0 & 0 \\ 0 & B_{22} \end{bmatrix}.$$

Since $PH = HP^*$, $PH(I-P)^* = P(I-P)H = (P-P^2)H = 0$, we have

$$\begin{aligned} H &= [P + (I-P)]H[P^* + (I-P)^*] = PHP^* + (I-P)H(I-P)^* \\ &= \begin{bmatrix} H_{11} & 0 \\ 0 & H_{22} \end{bmatrix}. \end{aligned}$$

In addition,

$$PCP^* - (I-P)C(I-P)^* = \begin{bmatrix} C_{11} & 0 \\ 0 & C_{22} \end{bmatrix}.$$

Since H and C are Hermitian positive definite matrices, we see that H_{11}, H_{22}, C_{11}, C_{22} are also Hermitian positive definite matrices. We note that

$$\begin{aligned} &APCP^*A^* + BPCP^*B^* \\ &\quad = \begin{bmatrix} A_{11}C_{11}A_{11}^* + (I-A_{11})C_{11}(I-A_{11})^* & 0 \\ 0 & 0 \end{bmatrix} = \begin{bmatrix} G_{11} & 0 \\ 0 & 0 \end{bmatrix}, \end{aligned}$$

$$\begin{aligned} &A(I-P)^*C(I-P)^*A^* + B(I-P)C(I-P^*)B^* \\ &\quad = \begin{bmatrix} 0 & 0 \\ 0 & B_{22}C_{22}B_{22}^* + (I-B_{22})C_{22}(I-B_{22})^* \end{bmatrix} = \begin{bmatrix} 0 & 0 \\ 0 & G_{22} \end{bmatrix}; \end{aligned}$$

moreover, G_{11} and G_{22} ($G_{11} = G_{11}^* > 0$, $G_{22} = G_{22}^* > 0$) are Hermitian positive definite matrices provided that C_{11} and C_{22} have the same properties. In the case $A+B = I$, the first matrix equation in (10.4.9) can be written as two equations

$$\begin{aligned} (I-A_{11})H_{11}(I-A_{11})^* &= G_{11} + A_{11}H_{11}A_{11}^*, \\ (I-B_{22})H_{22}(I-B_{22})^* &= G_{22} + B_{22}H_{22}B_{22}^*. \end{aligned} \tag{10.4.11}$$

Since H_{11}, G_{11}, H_{22}, and G_{22} are positive definite, $B_{11} = I - A_{11}$ and $A_{22} = I - B_{22}$ are nonsingular, and the equalities (10.4.11) can be written as two discrete Lyapunov equations, we conclude that

$$
\begin{aligned}
H_{11} &= [B_{11}^{-1}A_{11}]H_{11}[B_{11}^{-1}A_{11}]^* + B_{11}^{-1}G_{11}B_{11}^{-*}, \\
H_{22} &= [A_{22}^{-1}B_{22}]H_{22}[A_{22}^{-1}B_{22}]^* + A_{22}^{-1}G_{22}A_{22}^{-*}.
\end{aligned}
\tag{10.4.12}
$$

As was shown in §10.2, the positive definiteness of the matrices H_{ii} and G_{ii} in (10.4.12) means that the eigenvalues of $B_{11}^{-1}A_{11}$ and $A_{22}^{-1}B_{22}$ lie in the unit disk or, which is equivalent, the spectra of the pencils $\lambda B_{11} - A_{11}$ and $\lambda B_{22} - A_{22}$ lie inside and outside the unit disk, respectively. Thus,

$$P = \begin{bmatrix} I_{N_1} & 0 \\ 0 & 0 \end{bmatrix}$$

projects onto the space that reduces the pencil

$$\lambda B - A = \begin{bmatrix} \lambda B_{11} - A_{11} & 0 \\ 0 & \lambda B_{22} - A_{22} \end{bmatrix}$$

and corresponds to the eigenvalues lying in the unit disk. □

We note that the solutions H_{11} and H_{22} to the discrete Lyapunov equations (10.4.12) are unique. Consequently, a solution (the pair P and $H = H^* > 0$) to equations (10.4.9), (10.4.10) is unique.

We recall that under the above assumptions,

$$H = \begin{bmatrix} H_{11} & 0 \\ 0 & H_{22} \end{bmatrix}, \quad H_{11} > 0, \ H_{22} > 0.$$

As was shown, there exists a solution to equations (10.4.9), (10.4.10) such that $H = H^* > 0$ is represented as the integral (10.4.8). By the uniqueness of a solution, the only admissible solution (P, H) $(H = H^* > 0)$ to the problem (10.4.9), (10.4.10) is

$$
\begin{aligned}
P &= \frac{1}{2\pi}\int_0^{2\pi} (B - e^{i\varphi}A)^{-1}B\, d\varphi, \\
H &= \frac{1}{2\pi}\int_0^{2\pi} (A - e^{i\varphi}B)^{-1}(ACA^* + BCB^*)(A - e^{i\varphi}B)^{-*} d\varphi.
\end{aligned}
\tag{10.4.13}
$$

PROBLEM 10.4.1. Modify the proof for an arbitrary nonsingular matrix $A+B$.

For $B = rI$, when speaking about the dichotomy of the spectrum of the pencil $\lambda B - A$ by the unit circle $|\lambda| = 1$, we mean the circle dichotomy of the spectrum of the matrix A. In §9.1, we characterized this dichotomy by the norm $\|H_F\|$ of the matrix integral

$$H_F = \frac{1}{2\pi}\int_0^{2\pi} (A - re^{i\varphi}I)^{-1}F(A^* - re^{-i\varphi}I)^{-1}d\varphi, \tag{10.4.14}$$

where $F = F^* > 0$ is a Hermitian positive definite matrix. For H_F and the projection

$$P = \frac{1}{2\pi} \int_0^{2\pi} \left(I - \frac{e^{-i\varphi}}{r} A \right)^{-1} d\varphi \tag{10.4.15}$$

onto the invariant subspace of the matrix A corresponding to the eigenvalues in the disk of radius r we can write the following system of matrix equations:

$$\begin{gathered} r^2 H_F - A H_F A^* + F = PF + FP^*, \quad H_F = H_F^*, \\ P^2 = P, \quad PA = AP, \quad PH_F = H_F P^*. \end{gathered} \tag{10.4.16}$$

It suffices to give a detialed proof only for the first equality. By (10.4.1), for $B = rI$ the block B_2 is nonsingular and for T and Q the following relation holds:

$$T = rQ^{-1} \begin{bmatrix} I_{N_2} & 0 \\ 0 & B_2^{-1} \end{bmatrix}.$$

Using the last relation, we rewrite (10.4.6) with $Z = H$:

$$r^2 H - AHA^* = T \begin{bmatrix} D_{11} & 0 \\ 0 & -D_{22} \end{bmatrix} T^* = r^2 Q^{-1} \begin{bmatrix} D_{11} & 0 \\ 0 & -B_2^{-1} D_{22} B_2^{-1} \end{bmatrix} Q^{-*}.$$

We recall (see (10.4.4)) that

$$F = T \begin{bmatrix} D_{11} & D_{12} \\ D_{21} & D_{22} \end{bmatrix} T^* = r^2 Q^{-1} \begin{bmatrix} I_{N_2} & 0 \\ 0 & B_2^{-1} \end{bmatrix} \begin{bmatrix} D_{11} & D_{12} \\ D_{21} & D_{22} \end{bmatrix} \begin{bmatrix} I_{N_2} & 0 \\ 0 & B_2^{-*} \end{bmatrix} Q^{-*}$$

and deduce, using (10.4.2), the equality

$$PFP^* - (I-P)F(I-P^*) = r^2 Q^{-1} \begin{bmatrix} D_{11} & 0 \\ 0 & -B_2^{-1} D_{22} B_2^{-*} \end{bmatrix} Q^{-*}.$$

It is obvious that

$$r^2 H - AHA^* = PFP^* - (I-P)F(I-P^*) = -F + PF + FP^*.$$

The system (10.4.16) can be studied in the same way as the system (10.4.9), (10.4.10). The result is formulated in the following problem.

PROBLEM 10.4.2. If there exist matrices P and H_F satisfying (10.4.16) and, in addition, the matrix H_F is positive definite, then the matrix A has no eigenvalues of module r and the solution P and H_F can be computed by formulas (10.4.14) and (10.4.15). There are no other solutions to the system (10.4.16) with a positive definite matrix H_F.

As a criterion for the circle dichotomy of the spectrum of a matrix A by the circle $|\lambda| = r$ we can take the norms of various Hermitian matrices H_F. In some cases, it is convenient to use $\|H_F^{(1)}\|$ and in others, $\|H_F^{(2)}\|$, where

$$H_F^{(1)} = \frac{1}{2\pi} \int_0^{2\pi} (A - re^{i\varphi} I)^{-1} F (A^* - re^{-i\varphi})^{-1} d\varphi,$$

$$H_F^{(2)} = \frac{1}{2\pi} \int_0^{2\pi} (A - re^{i\varphi} I)^{-1} (r^2 F + AFA^*)(A^* - re^{-i\varphi})^{-1} d\varphi.$$

By (10.4.9) and (10.4.16), $H_F^{(1)}$, $H_F^{(2)}$, A, F, P (where P is a projection onto the invariant subspace of A corresponding to the eigenvalues in the unit disk) are connected by the following equations:

$$r^2 H_F^{(1)} - A H_F^{(1)} A^* + F - PF - FP^* = 0,$$
$$r^2 H_F^{(2)} - A H_F^{(2)} A^* + A[F - PF - FP^*]A^* + r^2[F - PF - FP^*] = 0.$$

The solvability conditions for positive definite matrices $H_F^{(1)}$ and $H_F^{(2)}$ coincide. It turns out that for a given matrix $H_F^{(1)}$ it is not hard to compute $H_F^{(2)}$. The converse assertion is also valid: for a given matrix $H_F^{(2)}$ and a projection P we can compute $H_F^{(1)}$. A rather simple and effective method for computing the integral

$$H = \frac{1}{2\pi} \int_0^{2\pi} (A - e^{i\varphi}B)^{-1}(AFA^* + BFB^*)(A^* - e^{-i\varphi}B^*)^{-1} d\varphi$$

will be described in §§14.3 and 14.4. It is obvious that $H_F^{(2)}$ is an integral of this form.

The proof of the formulas connecting $H_F^{(1)}$ and $H_F^{(2)}$ is left to the reader as an easy exercise.

PROBLEM 10.4.3. Prove the equalities

$$H_F^{(2)} = r^2 H_F^{(1)} + A H_F^{(1)} A^*, \quad H_F^{(1)} = \frac{1}{2}(H_F^{(2)} + F - PF - FP^*).$$

10.5. Linear and circle dichotomies

The definition of a Green matrix. The representation of a Green matrix. The case where the spectrum of a matrix lies to one side of the imaginary axis. An estimate for the growth of the norm of a Green matrix at infinity.

We will prove simple formulas connecting solutions to matrix equations that appear when we verify which kind of dichotomy, circle or linear, holds. In view of these formulas, we will restrict ourselves to the study of the circle dichotomy because the study of the linear dichotomy is reduced to the circle dichotomy (see Chapter 14).

The invariant subspace of a matrix D corresponding to the eigenvalues in the left half-plane coincides with the invariant subspace of the matrix exponential $A = e^{\tau D}$ $(\tau > 0)$ corresponding to the eigenvalues in the unit disk. Similarly, the invariant subspace of D corresponding to the eigenvalues in the right half-plane coincides with the invariant subspace of $A = e^{\tau D}$ $(\tau > 0)$ corresponding to the eigenvalues outside the unit disk. The projections onto these subspaces are denoted by P and $I - P$, respectively $(PD = DP,\ PA = AP,\ P^2 = P)$.

For a Hermitian positive definite matrix F $(F = F^* > 0)$ we introduce the Hermitian positive definite matrix

$$C = \int_0^{\tau} e^{sD^*} F e^{sD}\, ds.$$

It is convenient to use the canonical decompositions

$$D = Q^{-1}\begin{bmatrix} D_0 & 0 \\ 0 & D_1 \end{bmatrix} Q, \quad P = Q^{-1}\begin{bmatrix} I & 0 \\ 0 & 0 \end{bmatrix} Q,$$

$$A = e^{\tau D} = Q^{-1}\begin{bmatrix} A_0 & 0 \\ 0 & A_1 \end{bmatrix} Q = Q^{-1}\begin{bmatrix} e^{\tau D_-} & 0 \\ 0 & e^{\tau D_+} \end{bmatrix} Q,$$

where the eigenvalues of D_0, D_1, A_0, and A_1 lie inside the left half-plane, inside the right half-plane, inside the unit disk, and outside the unit disk, respectively.

By the *Green matrix* $G(t)$ we mean the matrix

$$G(t) = \begin{cases} Q^{-1}\begin{bmatrix} e^{tD_-} & 0 \\ 0 & 0 \end{bmatrix} Q, & t > 0, \\ \\ -Q^{-1}\begin{bmatrix} 0 & 0 \\ 0 & e^{tD_+} \end{bmatrix} Q, & t < 0. \end{cases} \tag{10.5.1}$$

It is bounded for all t; moreover, for $t \neq 0$ it satisfies the differential equation

$$\frac{d}{dt}G(t) - DG(t) = 0.$$

The limit from the right $G(+0)$ and the limit from the left $G(-0)$ of the Green matrix $G(t)$ at $t = 0$ are related by the equality

$$G(+0) - G(-0) = I.$$

The integral

$$X = \int_{-\infty}^{+\infty} G^*(t) F G(t)\, dt \tag{10.5.2}$$

can be represented as the sum

$$X = \sum_{k=1}^{\infty} e^{-k\tau D^*}(I - P^*)\left[\int_0^{\tau} e^{sD^*} F e^{sD}\, ds\right](I - P)e^{-k\tau D}$$

$$+ \sum_{k=0}^{\infty} e^{k\tau D^*} P^* \left[\int_0^{\tau} e^{sD^*} F e^{sD}\, ds\right] P e^{k\tau D}$$

$$= \sum_{k=1}^{\infty} (A^*)^{-k}(I - P^*)C(I - P)A^{-k} + \sum_{k=0}^{\infty}(A^*)^k P^* C P A^k = S_- + S_+,$$

where

$$S_- = \sum_{k=1}^{\infty}(A^*)^{-k}(I - P^*)C(I - P)A^{-k}, \quad S_+ = \sum_{k=0}^{\infty}(A^*)^k P^* C P A^k.$$

It is easy to check that

$$A^* S_- A - S_- = (I - P)^* C (I - P), \quad S_+ - A^* S_+ A = P^* C P.$$

Therefore, $X = S = S_- + S_+$ satisfies the equation

$$X - A^* X A = P^* C P - (I - P^*)C(I - P).$$

It is easy to check that $P^*X = XP = S_+$. Thus, if D has no purely imaginary eigenvalues, then, setting

$$A = e^{\tau D}, \quad C = \int_0^{\tau} e^{tD} F e^{tD^*}\, dt,$$

we can find the positive definite matrix (10.5.2) by solving the equations

$$X - A^*XA = P^*CP - (I - P^*)C(I - P),$$
$$X^* = X, \quad P^*X = XP, \quad PA = AP, \quad P^2 = P.$$

The last equations differ from equations (10.4.16) by notation only. We describe an algorithm of the solution of this problem in Chapter 14.

Having computed X and P, we obtain a solution to the following system of matrix equations (this system generalizes the classical Lyapunov equation to the case of the problem about the spectrum dichotomy by the imaginary axis (see §10.2)):

$$XD + D^*X + P^*FP - (I - P)^*F(I - P) = 0, \tag{10.5.3}$$

where $P^*X = XP$, $X = X^*$, $P^2 = P$, and $PD = DP$. Indeed, it suffices to check that the integral (10.5.2) satisfies the first equation of the above system. To show this, we represent X as the sum of integrals along the positive half-axis and along the negative half-axis, i.e., $X = X_+ + X_-$, where

$$X_+ = \int_0^{\infty} G^*(t)FG(t)\, dt, \quad X_- = \int_{-\infty}^{0} G^*(t)FG(t)\, dt; \tag{10.5.4}$$

moreover,

$$D^*X_+ + X_+D = \int_0^{\infty} \left(\frac{d}{dt}G^*(t)\right) FG(t)\, dt + \int_0^{\infty} G^*(t)F\left(\frac{d}{dt}G(t)\right) dt$$
$$= \int_0^{\infty} \frac{d}{dt}[G^*(t)FG(t)]\, dt = -G^*(+0)FG(+0),$$
$$D^*X_- + X_-D = \int_{-\infty}^{0} \frac{d}{dt}[G^*(t)FG(t)]dt = G^*(-0)FG(-0).$$

From (10.5.1) it follows that

$$G(+0) = Q^{-1}\begin{bmatrix} I & 0 \\ 0 & 0 \end{bmatrix} Q = P, \quad G(-0) = Q^{-1}\begin{bmatrix} 0 & 0 \\ 0 & -I \end{bmatrix} Q = -(I - P).$$

Therefore,

$$D^*X_+ + X_+D = -P^*FP,$$
$$D^*X_- + X_-D = (I - P^*)F(I - P),$$
$$D^*X + XD + P^*FP - (I - P^*)F(I - P) = 0.$$

We have proved (10.5.3).

REMARK 10.5.1. If the eigenvalues of a matrix A lie to one side of the imaginary axis, then either $X_+ = 0$ or $X_- = 0$ and either $P \equiv G(+0) = 0$ or $I - P \equiv -G(-0) = 0$.

To conclude the section, we deduce an important inequality that allows us to estimate the growth of the norm $\|G(t)\|$ as $t \to \pm\infty$ in terms of the norm $\|X\|$, i.e.,

$$\|G(t)\| \le \sqrt{2\|D\|\,\|F^{-1}\|\,\|X\|}\,e^{-|t|/(2\|X\|\,\|F^{-1}\|)}. \tag{10.5.5}$$

In the proof we will use the equalities

$$\begin{aligned} G(t) &= G(+0)G(t) = G(t)G(+0), \\ G(t+s) &= G(t)G(s) = G(s)G(t), \end{aligned}$$

for $t > 0$, $s > 0$,

$$\begin{aligned} G(t) &= -G(-0)G(t) = -G(t)G(-0), \\ G(t+s) &= -G(t)G(s) = -G(s)G(t), \end{aligned}$$

for $t < 0$, $s < 0$, and

$$\begin{gathered} DG(t) = G(t)D, \quad -\infty < t \le -0, \quad +0 \le t < +\infty, \\ G^*(+0)XG(+0) = X_+, \quad G^*(-0)XG(-0) = X_-. \end{gathered}$$

These equalities follow from (10.5.1) and (10.5.4). Let $X_+ \neq 0$, $s > 0$. Setting

$$Z(s) = \int_s^\infty G^*(t)FG(t)\,dt = G^*(s)\Big[\int_0^\infty G^*(t)FG(t)\,dt\Big]G(s) = G^*(s)X_+G(s)$$

and checking the equality

$$\begin{aligned} \frac{d}{ds}Z(s) &= D^*Z(s) + Z(s)D = G^*(s)[D^*X_+ + X_+D]G(s) \\ &= -G^*(s)[G^*(+0)FG(+0)]G(s), \end{aligned}$$

we arrive at the estimates ($t > 0$)

$$\begin{aligned} \frac{d}{ds}(Z(s)x, x) &\le -(G^*(+s)X_+G(+s)x, x)/(\|X\|\,\|F^{-1}\|) \\ &= -\frac{1}{\|X\|\,\|F^{-1}\|}(Z(s)x, x), \end{aligned}$$

$$\frac{d}{dt}[e^{t/(\|X\|\,\|F^{-1}\|)}(Z(t)x, x)] \le 0,$$

$$\begin{aligned} (XG(t)x, G(t)x) = (Z(t)x, x) &\le e^{-t/(\|X\|\,\|F^{-1}\|)}(Z(0)x, x) \\ &= e^{-t/(\|X\|\,\|F^{-1}\|)}(XG(+0)x, G(+0)x). \end{aligned}$$

We also use the elementary inequality

$$\|e^{tA}f\| \ge e^{-t\|A\|}\|f\|, \quad t > 0,$$

which follows from the relation

$$\|f\| = \|e^{-tD}e^{tD}f\| \le \|e^{-tD}\|\,\|e^{tD}f\| \le \|e^{-tD}\|\,\|e^{tD}f\| \le e^{t\|D\|}\|e^{tD}f\|.$$

This inequality will be used in the proof of the following estimates:

$$\begin{aligned}(XG(+0)y, G(+0)y) &= \int_0^\infty (Fe^{tD}G(+0)y, e^{tD}G(+0)y)\,dt \\ &\geq (G(+0)y, G(+0)y)\frac{1}{\|F^{-1}\|}\int_0^\infty e^{-2t\|D\|}dt \\ &= \frac{1}{2\|D\|\,\|F^{-1}\|}(G(+0)y, G(+0)y).\end{aligned}$$

Setting $y = G(t)x$ $(t > 0)$, we find

$$\begin{aligned}(G(t)x, G(t)x) &\leq 2\|D\|\,\|F^{-1}\|(XG(t)x, G(t)x) \\ &\leq e^{-t/(\|X\|\,\|F^{-1}\|)}(XG(+0)x, G(+0)x) \\ &= 2\|D\|\,\|F^{-1}\|e^{-t/(\|X\|\,\|F^{-1}\|)}(X_+x, x) \\ &\leq 2\|D\|\,\|F^{-1}\|e^{-t/(\|X\|\,\|F^{-1}\|)}(Xx, x) \\ &\leq 2\|D\|\,\|F^{-1}\|\,\|X\|e^{-t/(\|X\|\,\|F^{-1}\|)}(x, x),\end{aligned}$$

i.e., for $t > 0$ we obtain the estimate

$$\|G(t)\|^2 \leq 2\|D\|\,\|F^{-1}\|\,\|X\|e^{-t/(\|X\|\,\|F^{-1}\|)},$$

which coincides with the required inequality (10.5.5). The case $t < 0$ is considered in the same way.

10.6. Decomposition into invariant subspaces

Spectral spots and points. The maximum principle for analytic functions. The location of poles of the resolvent with respect to spectral spots. ε-spectrum of a matrix. Examples of ε-spectra of matrices. Illustrations.

We return to the study of the decomposition of an N-dimensional unitary space into invariant subspaces of a linear operator A acting in this space. It is natural to begin with the computation of the eigenvalues of a matrix A (in other words, the points of the spectrum of A) and to divide the eigenvalues into groups such that to each group some invariant subspace will correspond.

As was mentioned, eigenvalues of matrices cannot be always computed with sufficient accuracy. In particular, it is hard to distinguish a multiple eigenvalue from a group of eigenvalues that are close each other.

To divide the spectrum into such groups, it is natural to use the spectral portrait of the matrix A, i.e., the complex plane whose points are colored and the color of a point λ depends on the norm $\|(\lambda I - A)^{-1}\|$ of the resolvent. If $\varepsilon > 0$ is sufficiently small, then we can distinguish spectral spots, i.e., domains in which $\|(\lambda I - A)^{-1}\| > \dfrac{1}{\varepsilon\|A\|}$. It is obvious that every eigenvalue of A is contained in some of these spots. On the other hand, each spot contains at least one eigenvalue. The proof of this assertion is based on the following maximum principle for the norm of the resolvent $\|(\lambda I - A)^{-1}\| = \|R(\lambda)\|$: *The values of $\|R(\lambda)\|$ in a bounded closed* *domain do not exceed the maximal value of $\|R(\lambda)\|$ as λ runs over the boundary of the domain provided that the domain contains no eigenvalues of A.*

For the eigenvalues lying in a simply connected spectral spot $\|(\lambda I - A)^{-1}\| > \dfrac{1}{\varepsilon\|A\|}$ the corresponding invariant subspace is uniquely determined by the projection

$$P = \frac{1}{2\pi i} \oint\limits_{\gamma} (\lambda I - A)^{-1}\, d\lambda,$$

where γ is the boundary of the spot. The projection $I - P$ projects onto the complementary invariant subspace corresponding to the eigenvalues lying outside the spot.

For a fixed $\varepsilon > 0$, e.g., for $\varepsilon = 10^{-6}$ or $\varepsilon = 10^{-16}$, we consider the corresponding spectral spots, i.e., simply connected sets in the complex plane consisting of points λ (called points of the ε-spectrum of A (see §1.7)) such that $\|(\lambda I - A)^{-1}\| > \dfrac{1}{\varepsilon\|A\|}$. To each of these spots we associate the projection P_j ($P_j^2 = P_j$, $P_jA = AP_j$) onto the corresponding invariant subspace of A. Every point of the spectrum of A belongs to one of the spots of the ε-spectrum. Therefore, every invariant subspace is invariant under the action of one of the projections P_j. Consequently, the sum of the projections P_j coincides with the identity operator.

There is another approach to the construction of P_j. Using the Schur theorem (see §4.5) we can construct the maximal invariant subspace for every eigenvalue $\lambda_k(A)$. This subspace consists of all nonzero vectors x ($\|x\| \neq 0$) that are annihilated by the operator $(A - \lambda_k I)^{N-1}$. The linear span of all subspaces corresponding to the eigenvalues in a spot is the maximal invariant subspace for this spot.

We recall that under the perturbations of A the stability of the projection and the subspace determined by this projection depends on the depth of the "ravine" picked in the graph of $\|(\lambda I - A)^{-1}\|$. This "ravine" separates the spot under consideration from the remaining spectral spots.

As ε decreases, the spots of the ε-spectrum can shrink and split. This happens if the spot contains several eigenvalues. As ε increases, the spots of the ε-spectrum increase and can glue together. For example, for sufficiently large N and special b_j and λ_j the ε-spectrum of the matrix

$$A = \begin{bmatrix} \lambda_1 & b_2 & & & \Large 0 \\ & \lambda_2 & b_3 & & \\ & & \ddots & \ddots & \\ \Large 0 & & & \lambda_{N-1} & b_{N-1} \\ & & & & \lambda_N \end{bmatrix}$$

can look as in Figure 10.6.1 (examples of such matrices were constructed by Kostin).

We indicate another structure of a spectral portrait of a matrix which can be easily constructed using the above arguments (see Figure 10.6.2). We give the three-dimensional and plane representations.

For sufficiently small ε every ε-spectral spot contains only one eigenvalue. However, in practical situations this cannot be achieved for every matrix (see the example in §1.3).

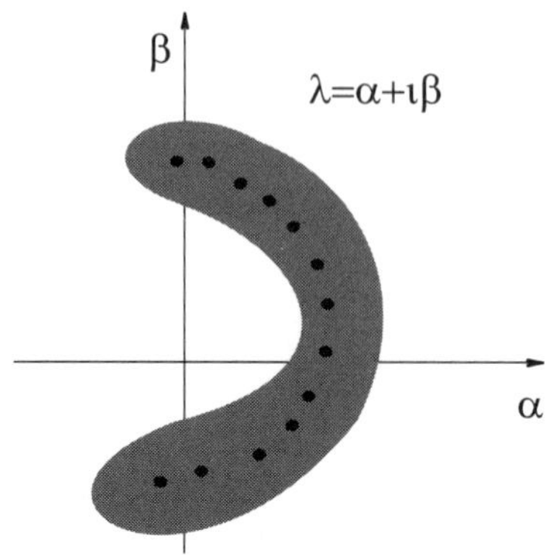

FIGURE 10.6.1

The maximal invariant subspaces corresponding to a single eigenvalue can have various structures. The reader can see it solving the following problem.

PROBLEM 10.6.1. Consider the matrices

$$A = \begin{bmatrix} 0 & 1 & 3 & 6 \\ 0 & 0 & 2 & 5 \\ 0 & 0 & 0 & 4 \\ 0 & 0 & 0 & 0 \end{bmatrix}, \quad B = \begin{bmatrix} 0 & 1 & 3 & 0 \\ 0 & 0 & 2 & 0 \\ 0 & 0 & 0 & 0 \\ 0 & 0 & 0 & 0 \end{bmatrix}$$

with zero eigenvalues. Show that the entire four-dimensional space can be represented as the direct sum of two invariant subspaces of the matrix B, whereas such a decomposition is impossible for the matrix A.

We do not discuss in detail possible structures of invariant subspaces corresponding to a root of the characteristic equation $\det(\tau E - A) = 0$. The description of such structures can be obtained by analyzing the so-called Jordan form of the matrix A (see any textbook on linear algebra, e.g., [**9**]).

If we want to decompose the space into invariant subspaces of a matrix A so that each invariant subspace corresponds to a spectral spot, then it is necessary to construct the spectral portrait of the matrix. This is not easy. However, we can often combine subspaces corresponding to the eigenvalues lying in some strip parallel to the imaginary axis or subspaces corresponding to the eigenvalues in some annulus. In this case we can construct simplified one-dimensional spectral portraits. Thus, if the line $z = a + it$, $-\infty < t < +\infty$, contains no eigenvalues of A, then the matrix integral

$$H(a) = \frac{1}{2\pi} \int_{-\infty}^{+\infty} (A^* - aI + itI)^{-1}(A - aI - itI)^{-1} dt$$

converges and the norm $\|H(a)\|$ characterizes the dichotomy quality for the spectrum of A by this line. Drawing the graph of the dependence of $\|H(a)\|$ on a (see Figure 10.6.3), we can single out the values of a for which the norm $\|H(a)\|$ is so large that its graphic representation has no room in the chosen scale of the picture. These values form one-dimensional spectral spots (see the dashed parts of

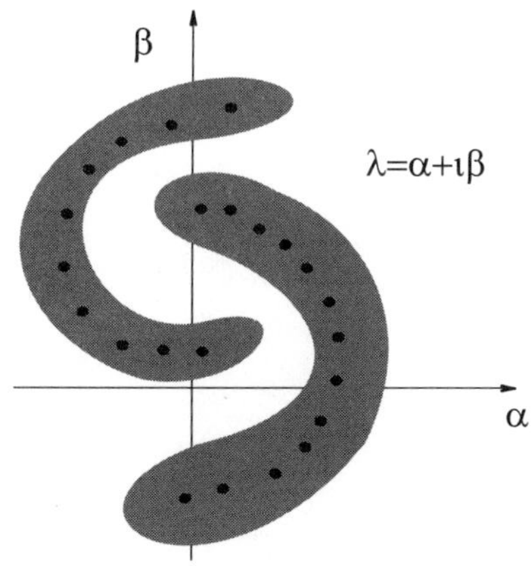

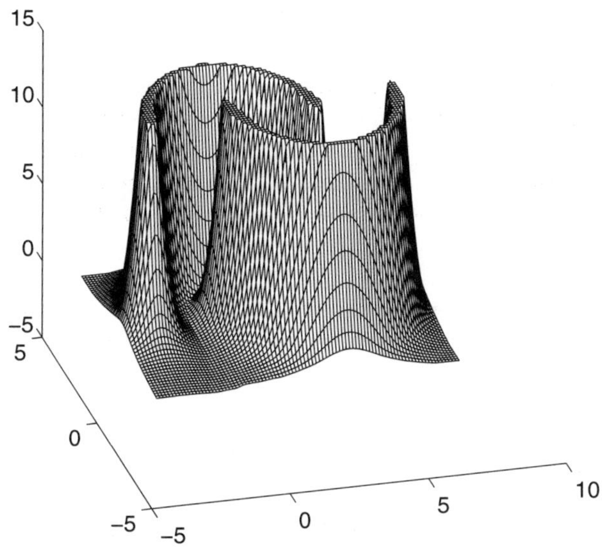

FIGURE 10.6.2

the a-axis in Figure 10.6.3). Later in this volume we will show that the graph of the dependence of $\|H(a)\|$ on a is a convex downward curve on every connected interval of the a-axis on which the norm $\|H(a)\|$ is finite.

For every a in such an interval the invariant subspaces corresponding to the eigenvalues lying to the left of the line $z = a + it$ coincide. Consequently, the projection $P = P(a)$ onto this invariant subspace is independent of the choice of a from the interval.

As we know (see §10.2), $H(a) > 0$ and $P(a)$ can be found as a solution to the system of matrix equations

$$P^2 = P, \quad PA = AP, \quad HP = P^*H, \quad H = H^*,$$
$$H(A - aI) + (A^* - aI)H + [P^*P - (I - P^*)(I - P)] = 0.$$

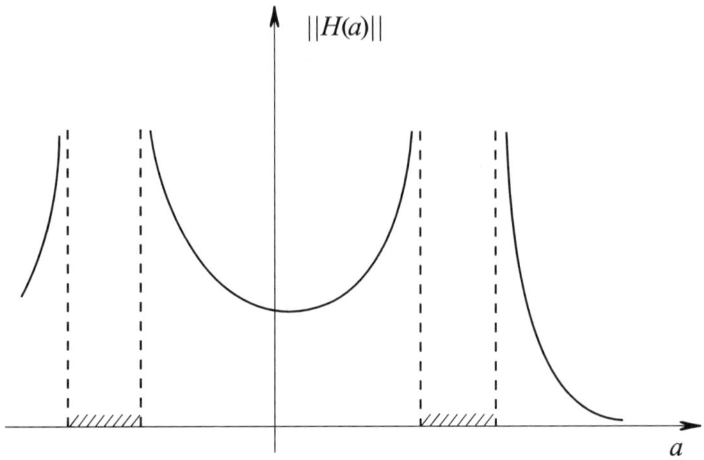

FIGURE 10.6.3

An effecient procedure for solving this system will be described in detail in Chapter 14. Realizing the procedure for a sufficiently dense sequence of points a, we draw the graph of the dependence of $\|H(a)\|$ on a, distinguish spectral spots, and compute the projections corresponding to the eigenvalues contained in these spots. Let a_1 and a_2 ($a_2 > a_1$) belong to neighboring intervals on which $H(a)$ is finite. For a spot separating these intervals, the corresponding projection can be found as the difference $P(a_2) - P(a_1)$.

The annular (or radial) spectral portrait of a matrix A is the graph of the dependence of the norm $\|H(r)\|$ on r, where H is a Hermitian matrix defined in the integral form

$$H = \frac{1}{2\pi} \int_0^{2\pi} [A^* - re^{-i\varphi}]^{-1}(A - re^{i\varphi})^{-1}\, d\varphi.$$

It follows from §10.4 that the matrix H and the projection $P = P(r)$ onto the invariant subspace of A corresponding to the eigenvalues $\lambda(A)$ in the disk $|\lambda(A)| < r$ satisfy the algebraic equations

$$PA = AP, \quad P^2 = P, \quad HP = P^*H, \quad H = H^* > 0,$$
$$r^2H - A^*HA = P^*P - (I - P^*)(I - P).$$

The condition $H = H^* > 0$ yields the uniqueness of a solution. These equations cannot be solved if the circle $z = re^{i\varphi}$ contains eigenvalues of the matrix A. In practice, the algorithm described in Chapter 14 does not work for sufficiently large $\|H(r)\|$, i.e., in the case where the circles $z = re^{i\varphi}$ are close to the eigenvalues of A. The graph of $\|H(r)\|$ (see Figure 10.6.4) constructed on the basis of this algorithm is similar to the above graph of $H(a)$.

The graph can be regarded as a radial spectral portrait of the matrix A. It is obvious that the projection $P(r)$, computed simultaneously with $H(r)$, can be used for constructing projections onto the invariant subspaces corresponding to some spectral spots. In the case under consideration, it is natural that the annuli $r_j < r < r_{j+1}$ containing the spectral spots are referred to as the *spectral annuli*.

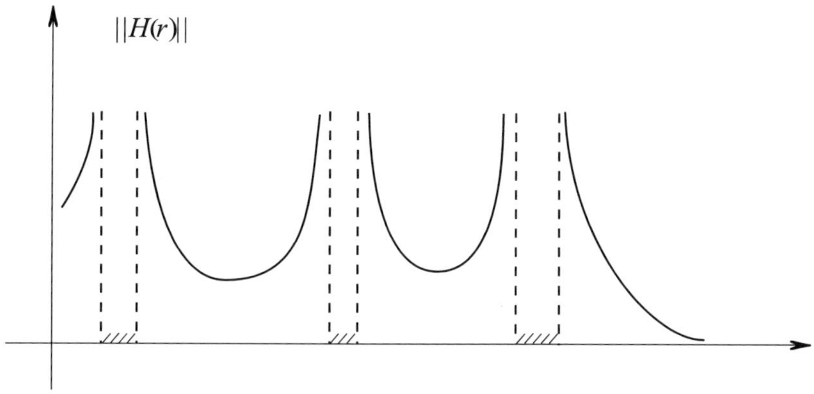

FIGURE 10.6.4

10.7. Remarks about criteria

Asymptotic representation of a solution to the Cauchy problem. An estimate for the remainder. Operator norms. Characterization of the separation of spectra of diagonal blocks in the canonical representations of a matrix. Comparison with the characteristic $\kappa(A)$ of the spectrum dichotomy by the imaginary axis and with other spectral characteristics.

In applications of matrix theory, eigenvalues are often used in asymptotic estimates for solutions to differential or difference (discrete) equations. For example, if a simple eigenvalue $\lambda_0(A)$ of a matrix A has the maximal real part $\alpha = \operatorname{Re}\lambda_0$, whereas the remaining eigenvalues $\lambda_j(A)$ have the real part less than α ($\operatorname{Re}\lambda_j < \operatorname{Re}\lambda_0 = \alpha$, $j \neq 0$), then the solution $x(t)$ to the Cauchy problem

$$\frac{dx}{dt} = Ax, \quad x(0) = f,$$

admits the asymptotic representation

$$x(t) = e^{\lambda_0 t}Pf + o(e^{\alpha t}), \quad t \to \infty, \tag{10.7.1}$$

where P is the projection ($PA = AP$, $P^2 = P$) of rank 1 onto the invariant subspace (eigenvector) corresponding to the eigenvalue λ_0. To obtain an estimate for the remainder in this representation, we need more detailed information about the location of the remaining eigenvalues and about the possibility to surround the point λ_0 at which $\|(\lambda_0 I - A)^{-1}\| = \infty$ by a contour γ such that the length of γ and the norm $\|(\lambda I - A)^{-1}\|$ for all $\lambda \in \gamma$ are not too large.

If the eigenvalues with the largest real part cannot be isolated, or it is hard to isolate every such eigenvalue separately, but it is possible to isolate a group of eigenvalues (a cluster) lying in the right spot of the ε-spectrum ($\varepsilon \ll 1$), then the study of the asymptotic behavior of $x(t)$ as $t \to +\infty$ is reduced to the study of the representation $x(t) = e^{tPA}f$ (see (10.7.1)) plus the remainder. In this representation, P is the projection onto the invariant subspace of A corresponding to the eigenvalues in the right cluster and the remainder coincides with $e^{t(I-P)A}f$.

Suppose that the imaginary axis contains no eigenvalues of A. Let P be the projection onto the invariant subspace corresponding to the eigenvalues in the right half-plane. By the Schur theorem (see §4.2), the matrix A admits the canonical decomposition

$$A = U^* \begin{bmatrix} B & D \\ 0 & C \end{bmatrix} U, \tag{10.7.2}$$

where U is a unitary matrix ($UU^* = I_N$), the eigenvalues of the $N_1 \times N_1$-block B lie in the right half-plane, and the eigenvalues of the $N_2 \times N_2$-block C lie in the left half-plane; here $N = N_1 + N_2$. Hence we can choose a rectangular matrix M such that

$$\begin{bmatrix} B & D \\ 0 & C \end{bmatrix} = \begin{bmatrix} I_{N_1} & -M \\ 0 & I_{N_2} \end{bmatrix} \begin{bmatrix} B & 0 \\ 0 & C \end{bmatrix} \begin{bmatrix} I_{N_1} & M \\ 0 & I_{N_2} \end{bmatrix}. \tag{10.7.3}$$

The matrix M must be a solution to the matrix Sylvester equation

$$BM - MC = D. \tag{10.7.4}$$

Equation (10.7.4) is solvable because the matrices B and C have no common eigenvalues. Consequently, there exists a linear operator $\mathfrak{L}^{-1}$ that is inverse to $\mathfrak{L} : M \to D$ and, in accordance with (10.7.4), transforms M to D where M and D are taken from the canonical decomposition (10.7.2) of A.

It is important that M occurs in the matrix

$$Q = \begin{bmatrix} I_{N_1} & M \\ 0 & I_{N_2} \end{bmatrix} U$$

which reduces A to the block diagonal canonical form

$$A = Q^{-1} \begin{bmatrix} B & 0 \\ 0 & C \end{bmatrix} Q.$$

The projections P and $I - P$ onto the invariant subspaces of A corresponding to the eigenvalues in the right and left half-planes,[1] respectively, can be written as follows:

$$P = Q^{-1} \begin{bmatrix} 0 & 0 \\ 0 & I_{N_2} \end{bmatrix} Q = U^* \begin{bmatrix} 0 & -M \\ 0 & I_{N_2} \end{bmatrix} U,$$

$$I - P = Q^{-1} \begin{bmatrix} I_{N_1} & 0 \\ 0 & 0 \end{bmatrix} Q = U^* \begin{bmatrix} I_{N_1} & M \\ 0 & 0 \end{bmatrix} U.$$

It is easy to check (see §8.1) that $\|P\| = \|I - P\| = \sqrt{1 + \|M\|^2}$. We recall that the distance between the spectra of the diagonal blocks B and C can be characterized (see §4.3) by the parameter $\operatorname{sep}[B, C] = \|\mathfrak{L}^{-1}\|_{\text{op}}$, where $\|\mathfrak{L}^{-1}\|_{\text{op}}$ denotes the operator norm of the operator $\mathfrak{L}^{-1}$ that acts in the unitary space of rectangular matrices ($D = \mathfrak{L}M$, $M = \mathfrak{L}^{-1}D$). The vector norm is defined as follows:

$$\|D\|_E = \sqrt{\operatorname{tr} D^* D} \equiv \sqrt{\sum_{ij} |D_{ij}|^2} = \sqrt{\sum_k \sigma_k^2(D)},$$

$$\|M\|_E = \sqrt{\operatorname{tr} M^* M} = \sqrt{\sum_k \sigma_k^2(M)}.$$

[1] Contrary to §§8.1 and 10.5, here P denotes not a projection onto the invariant subspace of the matrix A corresponding to the eigenvalues in the left half-plane, but the projection onto the complementary subspace which corresponds to the eigenvalues in the right half-plane.

The norm introduced is often called the *Euclidean* (or, sometimes, *Frobenius*) *norm of a matrix*, which explains the notation $\|\cdot\|_E$. The Euclidean norm should not be confused with the operator norm of a matrix because the operator norm is equal to the maximal singular value ($\sigma_{\max}(D)$ or $\sigma_{\max}(M)$). It is obvious that

$$\|D\| \leq \|D\|_E \leq \sqrt{\min\{N_1, N_2\}}\|D\|.$$

The operators $\mathfrak{L}$ and $\mathfrak{L}^{-1}$ act in the space of rectangular matrices with the norm defined in a usual way, i.e.,

$$\|\mathfrak{L}^{-1}\|_{\text{op}} = \sup_{X \neq 0} \frac{\|\mathfrak{L}^{-1}X\|_E}{\|X\|_E}.$$

It is clear that $\|M\| \leq \|M\|_E \leq \|\mathfrak{L}^{-1}\|_{\text{op}}\, \|D\|_E$. Therefore,

$$\|P\| = \|I - P\| = \sqrt{1 + \|M\|^2} \leq \sqrt{1 + \text{sep}^2[B, C]\, \min\{N_1, N_2\}\, \|D\|_E^2}.$$

Thus, to estimate the norm of the projection it is not sufficient to know only the characteristic sep $[B, C]$. We also need to use the norm $\|D\|_E$. Moreover, sep $[B, C]$ gives no information about the location of the spectra of B and C with respect to the imaginary axis. However, necessary information can be obtained from the analysis of solutions to the system generalizing the classical Lyapunov equation (see §10.2)[2]

$$\begin{aligned} &H_F A + A^* H_F + (I - P^*)F(I - P) - P^* F P = 0, \\ &PA = AP, \quad P^2 = P, \quad P^* H_F = H_F P. \end{aligned} \tag{10.7.5}$$

Recall that for a Hermitian positive definite matrix $F = F^* > 0$, a solution H_F to (10.7.5) is a Hermitian positive definite matrix ($H_F = H_F^* > 0$) if and only if A has no eigenvalues on the imaginary axis and P is the projection onto the invariant subspace of A corresponding to the eigenvalues in the right half-plane.

As a criterion for the quality of the spectrum dichotomy of a matrix A by the imaginary axis we take the parameter $\kappa(A) = 2\|A\|\, \|H\|$ that is proportional to the norm $\|H\|$ of a solution $H \equiv H_I$ to the system (10.7.5) with $F = I (\equiv I_N)$. Such a choice allows us to estimate the distance between the eigenvalues of B, C and the imaginary axis as well as the norm of M in the canonical decomposition (10.7.3). The norms $\|P\| = \|I - P\|$ are estimated in terms of the norm $\|H\|$ (more exactly, in terms of $\kappa(A) = 2\|A\|\, \|H\|$).

In the case

$$A = Q^{-1} \begin{bmatrix} B & 0 \\ 0 & C \end{bmatrix} Q, \quad \text{Re}\, \lambda_j(B) > 0, \quad \text{Re}\, \lambda_j(C) < 0,$$

the Green matrix $G(t)$ (see §10.5) constructed for A has the form

$$G(t) = \begin{cases} Q^{-1} \begin{bmatrix} 0 & 0 \\ 0 & e^{tC} \end{bmatrix} Q, & t > 0, \\[2ex] -Q^{-1} \begin{bmatrix} e^{tB} & 0 \\ 0 & 0 \end{bmatrix} Q, & t < 0. \end{cases}$$

---

[2]We emphasize the difference between the notation of projections adopted here and in §10.2.

The Green matrix $G(t)$ satisfies the estimate (see (10.5.5))

$$\|G(t)\| \leq \sqrt{2\|A\|\,\|H\|}e^{-(|t|\,\|A\|)/(2\|A\|\,\|H\|)} = \sqrt{\kappa(A)}e^{-(|t|\,\|A\|)/(\kappa(A))}.$$

Consequently,

$$\|G(+0)\| = \|G(-0)\| \leq \sqrt{\kappa(A)} = \sqrt{2\|A\|\,\|H\|}.$$

Since $P = G(-0)$ and $I - P = -G(+0)$, we have

$$\|P\| = \|I - P\| \leq \sqrt{\kappa(A)} = \sqrt{2\|A\|\,\|H\|}.$$

Recalling the representations

$$A = U^* \begin{bmatrix} B & D \\ 0 & C \end{bmatrix} U,$$

$$\begin{bmatrix} B & D \\ 0 & C \end{bmatrix} = \begin{bmatrix} I_{N_1} & -M \\ 0 & I_{N_2} \end{bmatrix} \begin{bmatrix} B & 0 \\ 0 & C \end{bmatrix} \begin{bmatrix} I_{N_1} & M \\ 0 & I_{N_2} \end{bmatrix},$$

$$I - P = U^* \begin{bmatrix} 0 & -M \\ 0 & I_{N_2} \end{bmatrix} U, \quad P = U^* \begin{bmatrix} I_{N_1} & M \\ 0 & 0 \end{bmatrix} U,$$

we find that $\sqrt{1 + \|M\|^2} = \|P\| = \|I - P\| \leq \sqrt{\kappa(A)}$, $\|M\| \leq \sqrt{\kappa(A)}$, i.e., the norm $\|M\|$ can also be estimated in terms of $\kappa(A) = 2\|A\|\,\|H\|$. Regarding the solution $x(t) = e^{tA}f$ to the Cauchy problem $x' = -Ax$, $x(0) = f$, we note that for $t > 0$

$$e^{tA} = Q^{-1} \begin{bmatrix} e^{tB} & 0 \\ 0 & e^{tC} \end{bmatrix} Q = G(t) - G(-t),$$

$$x(t) = e^{tA}f = -G(-t)f + G(t)f = e^{tA}Pf + G(t)f,$$

$$G(t)f = e^{tA}(I - P)f,$$

and the inequality (10.5.5) rewritten in the form

$$\|G(t)\| \leq \sqrt{\kappa(A)}e^{-(|t|\,\|A\|)/(\kappa(A))}, \tag{10.7.6}$$

leads to the following estimate for the remainder $e^{tA}(I - P)f$ in the asymptotical representation $x(t) \sim e^{tA}Pf + \delta(t)$:

$$\|\delta(t)\| = \|e^{tA}(I - P)f\| \leq \sqrt{\kappa(A)}e^{-t\|A\|/\kappa(A)}\|f\|, \quad t > 0.$$

PROBLEM 10.7.1. Using the estimate (10.7.6), prove that the matrix A has no eigenvalues, in the strip $|\operatorname{Re}\lambda| < \|A\|/\kappa(A)$, which contains the imaginary axis.

The above estimates show that the criteria for dichotomy based on solutions to matrix Lyapunov equations often turn out to be more informative than the criterion $\operatorname{sep}[B, C]$ characterizing the distance between parts of the spectrum of the operator A.

Similar estimates based on the criterion for the circle dichotomy can be used in the study of discrete equations $x_{n+1} = Ax_n$. We will return to this question in Chapter 14.

CHAPTER 11

The Hausdorff Set of a Matrix

11.1. The simplest properties of the Hausdorff set

The definition of the Hausdorff set of a square matrix. The distance between sets. An estimate for the distance between the Hausdorff sets. Properties of the Hausdorff set. The Hausdorff theorem.

Studying the variational principles for quadratic and Hermitian forms, we discussed properties of the Rayleigh quotient $\frac{(Au,u)}{(u,u)}$ in detail. Now we drop the condition $A = A^*$ and consider an arbitrary matrix A. Taking different nonzero vectors u, we obtain different values of the Rayleigh quotient $\frac{(Au,u)}{(u,u)}$. Even if every element of A is real, some values of the Rayleigh quotient $\frac{(Au,u)}{(u,u)}$ can be complex because we do not suppose the components of u to be real.

The set of all values of the Rayleigh quotient $\frac{(Au,u)}{(u,u)}$ is called the *range* or the *Hausdorff set* of the operator (matrix) A.

If matrices A_1 and A_2 are unitarily similar, i.e., there exists a unitary matrix U such that $A_1 = UA_2U^*$, then the Hausdorff sets of the matrices A_1 and A_2 coincide. By the Schur theorem (see §4.2), if we want to construct the Hausdorff set of an arbitrary matrix, we may assume that the matrix is upper triangular so that the diagonal elements coincide with the eigenvalues.

It is obvious that $|(Au,u)| \leq \|A\|\,\|u\|^2$. Therefore, the points $z = x + iy = \frac{(Au,u)}{(u,u)}$ lie in the disk $|z| = \sqrt{x^2+y^2} \leq \|A\|$.

We emphasize that under small perturbations of the matrix the change of the Hausdorff set of a matrix is small. To give a precise formulation, we introduce the notion of the distance $\rho(\mathcal{A},\mathcal{B})$ between bounded sets $\mathcal{A}$ and $\mathcal{B}$ in the complex plane.

By the *distance between sets $\mathcal{A}$ and $\mathcal{B}$* in the complex plane we mean the number

$$\rho(\mathcal{A},\mathcal{B}) = \max\{\sup_{\alpha\in\mathcal{A}}[\inf_{\beta\in\mathcal{B}}|\alpha-\beta|], \sup_{\beta\in\mathcal{B}}[\inf_{\alpha\in\mathcal{A}}|\alpha-\beta|]\}. \tag{11.1.1}$$

Let us explain the meaning of formula (11.1.1). Fix a point $\alpha \in \mathcal{A}$, take an arbitrary point $\beta \in \mathcal{B}$, and find the lower bound of $|\alpha - \beta|$, i.e., the lower bound of the distances between α and β. The upper bound $\sup_{\alpha\in\mathcal{A}}[\inf_{\beta\in\mathcal{B}}|\alpha-\beta|]$ of distances between $\alpha \in \mathcal{A}$ and the nearest points $\beta \in \mathcal{B}$ is the distance at which points of $\mathcal{A}$ may be located from the nearest point of $\mathcal{B}$. This distance does not necessarily coincide with the number $\sup_{\beta\in\mathcal{B}}[\inf_{\alpha\in\mathcal{A}}|\beta-\alpha|]$. In the above definition, suggested by Hausdorff, the greatest "unilateral" distance is taken for the distance between $\mathcal{A}$ and $\mathcal{B}$.

We illustrate the definition by an example. Let $\mathcal{A} = [0,2]$ and $\mathcal{B} = [1,4]$. Then

$$\sup_{\alpha\in\mathcal{A}}[\inf|\alpha-\beta|] = \max_{0\le\alpha\le2}\Big[\min_{1\le\beta\le4}|\alpha-\beta|\Big] = 1,$$
$$\sup_{\beta\in\mathcal{B}}[\inf|\alpha-\beta|] = \max_{1\le\beta\le4}\Big[\min_{0\le\alpha\le2}|\alpha-\beta|\Big] = 2.$$

Consequently, $\rho(\mathcal{A},\mathcal{B}) = 2$.

For closed sets $\mathcal{A}$ and $\mathcal{B}$ the lower and upper bounds can be replaced by maximum and minimum:

$$\rho(\mathcal{A},\mathcal{B}) = \max\{\max_{\alpha\in\mathcal{A}}[\min_{\beta\in\mathcal{B}}|\alpha-\beta|], \max_{\beta\in\mathcal{B}}[\min_{\alpha\in\mathcal{A}}|\alpha-\beta|]\}.$$

PROBLEM 11.1.1. Prove that the Hausdorff set of an $N\times N$-matrix A is closed.

Let $\mathcal{A}$ and $\mathcal{B}$ be the Hausdorff sets of $N\times N$-matrices A and B. For $\alpha\in\mathcal{A}$ there exists a vector $u\neq 0$ such that $\alpha = \dfrac{(Au,u)}{(u,u)}$. For $\beta = \dfrac{(Bu,u)}{(u,u)}$ we have

$$|\alpha-\beta| = \left|\frac{(Au,u)-(Bu,u)}{(u,u)}\right| = \left|\frac{((A-B)u,u)}{(u,u)}\right| \le \|A-B\|.$$

Consequently,

$$\min_{\beta\in\mathcal{B}}|\alpha-\beta| \le \|A-B\|, \quad \max_{\alpha\in\mathcal{A}}\min_{\beta\in\mathcal{B}}|\alpha-\beta| \le \|A-B\|.$$

Similarly, we can prove the inequality

$$\max_{\beta\in\mathcal{B}}\min_{\alpha\in\mathcal{A}}|\beta-\alpha| \le \|A-B\|.$$

Thus, the following theorem holds.

THEOREM 11.1.1. *The distance $\rho(\mathcal{A},\mathcal{B})$ between sets $\mathcal{A}$ and $\mathcal{B}$ is estimated as follows:*

$$\rho(\mathcal{A},\mathcal{B}) \le \|A-B\|.$$

For $\alpha\in\mathcal{A}$ a vector u such that $\|u\| = \sqrt{(u,u)} = 1$ and $\alpha = (Au,u) = \dfrac{(Au,u)}{(u,u)}$ is not unique. For any vector $v = e^{i\varphi}u$ ($\|v\| = \|u\| = 1$) we obtain the same value $\alpha = (Av,v) = (Au,u)$.

Let the Hausdorff set of a matrix A contain the points $\alpha_1 = 1 = (Au,u)$ and $\alpha_0 = 0 = (Av,v)$ ($\|u\| = \|v\| = 1$). We note that the vectors u and v are not collinear (for $\gamma\neq0$ from $v=\gamma u$ it follows that $\alpha_0 = (Av,v) = |\gamma|^2(Au,u) = |\gamma|^2 \neq 0$). Therefore, for any real φ and s ($0\le\varphi\le2\pi, 0\le s\le1$) the linear combination $w(s) = e^{i\varphi}(1-s)u + sv$ of u and v does not vanish and the scalar square of $w(s)$ is real and strictly positive, i.e.,

$$(w(s),w(s)) = \|e^{i\varphi}(1-s)u + sv\|^2 > 0.$$

We choose φ so that $e^{i\varphi}(Au,v) + e^{-i\varphi}(Av,u)$ is real. Let us show that such a choice is possible. Let $(Au,v) = e^{i\theta_1}a_1$ and $(Av,u) = e^{i\theta_2}a_2$, where θ_1, θ_2, a_1, a_2

are real. It is obvious that

$$\begin{aligned}
&\operatorname{Im}[e^{i\varphi}(Au,v)+e^{-i\varphi}(Av,u)] = \operatorname{Im}[e^{i(\varphi+\theta_1)}a_1 + e^{i(-\varphi+\theta_2)}a_2] \\
&\quad = a_1\sin(\varphi+\theta_1) - a_2\sin(\varphi-\theta_2) \\
&\quad = (a_1\cos\theta_1 - a_2\cos\theta_2)\sin\varphi + (a_1\sin\theta_1 - a_2\sin\theta_2)\cos\varphi \\
&\quad = \begin{cases} 0 & \text{if } (a_1\cos\theta_1 + a_2\cos\theta_2)^2 + (a_1\sin\theta_1 - a_2\sin\theta_2)^2 = 0, \\ a\sin(\varphi+\theta) & \text{otherwise,} \end{cases}
\end{aligned}$$

where

$$a = \sqrt{(a_1\cos\theta_1 - a_2\cos\theta_2)^2 + (a_1\sin\theta_1 + a_2\sin\theta_2)^2},$$

$$e^{i\theta} = \frac{(a_1\cos\theta_1 - a_2\cos\theta_2) + i(a_1\sin\theta_1 + a_2\sin\theta_2)}{a}.$$

Therefore, we can assume that the imaginary part is equal to zero; otherwise, we set $\varphi = -\theta$.

Consider the Rayleigh quotient $\dfrac{(Aw,w)}{(w,w)}$, where $w(s)$ is the above function of the real parameter s which changes from 0 to 1, i.e., $w(s) = e^{i\varphi}(1-s)u + sv$, and φ is such that the number $e^{i\varphi}(Au,v) + e^{-i\varphi}(Av,u)$ is real. Furthermore,

$$\begin{aligned}
\frac{(Aw(s),w(s))}{(w(s),w(s))} &= \frac{(1-s)^2(Au,u) + s(1-s)[e^{i\varphi}(Au,v)+e^{-i\varphi}(Av,u)] + s^2(Av,v)}{(w(s),w(s))} \\
&= \frac{(1-s)^2 + s(1-s)[e^{i\varphi}(Au,v)+e^{-i\varphi}(Av,u)]}{\|w(s)\|^2},
\end{aligned}$$

where we used the equalities $(Au,u) = 1$ and $(Av,v) = 0$. By this formula and the above choice of φ, the Rayleigh quotient $\dfrac{(Aw,w)}{(w,w)}$ is a continuous real function of s and takes the values 1 and 0 at $s = 0$ and $s = 1$, respectively. Therefore, in view of the well-known theorem from analysis, it takes any intermediate value between 0 and 1.

Thus, if the points $\alpha_1 = 0$ and $\alpha_2 = 1$ belong to the Hausdorff set of a matrix A, then the Hausdorff set contains the entire segment joining α_1 and α_2.

Let z_1 and z_2 ($z_1 \neq z_2$) belong to the Hausdorff set of a matrix Z. In other words, assume that there exist vectors $u^{(1)}$ and $u^{(2)}$ such that

$$\frac{(Zu^{(1)},u^{(1)})}{(u^{(1)},u^{(1)})} = z_1, \qquad \frac{(Zu^{(2)},u^{(2)})}{(u^{(2)},u^{(2)})} = z_2.$$

Consider the matrix

$$A = \frac{1}{z_1 - z_2}Z - \frac{z_2}{z_1 - z_2}I$$

such that

$$\frac{(Au^{(1)},u^{(1)})}{(u^{(1)},u^{(1)})} = 1, \qquad \frac{(Au^{(2)},u^{(2)})}{(u^{(2)},u^{(2)})} = 0.$$

As was shown, for any $s \in [0,1]$ there exists a vector $u(s)$ such that $\dfrac{(Au(s),u(s))}{(u(s),u(s))} = s$. Since $Z = (z_1 - z_2)A + z_2 I$, it is obvious that $(Zu(s),u(s)) = (z_1 - z_2)s + z_2 = z_1 s + z_2(1-s)$. Thus, we have proved the following important assertion.

THEOREM 11.1.2 (Hausdorff). *If the Hausdorff set of a matrix contains two different points of the complex plane, then it also contains the segment joining these points. In other words, the Hausdorff set of any matrix is convex.*

This theorem, proved by Hausdorff in 1919, was a reason for introducing the term "Hausdorff set." In his work, Hausdorff continued the investigations of Toeplitz (1918) who proved that the exterior boundary of the range is a convex curve. However, the following question remained open: Is it true that every interior point of the domain bounded by the convex curve is a value of the Rayleigh quotient $\dfrac{(Au,u)}{(u,u)}$?

We note the following obvious fact. Every eigenvalue of a matrix A belongs to the Hausdorff set of A, i.e., the Hausdorff set of A contains the convex hull of all eigenvalues of A.

11.2. The Hausdorff set of a second order matrix

Examples of the Hausdorff sets. An example for the simplest diagonal matrix. An example for an upper triangular real matrix. An example for an arbitrary upper triangular matrix. The Hausdorff set of a second order matrix.

In this section, we give a complete description of the Hausdorff sets of square second order matrices. We begin with examples.

EXAMPLE 11.2.1. Consider the simplest diagonal matrix

$$A = \begin{bmatrix} -1 & 0 \\ 0 & 1 \end{bmatrix}.$$

For a nonzero vector u with components u_1 and u_2 we have

$$(Au,u) = -u_1\overline{u}_1 + u_2\overline{u}_2 = |u_2|^2 - |u_1|^2, \quad (u,u) = |u_2|^2 + |u_1|^2 \neq 0.$$

Therefore,

$$\frac{(Au,u)}{(u,u)} = \frac{|u_2|^2 - |u_1|^2}{|u_2|^2 + |u_1|^2} = 1 - \frac{2|u_1|^2}{|u_1|^2 + |u_2|^2}.$$

Since u_1 and u_2 are subject to only the condition $|u_1|^2 + |u_2|^2 \neq 0$, the ratio $\dfrac{|u_1|^2}{|u_1|^2 + |u_2|^2}$ may take any value between 0 and 1. The set of all possible values of $\dfrac{(Au,u)}{(u,u)}$, i.e., the Hausdorff set of A, coincides with the segment $[-1,+1]$.

The assertion from Example 11.2.1 is not new to us because A is a Hermitian matrix ($A = A^*$). Recall that for Hermitian (in particular, real symmetric) matrices the number $\dfrac{(Au,u)}{(u,u)}$ is the Rayleigh quotient whose values (see §§7.1–7.3) fill out the entire segment on the real axis that joins the largest eigenvalue and the smallest eigenvalue of A. Indeed, for $A = A^*$ we can choose a unitary transformation U such that the matrix U^*AU is diagonal with real diagonal elements λ_j ($\lambda_1 \leq \lambda_2 \leq \cdots \leq$

λ_N). In addition, for any vector x and $y = Ux$ we have

$$\frac{(Ax,x)}{(x,x)} = \frac{\sum\limits_{j=1}^{N} \lambda_j |y_j|^2}{\sum\limits_{j=1}^{N} |y_j|^2}. \tag{11.2.1}$$

As was mentioned in Chapter 7, (11.2.1) implies that the Rayleigh quotient $\frac{(Ax,x)}{(x,x)}$ is real and the inequality $\lambda_N \geq \frac{(Ax,x)}{(x,x)} \geq \lambda_1$ holds. Choosing $y = y(s)$ with the components $y_N = \cos s$, $y_{N-1} = y_{N-2} = \cdots = y_2 = 0$, $y_1 = \sin s$ and setting $x(s) = U^* y(s)$, we find that

$$\frac{(A(x(s)), x(s))}{(x(s), x(s))} = \lambda_N \cos^2 s + \lambda_1 \sin^2 s.$$

Choosing a suitable s, we see that the Rayleigh quotient takes any given value in the segment $[\lambda_1, \lambda_N]$.

EXAMPLE 11.2.2. Consider a slightly more complicated example of the real upper triangular matrix

$$A = \begin{bmatrix} -1 & 2q \\ 0 & 1 \end{bmatrix}, \qquad q > 0.$$

Consider the curve parametrized by the Rayleigh quotient $\frac{(Au,u)}{(u,u)}$, where the parameter t in the componentwise representation of the vector

$$u(t) = \begin{pmatrix} v(t) \\ w(t) \end{pmatrix} = \begin{pmatrix} q\cos t - i\sqrt{1+q^2}\sin t \\ \cos t + \sqrt{1+q^2} \end{pmatrix}$$

runs over the segment $0 \leq t \leq 2\pi$. For such $u(t)$ we have

$$\begin{aligned} (u(t), u(t)) &= \|u(t)\|^2 = |v(t)|^2 + |w(t)|^2 = 2\sqrt{1+q^2}(\sqrt{1+q^2} + \cos t), \\ (Au(t), u(t)) &= 2\sqrt{1+q^2}(\cos t + \sqrt{1+q^2})(\sqrt{1+q^2}\cos t + iq\sin t). \end{aligned}$$

Therefore,

$$\frac{(Au(t), u(t))}{(u(t), u(t))} = z(t) = \sqrt{1+q^2}\cos t + iq\sin t.$$

A point $z(t) = x(t) + iy(t)$ describes a closed curve, namely, the ellipse

$$\frac{x^2}{1+q^2} + \frac{y^2}{q^2} = 1$$

with the major semiaxis $\sqrt{1+q^2}$, minor semiaxis q, and foci at $x = +1, y = 0$ and $x = -1, y = 0$.

The sum of the distances between an arbitrary point (x, y) of the ellipse and the foci is equal to the major axis $2\sqrt{1+q^2}$:

$$\sqrt{(x-1)^2 + y^2} + \sqrt{(x+1)^2 + y^2} = 2\sqrt{1+q^2}.$$

By Theorem 11.1.2, the Hausdorff set of the matrix A contains not only the ellipse itself but also its interior

$$\frac{x^2}{1+q^2} + \frac{y^2}{q^2} \leq 1.$$

So far we do not know whether there exist points of the Hausdorff set outside the ellipse.

Choose an arbitrary t_0 and construct the matrix

$$\widehat{A} = (q\cos t_0 - i\sqrt{1+q^2}\sin t_0)A - q\sqrt{1+q^2}I.$$

It is easy to check that the Hermitian matrix

$$\widehat{B} = \widehat{B}^* = \frac{1}{2}(\widehat{A} + \widehat{A}^*) = q\begin{bmatrix} -(\sqrt{1+q^2}+\cos t_0) & q\cos t_0 - i\sqrt{1+q^2}\sin t_0 \\ q\cos t_0 + i\sqrt{1+q^2}\sin t_0 & -\sqrt{1+q^2}+\cos t_0 \end{bmatrix}$$

has eigenvalues $\lambda_1(\widehat{B}) = 0$, $\lambda_2(\widehat{B}) = -2q\sqrt{1+q^2}$. Therefore, the quadratic form $(\widehat{B}u, u)$ is negative semidefinite, i.e., $(\widehat{B}u, u) \leq 0$. We set $\widehat{C} = \dfrac{1}{2i}(\widehat{A} - \widehat{A}^*)$ and note that the quadratic forms $(\widehat{B}u, u)$ and $(\widehat{C}u, u)$ are real because $\widehat{B}$ and $\widehat{C}$ are Hermitian matrices. We verify that all possible values of the Rayleigh quotient $\dfrac{(\widehat{A}u, u)}{(u, u)} = \dfrac{(\widehat{B}u, u) + i(\widehat{C}u, u)}{\|u\|^2}$ lie in the closed left half-plane.

If for a nonzero vector u ($\|u\| \neq 0$) we have

$$\frac{(Au, u)}{(u, u)} = x_0 + iy_0 = r(\sqrt{1+q^2}\cos t_0 + iq\sin t_0), \quad r > 1,$$

i.e., if some of the values of the Rayleigh quotient $\dfrac{(Au, u)}{(u, u)}$ lie outside the ellipse, then

$$\begin{aligned}
\operatorname{Re}&\left[\frac{(\widehat{A}(t_0)u, u)}{(u, u)}\right] \\
&= \operatorname{Re}\left[(q\cos t_0 - i\sqrt{1+q^2}\sin t_0)\frac{(Au, u)}{(u, u)} - q\sqrt{1+q^2}\right] \\
&= \operatorname{Re}\left[(q\cos t_0 - i\sqrt{1+q^2}\sin t_0)(x_0 + iy_0) - q\sqrt{1+q^2}\right] \\
&= \operatorname{Re}\left[(q\cos t_0 - i\sqrt{1+q^2}\sin t_0)\cdot r(\sqrt{1+q^2}\cos t_0 + iq\sin t_0) - q\sqrt{1+q^2}\right] \\
&= (r-1)q\sqrt{1+q^2} > 0,
\end{aligned}$$

which contradicts the inequality

$$\operatorname{Re}\left[\frac{(\widehat{A}(t_0)u, u)}{(u, u)}\right] \leq 0$$

for any t_0, $u \neq 0$. This means that no point of the Hausdorff set is located outside the ellipse.

To explain the idea of the above proof, we first note that the transformation

$$\widehat{x} + i\widehat{y} = (q\cos t_0 - i\sqrt{1+q^2}\sin t_0)(x + iy) - q(1+q^2)$$

takes the tangent line to the ellipse $\dfrac{x^2}{(1+q^2)} + \dfrac{y^2}{q^2} = 1$ given by the equation

$$\begin{aligned} &\operatorname{Re}\left[(x+iy)(q\cos t_0 - i\sqrt{1+q^2}\sin t_0) - q\sqrt{1+q^2}\right] \\ &= xq\cos t_0 + y\sqrt{1+q^2}\sin t_0 - q\sqrt{1+q^2} = 0, \end{aligned} \tag{11.2.2}$$

to the vertical line $\widehat{x} = 0$. Moreover, under this transformation, the image of the ellipse lies in the left half-plane $\widehat{x} \le 0$. Under the transformation of the matrices $A \to \widehat{A}$, the Hausdorff set is also transformed: $x + iy \to \widehat{x} + i\widehat{y}$. In the above proof, every point that lies to one side of the tangent line to the ellipse at the point

$$\frac{x_0 + iy_0}{r} = \sqrt{1+q^2}\cos t_0 + iq\sin t_0$$

goes to a point in the left half-plane. Since the center of the ellipse $x = 0$, $y = 0$ goes to the point $\widehat{x} = -q(1+q^2)$, $\widehat{y} = 0$ in the left half-plane, we conclude that all interior points of the ellipse go to points of the left half-plane. The equality (11.2.2) is just the tangent line to the ellipse at the point (x_0, y_0).

Thus, the Hausdorff set of the real matrix

$$A = \begin{bmatrix} -1 & 2q \\ 0 & 1 \end{bmatrix} \quad (q > 0)$$

coincides with the ellipse

$$\frac{x^2}{1+q^2} + \frac{y^2}{q^2} \le 1.$$

The case $q = 0$ was considered earlier. Recall that in this case the Hausdorff set is the segment $[-1, 1]$ of the real axis which can be regarded as an infinitely thin ellipse with the semiaxes $q = 0$ and $1 + q^2 = 1$. Thereby, we can expand the description of the range of the matrix A obtained for $q > 0$ to the limit case $q = 0$.

It turns out that the Hausdorff sets of matrices of the form

$$A = \begin{bmatrix} -1 & 2e^{i\varphi}q \\ 0 & 1 \end{bmatrix}$$

is the same for all real φ. Indeed, for

$$A = \begin{bmatrix} -1 & 2e^{i\varphi_1}q \\ 0 & 1 \end{bmatrix}, \quad B = \begin{bmatrix} -1 & 2e^{i\varphi_2}q \\ 0 & 1 \end{bmatrix}$$

we set

$$u = \begin{pmatrix} u_1 \\ e^{-i\varphi_1}u_2 \end{pmatrix} \quad v = \begin{pmatrix} u_1 \\ e^{-i\varphi_2}u_2 \end{pmatrix}$$

and obtain

$$(Au, u) = -u_1\overline{u}_1 + 2qu_2\overline{u}_1 + u_2\overline{u}_2 = (Bv, v),$$

$$(u, u) = u_1\overline{u}_1 + u_2\overline{u}_2 = (v, v), \quad \frac{(Au, u)}{(u, u)} = \frac{(Bv, v)}{(v, v)}.$$

It is clear that every nonzero vector u can be represented as above. Consequently, for every value of $\dfrac{(Au, u)}{(u, u)}$ there is a vector v such that $\dfrac{(Au, u)}{(u, u)} = \dfrac{(Bv, v)}{(v, v)}$.

Thus, the Hausdorff set of the matrix B is contained in the Hausdorff set of the matrix A. The converse assertion is obvious. We have proved that the Hausdorff sets of the matrices A and B coincide.

The Hausdorff set of the matrix

$$A = \begin{bmatrix} -1 & q \\ 0 & 1 \end{bmatrix}$$

is the ellipse

$$\frac{x^2}{1+|q|^2} + \frac{y^2}{|q|^2} \le 1$$

if $|q| > 0$ and, in the case $q = 0$, is the segment $y = 0$, $-1 \le x \le 1$, which can be regarded as an infinitely thin ellipse.

EXAMPLE 11.2.3. Consider a triangular matrix

$$A = \begin{bmatrix} 0 & e^{i\varphi} \\ 0 & 0 \end{bmatrix}.$$

Setting $u = \begin{pmatrix} u_1 \\ u_2 \end{pmatrix}$, we see that

$$(Au, u) = e^{i\varphi} u_2 \overline{u}_1, \quad (u, u) = |u_1|^2 + |u_2|^2,$$

$$\frac{(Au, u)}{(u, u)} = e^{i\varphi} \frac{u_2 \overline{u}_1}{|u_1|^2 + |u_2|^2}.$$

Since

$$|e^{i\varphi} u_2 \overline{u}_1| = |u_2|\,|u_1| \le \frac{1}{2}(|u_2|^2 + |u_1|^2),$$

we have

$$\left| \frac{(Au, u)}{(u, u)} \right| \le \frac{1}{2}.$$

In other words, the Hausdorff set of A lies in the ellipse $|z| = \sqrt{x^2 + y^2} \le 1/2$. Actually, this ellipse coincides with the Hausdorff set of A. Indeed, any point $z = (1/2)e^{i\omega} \sin\gamma$ in the ellipse admits the representation $z = e^{i\varphi} u_2 \overline{u}_1 / (|u_1|^2 + |u_2|^2)$. It suffices to put $u_1 = e^{i\varphi} \cos(\gamma/2)$ and $u_2 = e^{i\omega} \sin(\gamma/2)$.

An arbitrary upper triangular second order matrix

$$A = \begin{bmatrix} \lambda_1 & \alpha \\ 0 & \lambda_2 \end{bmatrix}$$

with complex elements is represented by one of the following formulas:

$$A = \frac{\lambda_2 - \lambda_1}{2} \begin{bmatrix} -1 & 0 \\ 0 & 1 \end{bmatrix} + \frac{\lambda_2 + \lambda_1}{2} \begin{bmatrix} 1 & 0 \\ 0 & 1 \end{bmatrix}, \quad \alpha = 0,$$

$$A = \frac{\lambda_2 - \lambda_1}{2} \begin{bmatrix} -1 & \dfrac{2\alpha}{\lambda_2 - \lambda_1} \\ 0 & 1 \end{bmatrix} + \frac{\lambda_2 + \lambda_1}{2} \begin{bmatrix} 1 & 0 \\ 0 & 1 \end{bmatrix}, \quad \alpha \ne 0,\ \lambda_2 \ne \lambda_1,$$

$$A = |\alpha| \begin{bmatrix} 0 & e^{i\varphi} \\ 0 & 0 \end{bmatrix} + \lambda \begin{bmatrix} 1 & 0 \\ 0 & 1 \end{bmatrix}, \quad \lambda_2 = \lambda_1 = \lambda.$$

In view of these representations and the above examples, it is easy to show that the Hausdorff set of a triangular second order matrix is an ellipse with foci located at eigenvalues. If A is a diagonal matrix, the ellipse degenerates into a segment joining points of the spectrum of A. If these points coincide, then the ellipse becomes a circle. In particular, the radius of the circle may be zero. This happens if $\lambda_1 = \lambda_2 = \lambda$ and $\alpha = 0$. In this case, the Hausdorff set consists of a single point λ.

If an ellipse is the Hausdorff set of a matrix, the semiaxes are computed from λ_2, λ_1, α by the following rule:

for the major semiaxis,

$$a = \frac{|\lambda_2 - \lambda_1|}{2}\sqrt{1 + \left|\frac{\alpha}{\lambda_2 - \lambda_1}\right|^2} = \frac{1}{2}\sqrt{|\lambda_2 - \lambda_1|^2 + |\alpha|^2};$$

for the minor semiaxis,

$$b = \frac{|\lambda_2 - \lambda_1|}{2}\frac{|\alpha|}{|\lambda_2 - \lambda_1|} = \frac{1}{2}|\alpha|.$$

The formulas $a = \frac{1}{2}\sqrt{|\lambda_2 - \lambda_1|^2 + |\alpha|^2}$ and $b = \frac{1}{2}|\alpha|$ for the semiaxes can be applied in the limit cases $\alpha = 0$ and $\lambda_2 = \lambda_1$.

Any matrix can be transformed to a triangular matrix by a unitary transformation in such a way that the Hausdorff set does not change. Therefore, the Hausdorff set of an arbitrary second order matrix is an ellipse with foci located at eigenvalues of the matrix.

We note yet another interesting fact. The ratio of the distance between the foci to the major axis $2a$ is equal to the sine of the angle θ between eigenvectors of the matrix A. This assertion can be formulated in another way. The sum of the distances from an arbitrary point of the ellipse to the foci is constant and is equal to $2a$. The sine of the angle between eigenvectors can be computed as the ratio of the distance between the foci to the sum of the distances from a point of the ellipse to the foci. If an ellipse degenerates into a segment (and only in this case), then the sine of the angle is equal to 1, which is equivalent to the existence of two orthogonal eigenvectors. Only under this condition, the eigenvalues (the foci of the ellipse) can lie on the boundary of the Hausdorff set.

To prove a formula relating θ and the parameters characterizing an ellipse, it suffices to consider the upper triangular matrix

$$A = \begin{bmatrix} \lambda_1 & \alpha \\ 0 & \lambda_2 \end{bmatrix}$$

with the eigenvectors $u = \begin{pmatrix} 1 \\ 0 \end{pmatrix}$ and $v = \begin{pmatrix} \alpha \\ \lambda_2 - \lambda_1 \end{pmatrix}$. For $\lambda_2 = \lambda_1$ the vectors u and v are collinear. No eigenvector is collinear to u or v if $\alpha \neq 0$.

The cosine and sine of the angle θ between the vectors u and v are found from the known formulas

$$\cos\theta = \frac{|(u,v)|}{\|u\|\,\|v\|} = \frac{|\alpha|}{|\alpha|^2 + |\lambda_2 - \lambda_1|^2},$$

$$\sin\theta = \sqrt{1 - \cos^2\theta} = \frac{|\lambda_2 - \lambda_1|}{\sqrt{|\lambda_2 - \lambda_1|^2 + |\alpha|^2}} = \frac{|\lambda_2 - \lambda_1|}{2a}.$$

If $\lambda_2 = \lambda_1$, then $\sin\theta = 0$. However, in this case the above formulas are not applicable. If $\lambda_2 = \lambda_1$, but $\alpha \neq 0$, then A has only one eigenvector. If, in addition, $\alpha = 0$, then any vector is an eigenvector. Hence we can find a pair of orthogonal vectors.

PROBLEM 11.2.1. Show that the formula connecting $\sin\theta$ with the parameters characterizing an ellipse (in other words, parameters that define the form of the Hausdorff set) holds for any second order matrix, not only for an upper triangular one.

As was mentioned in §11.1, the Hausdorff set of a matrix A lies in the disk of center at $z = 0$ and radius $\|A\|$. Actually, the following assertion holds.

PROPOSITION 11.2.1. *The Hausdorff set of a matrix A contains points that are at a distance at most $\frac{1}{2}\|A\|$ from the boundary of the enveloping disk, i.e., there exists a vector $v \neq 0$ such that*

$$\left|\frac{(Av,v)}{(v,v)}\right| \geq \frac{1}{2}\|A\|.$$

PROOF. We consider second order matrices. As above, it suffices to consider the upper triangular matrix

$$A = \begin{bmatrix} \lambda_1 & \alpha \\ 0 & \lambda_2 \end{bmatrix}$$

because any matrix is reduced to such a matrix by a similar unitary transformation. Moreover, we can assume that $|\lambda_1| \geq |\lambda_2|$. Hereinafter, this inequality is assumed to hold. For such a matrix and a vector $v = (v_1, v_2)$ we have

$$\left|\frac{(Av,v)}{(v,v)}\right| = \left|\frac{\lambda_1|v_1|^2 + \alpha v_2\overline{v}_1 + \lambda_2|v_2|^2}{|v_1|^2 + |v_2|^2}\right|.$$

Choose v so that the argument of the product $\alpha v_2\overline{v}_1$ coincides with the argument of the eigenvalue λ_1 with the largest module. For such vectors

$$\left|\frac{\lambda_1|v_1|^2 + \alpha v_2\overline{v}_1 + \lambda_2|v_2|^2}{|v_1|^2 + |v_2|^2}\right| \geq \frac{|\lambda_1|\,|v_1|^2 + |\alpha|\,|v_2|\,|v_1| - |\lambda_2|\,|v_2|^2}{|v_1|^2 + |v_2|^2}.$$

The largest value of the ratio

$$\frac{|\lambda_1|\xi^2 + |\alpha|\xi\eta - |\lambda_2|\eta^2}{\xi^2 + \eta^2},$$

coincides with the maximal eigenvalue

$$\mu = \frac{1}{2}(|\lambda_1| - |\lambda_2| + \sqrt{(|\lambda_1| + |\lambda_2|)^2 + |\alpha|^2})$$

of the symmetric matrix

$$S = \begin{bmatrix} |\lambda_1| & \frac{1}{2}|\alpha| \\ \frac{1}{2}|\alpha| & -|\lambda_2| \end{bmatrix}.$$

To μ we associate the eigenvector of the matrix S with the components

$$\xi_0 = |\alpha| \geq 0, \quad \eta_0 = \sqrt{(|\lambda_1| + |\lambda_2|)^2 + |\alpha|^2} - |\lambda_1| - |\lambda_2|.$$

Choosing v_1 and v_2 such that $|v_1| = \xi_0$, $|v_2| = \eta_0$ and the arguments of λ_1 and $\alpha v_2\overline{v}_1$ coincide, we find a vector v such that

$$\frac{|\lambda_1|\,|v_1|^2 + |\alpha|\,|v_2|\,|v_1| - |\lambda_2|\,|v_2|^2}{|v_1|^2 + |v_2|^2} = \frac{|\lambda_1| - |\lambda_2| + \sqrt{(|\lambda_1| + |\lambda_2|)^2 + |\alpha|^2}}{2}.$$

Consequently, for v we have

$$\begin{aligned}\left|\frac{(Av,v)}{(v,v)}\right| &\geq \frac{|\lambda_1|\,|v_1|^2+|\alpha|v_2\overline{v}_1-|\lambda_2|\,|v_2|^2}{|v_1|^2+|v_2|^2}\\ &= \frac{|\lambda_1|-|\lambda_2|+\sqrt{(|\lambda_1|+|\lambda_2|)^2+|\alpha|^2}}{2}\\ &\geq \frac{1}{2}\sqrt{|\lambda_1|^2+|\lambda_2|^2+|\alpha|^2}=\frac{1}{2}\sqrt{\sigma_2^2(A)+\sigma_1^2(A)}\geq\frac{\sigma_2(A)}{2}=\frac{\|A\|}{2}.\end{aligned}$$

We used the fact that for an arbitrary matrix the sum of squares of modules of elements coincides with the sum of squares of singular values, i.e.,

$$\sum_{i.j=1}^{N}|a_{ij}|^2=\operatorname{tr}A^*A=\sum_{j=1}^{N}\sigma_j^2(A).$$

For the matrix A we have

$$\sum_{i,j=1}^{2}|a_{ij}|^2=|\lambda_1|^2+|\lambda_2|^2+|\alpha|^2=\sigma_1^2(A)+\sigma_2^2(A).$$

□

Recall that the Hausdorff set of the matrix $\begin{bmatrix}0&1\\0&0\end{bmatrix}$ is the disk $|z|\leq 1/2$. Therefore, in the assertion that there exists a vector v such that

$$\left|\frac{(Av,v)}{(v,v)}\right|\geq\frac{1}{2}\|A\|,$$

the coefficient on the right-hand side cannot be larger than 1/2 if no additional condition is imposed on A.

11.3. Geometry of Hausdorff sets and invariant subspaces

The location of eigenvalues of a matrix with respect to the Hausdorff set. Connection with invariant subspaces. The case of multiple eigenvalues. An estimate for the angle between invariant subspaces.

As was mentioned in §11.1, the Hausdorff set of an $N\times N$-matrix A is convex and lies in the disk $|z|\leq\|A\|$. As was shown in §11.2, the Hausdorff set of a second order matrix cannot lie too "deeply" in the interior of this disk. At least one point z of this set is such that $|z|\geq\frac{1}{2}\|A\|$. We show that this assertion remains true for a matrix of an arbitrary order $N\geq 2$. (It is obvious that for $N=1$ the Hausdorff set consists of a single point $z=a_{11}$; moreover, $|z|=|a_{11}|=\|A\|\geq\|A\|/2$.)

Let the normalized singular vectors v and w ($\|v\|=\|w\|=1$) of a matrix A be such that $Av=\sigma_N w$ ($\sigma_N=\|A\|$). Let a two-dimensional subspace H_2 contain these vectors. The orthogonal projection onto H_2 is denoted by Π ($\Pi^2=\Pi$, $\Pi^*=\Pi$, $\operatorname{tr}\Pi=2$, $\Pi H_2=H_2$). It is obvious that $\Pi v=v$ and $\Pi w=w$. Therefore,

$$\Pi^*A\Pi v=\Pi Av=\Pi\sigma_N\Pi w=\sigma_N\Pi w=\sigma_N w,$$
$$\|\Pi^*A\Pi v\|=\sigma_N\|w\|=\sigma_N=\sigma_N\|v\|.$$

Consequently, $\|\Pi^*A\Pi\|\geq\sigma_N$. In fact, $\|\Pi^*A\Pi\|=\sigma_N=\|A\|$ because $\|\Pi^*A\Pi\|\leq\|A\|\,\|\Pi\|^2=\|A\|$ (as was shown in §8.1, the norm of a nonzero orthogonal projection is equal to 1, hence $\|\Pi\|=1$).

It is obvious that the Hausdorff set of A contains all $z = \dfrac{(Au,u)}{(u,u)}$, where u runs over H_2; moreover,

$$\frac{(Au,u)}{(u,u)}\bigg|_{u\neq 0, u\in H_2} = \frac{(A\Pi u,\Pi u)}{(\Pi u,\Pi u)}\bigg|_{\Pi u\neq 0, u=\Pi u} = \frac{(\Pi^* A\Pi u,u)}{(u,u)}\bigg|_{u=\Pi u\neq 0},$$

i.e., these values fill the Hausdorff set of the operator $\Pi^* A\Pi$, which can be regarded as an operator acting in the two-dimensional space H_2. Therefore, there is $z = \dfrac{(Au,u)}{(u,u)}$ such that

$$|z| \geq \frac{1}{2}\|\Pi^* A\Pi\| = \frac{1}{2}\|A\|.$$

The above arguments admit the following obvious generalization. If U is an $N\times k$-matrix with orthonormal columns, then the Hausdorff set $W(U^*AU)$ of the matrix U^*AU is contained in the Hausdorff set $W(A)$ of the matrix A. As was mentioned earlier (see §11.1), the Hausdorff set of a matrix A contains all eigenvalues $\lambda_j(A)$ of A and, consequently, their convex hull.

Suppose that an eigenvalue $\lambda_0(A)$ of multiplicity k lies on the boundary of the Hausdorff set. Without loss of generality, we can assume that the matrix A is triangular, i.e.,

$$A = \begin{bmatrix} \lambda_0 & a_{12} & \dots & a_{1k} & a_{1\,k+1} & \dots & a_{1N} \\ & \lambda_0 & \dots & a_{2k} & a_{2\,k+1} & \dots & a_{2N} \\ & & \ddots & \dots & \dots & & \vdots \\ & & & \lambda_0 & a_{k\,k+1} & \dots & a_{kN} \\ & & & & \lambda_{k+1} & \dots & a_{k+1,N} \\ & \mathbf{0} & & & & \ddots & \vdots \\ & & & & & & \lambda_N \end{bmatrix}.$$

Consider the Rayleigh quotient $\dfrac{(Au,u)}{(u,u)}$ for vectors u with only two nonzero components: $u = (0,\dots,0,u_i,0,\dots,0,u_j,0\dots,0)^T$, $1\leq i\leq k$, $i\leq j\leq N$. Furthermore,

$$\frac{(Au,u)}{(u,u)} = \frac{\left(\begin{bmatrix}\lambda_0 & a_{ij}\\ 0 & \lambda_j\end{bmatrix}\begin{pmatrix}u_i\\u_j\end{pmatrix},\begin{pmatrix}u_i\\u_j\end{pmatrix}\right)}{|u_i|^2+|u_j|^2},$$

where as usual $\lambda_i = \lambda_0$ and, in addition, $\lambda_j = \lambda_0$ for $j\leq k$. It is obvious that all possible values of the Rayleigh quotient form a set that coincides with the Hausdorff set of the second order matrix $\begin{bmatrix}\lambda_i & a_{ij}\\ 0 & \lambda_j\end{bmatrix}$.

As was proved in §11.2, for $a_{ij}\neq 0$ the Hausdorff set of this second order matrix, being an ellipse, does not degenerate into a segment (for $\lambda_0 = \lambda_j$ and $a_{ij}\neq 0$ the Hausdorff set is a disk of nonzero radius). It is obvious that this set is a part of the Hausdorff set of the original $N\times N$-matrix A, i.e., for $a_{ij}\neq 0$ the foci λ_i and λ_j are interior points of the Hausdorff set.

The above arguments show that an eigenvalue $\lambda_j(A)$ lies on the boundary of the Hausdorff set of a matrix A only if the elements a_{ij}, $j = i+1,\dots,N$, of the ith row of the matrix $\widehat{A}$ vanish. In particular, the assumption that an eigenvalue

$\lambda_0 = \lambda_0(A)$ of multiplicity k lies on the boundary of the Hausdorff set leads to the representation of A in the form

$$A = U \left[\begin{array}{cc} \boxed{\begin{matrix} \lambda_0 & & & 0 \\ & \lambda_0 & & \\ & & \ddots & \\ 0 & & & \lambda_0 \end{matrix}} & 0 \\ 0 & \boxed{\begin{matrix} \lambda_{k+1} & \cdots & a_{k+1\,N} \\ & \ddots & \vdots \\ 0 & & \lambda_N \end{matrix}} \end{array}\right] U^*, \tag{11.3.1}$$

where U is a unitary matrix that reduces A to an upper triangular matrix. In view of this representation, in the k-dimensional invariant subspace corresponding to the eigenvalue λ_0 we can choose an orthonormal basis consisting of eigenvectors; moreover, the orthogonal complement to this subspace is also an invariant subspace corresponding to the eigenvalues $\lambda_j(A)$ different from $\lambda_0(A)$.

If there are two orthogonal complementary invariant subspaces of A, then the following representation holds:

$$A = U \begin{bmatrix} A_1 & 0 \\ 0 & A_2 \end{bmatrix} U^*;$$

moreover, the Hausdorff set of A coincides with the Hausdorff set of the block diagonal matrix

$$\begin{bmatrix} A_1 & 0 \\ 0 & A_2 \end{bmatrix},$$

i.e., it consists of the values of the ratio

$$\frac{(A_1 u_1, u_1) + (A_2 u_2, u_2)}{\|u_1\|^2 + \|u_2\|^2},$$

where u_1 and u_2 are such that $\|u_1\|^2 + \|u_2\|^2 \neq 0$. In particular, under the assumption that $u_1 \neq 0$, $u_2 = 0$, we prove that the Hausdorff set $W(A_1)$ of the block A_1 is contained in the Hausdorff set $W(A)$ of the matrix A. Similarly, $W(A_2) \subset W(A)$. Since $W(A)$ is convex, it contains the convex hull of $W(A_1)$ and $W(A_2)$. In fact, $W(A)$ coincides with the convex hull of $W(A_1)$ and $W(A_2)$. Indeed, let

$$\frac{(Au, u)}{(u, u)} = \frac{(A_1 u_1, u_1) + (A_2 u_2, u_2)}{\|u_1\|^2 + \|u_2\|^2} = z,$$

where $\|u_1\| \neq 0$ and $\|u_2\| \neq 0$. It is obvious that

$$z = \left(1 - \frac{\|u_2\|^2}{\|u_1\|^2 + \|u_2\|^2}\right) z_1 + \frac{\|u_2\|^2}{\|u_1\|^2 + \|u_2\|^2} z_2,$$

where

$$z_1 = \frac{(Au_1, u_1)}{(u_1, u_1)}, \quad z_2 = \frac{(Au_2, u_2)}{(u_2, u_2)},$$

i.e., z belongs to the convex hull of $W(A_1)$ and $W(A_2)$. If $\|u_1\| = 0$, $\|u_2\| \neq 0$ or $\|u_2\| = 0$, $\|u_1\| \neq 0$, then z lies in the Hausdorff set $W(A_1)$ of the block A_1 or in the Hausdorff set $W(A_2)$ of the block A_2. Then z also belongs to the convex hull of $W(A_1)$ and $W(A_2)$. In particular, if an eigenvalue $\lambda_0(A)$ of multiplicity k of the matrix (11.3.1) lies on the boundary of the Hausdorff set $W(A)$, then $W(A)$ coincides with the convex hull of the point λ_0 and the Hausdorff set $W(A_2)$ of the block

$$A_2 = \begin{bmatrix} \lambda_{k+1} & a_{k+1\,k+1} & \dots & a_{k+1\,N} \\ & \lambda_{k+2} & & \vdots \\ & & \ddots & \\ \mathbf{0} & & & \lambda_N \end{bmatrix},$$

which can be associated with the action of A in the invariant subspace corresponding to the eigenvalues $\lambda_j(A)$ different from λ_0. The value λ_0 either belongs to this set (e.g., it lies on the boundary) or does not belong to it. If λ_0 does not belong to this set, then $W(A)$ is the union of all segments joining λ_0 and points of $W(A_2)$.

If the boundary of the Hausdorff set $W(A)$ of a matrix A contains two different eigenvalues λ_0' and λ_0'' ($\lambda_0' \neq \lambda_0''$), then the invariant subspaces corresponding to λ_0' and λ_0'' are orthogonal. This follows from the above arguments. Moreover, in the invariant subspace R_0 corresponding to an eigenvalue λ_0 lying on the boundary of the Hausdorff set $W(A)$, the action of A is reduced to the multiplication by λ_0. It is obvious that the set $z = \dfrac{(Au,u)}{(u,u)}$, where u runs over R_0, is a single point $z = \lambda_0$. In other words, if λ_0 lies on the boundary of the Hausdorff set of A, then the Hausdorff set of the restriction of a matrix A to R_0 is a point even in the case of a multiple eigenvalue λ_0. This assertion has useful generalizations.

We replace the assumption that an eigenvalue of multiplicity k lies on the boundary of the Hausdorff set $W(A)$ with the assumption that several eigenvalues $\lambda_1(A), \lambda_2(A), \dots, \lambda_k(A)$ lie near a boundary point λ_0 (here λ_0 does not necessarily belong to the spectrum of A). We characterize the closeness between the group of eigenvalues and the boundary point λ_0 by the inequality $|\lambda_j(A) - \lambda_0| \leq \nu$. Denote by A_0 the restriction of A to the invariant subspace corresponding to the eigenvalues $\lambda_1(A), \lambda_2(A), \dots, \lambda_k(A)$. In an orthonormal basis for this subspace A_0 can be written as a triangular matrix:

$$A_0 = \begin{bmatrix} \lambda_1 & a_{12} & \dots & a_{1k} \\ & \lambda_2 & \dots & a_{2k} \\ & & \ddots & \vdots \\ \mathbf{0} & & & \lambda_k \end{bmatrix} = \lambda_0 I + \begin{bmatrix} \lambda_1 - \lambda_0 & a_{12} & \dots & a_{1k} \\ & \lambda_2 - \lambda_0 & \dots & a_{2k} \\ & & \ddots & \vdots \\ & \mathbf{0} & & \lambda_k - \lambda_0 \end{bmatrix}.$$

The point $z_0 = \lambda_0$ lies on the boundary of $W(A)$. Since $W(A_0) \subset W(A)$, z_0 lies either on the boundary of $W(A_0)$ or outside $W(A_0)$. Therefore, there is a line passing through z_0 such that the points of $W(A_0)$ are located to one side of this line and, possibly, on this line. Let the line be described by the equation $\operatorname{Re}(e^{i\varphi}(z - \lambda_0)) = 0$. For $z \in W(A_0)$ we have $\operatorname{Re}(e^{i\varphi}(z - \lambda_0)) \leq 0$ or $\operatorname{Re}(e^{i\varphi}(z - \lambda_0)) \geq 0$. The second case is reduced to the first case by replacing φ with $\varphi + \pi$. Therefore, we can consider only the case where no point of the

Hausdorff set of the matrix

$$B = e^{i\varphi}(A_0 - \lambda_0 I) = \begin{bmatrix} e^{i\varphi}(\lambda_1 - \lambda_0) & e^{i\varphi}a_{12} & \dots & e^{i\varphi}a_{1k} \\ & e^{i\varphi}(\lambda_2 - \lambda_0) & \dots & e^{i\varphi}a_{2k} \\ & & \ddots & \vdots \\ \mathbf{0} & & & e^{i\varphi}(\lambda_k - \lambda_0) \end{bmatrix}$$

is contained in the right half-plane. Points of the Hausdorff set of any block of the form

$$D_{(p,q)} = \begin{bmatrix} e^{i\varphi}(\lambda_p - \lambda_0) & e^{i\varphi}a_{pq} \\ 0 & e^{i\varphi}(\lambda_q - \lambda_0) \end{bmatrix}$$

lie to the left of the imaginary axis or on it. This happens if the Hermitian matrices

$$\begin{aligned} B_{(p,q)} &= \frac{1}{2}[D_{(p,q)} + D^*_{(p,q)}] \\ &= \begin{bmatrix} \dfrac{e^{i\varphi}(\lambda_p - \lambda_0) + e^{-i\varphi}(\overline{\lambda}_p - \overline{\lambda}_0)}{2} & \dfrac{1}{2}e^{i\varphi}a_{pq} \\ \dfrac{1}{2}e^{-i\varphi}\overline{a}_{pq} & \dfrac{e^{i\varphi}(\lambda_q - \lambda_0) + e^{-i\varphi}(\overline{\lambda}_q - \overline{\lambda}_0)}{2} \end{bmatrix} \end{aligned}$$

have no positive eigenvalues, which is possible only if $\det B_{(p,q)} \geq 0$. Since

$$\begin{aligned} \det B_{(p,q)} &= \left[\operatorname{Re} e^{i\varphi}(\lambda_p - \lambda_0)\right]\left[\operatorname{Re} e^{i\varphi}(\lambda_q - \lambda_0)\right] - \frac{1}{4}|a_{pq}|^2 \\ &\leq |\lambda_p - \lambda_0|\,|\lambda_q - \lambda_0| - \frac{1}{4}|a_{pq}|^2 \leq \nu^2 - \frac{1}{4}|a_{pq}|^2, \end{aligned}$$

from our assumptions it follows that $|a_{pq}| \leq 2\nu$,

$$\begin{aligned} \|A_0 - \lambda_0 I\| = \|B\| &\leq \sqrt{\sum_{j=1}^{k} |\lambda_j - \lambda_0|^2 + \sum_{1 \leq p < q \leq k} |a_{pq}|^2} \\ &\leq \sqrt{k\nu^2 + \frac{k(k-1)}{2}(2\nu)^2} \leq \sqrt{2}k\nu. \end{aligned} \tag{11.3.2}$$

By (11.3.2), the Hausdorff set $W(A_0)$ of the restriction A_0 of A to the invariant subspace lies in the disk of radius $\sqrt{2}k\nu$ and center λ_0. It is obvious that the diameter of $W(A_0)$ does not exceed $2\sqrt{2}k\nu$. Thus, if certain k eigenvalues of A are close to a boundary point λ_0 of the Hausdorff set $W(A)$ (more exactly, lie in the ν-neighborhood of λ_0) and A_0 is the restriction of A to the k-dimensional invariant subspace of A corresponding to these k eigenvalues, then the Hausdorff set $W(A_0)$ of A_0 is a spot of diameter at most $2\sqrt{2}k\nu$.

PROBLEM 11.3.1. For the matrix

$$A = \begin{bmatrix} 0 & 2\nu & 2\nu & \dots & 2\nu \\ & 0 & 2\nu & \dots & 2\nu \\ & & \ddots & & \vdots \\ & & & 0 & 2\nu \\ & & & & 0 \end{bmatrix}$$

of order k, prove that the ratio of the radius of the circle centered at $\lambda_0 = 0$ which circumscribes $W(A)$ to the radius of the inscribed circle centered at the same point

$\lambda_0 = 0$ is at least $k-1$. Thereby, the dependence on the dimension k in the estimate for the diameter $2\sqrt{2}k\nu$ of the Hausdorff set $W(A)$ cannot be removed.

Hint. Find the eigenvalues of the symmetric matrix $B = \frac{1}{2}(A + A^*)$. The eigenvector with equal components corresponds to the eigenvalue $(k-1)\nu$. The matrix $B + \nu I$ is nonsingular.

Let $\lambda_1 = \lambda_1(A)$ and $\lambda_2 = \lambda_2(A)$ be different eigenvalues of a matrix A so that $|\lambda_1(A) - \lambda_2(A)| > 0$. Let z run over the boundary of the Hausdorff set $W(A)$. Find the length $L = \min\limits_{z \in \partial W(A)} [|\lambda_1 - z| + |z - \lambda_2|]$ of the shortest path between $\lambda_1(A)$ and $\lambda_2(A)$ passing through a point of $\partial W(A)$. The inequality $|\lambda_1 - \lambda_2|/L \leq 1$ can be used to estimate the angle φ between the eigenvectors u_1 and u_2 corresponding to the eigenvalues λ_1 and λ_2. Namely, the following inequality holds:

$$\sin\varphi \geq |\lambda_1 - \lambda_2|/L. \tag{11.3.3}$$

For a square second order matrix A this assertion was proved in §11.2. For a matrix of arbitrary order it suffices to consider the Rayleigh quotient $\dfrac{(Au, u)}{(u, u)}$ for vectors u of the two-dimensional invariant subspace spanned by the eigenvectors u_1 and u_2. As was shown above, the values of the Rayleigh quotient $\dfrac{(Au, u)}{(u, u)}$ form an ellipse with the foci λ_1 and λ_2 and the sum of the focal radii L'. The angle φ between u_1 and u_2 can be found from the equality $\sin\varphi = |\lambda_1 - \lambda_2|/L'$. The ellipse lies inside the Hausdorff set $W(A)$. Therefore, $L \geq L'$, which proves (11.3.3).

Let R_1 and R_2 be invariant subspaces of a matrix A such that their intersection contains only the zero vector. Denote by A_1 and A_2 the restrictions of A to the subspaces R_1 and R_2. Let $W(A)$, $W(A_1)$, and $W(A_2)$ be the Hausdorff sets of the matrices A, A_1, and A_2, respectively. Assume that the intersection of $W(A_1)$ and $W(A_2)$ is empty. If the distance between $W(A_1)$ and $W(A_2)$ is known and we know how "deeply" the sets $W(A_1)$ and $W(A_2)$ lie inside the set $W(A)$, then we can estimate the minimal angle between a vector of R_1 and a vector of R_2. Such an estimate is a natural generalization of the above estimate for the angle between the eigenvectors.

Let $v^{(1)}, v^{(2)}, \dots, v^{(m)}$ be an orthonormal basis for R_1 and $w^{(1)}, w^{(2)}, \dots, w^{(n)}$ an orthonormal basis for R_2. The matrices $V = [v^{(1)}, v^{(2)}, \dots, v^{(m)}]$ and $W = [w^{(1)}, w^{(2)}, \dots, w^{(n)}]$ whose columns are the basis vectors $v^{(i)}$ and $w^{(i)}$, satisfy the equalities $V^*V = I_m$ and $W^*W = I_n$ because the bases are orthonormal. The angle φ $(0 \leq \varphi \leq \pi/2)$ between arbitrary vectors

$$v = \sum_{i=1}^{m} \xi_i v^{(i)} \in R_1, \quad w = \sum_{j=1}^{n} \eta_j w^{(j)}$$

can be found from the equality

$$\cos\varphi = \left|\frac{(v, w)}{\|v\|\,\|w\|}\right| = \frac{\left|\sum\limits_{i,j} \xi_i \bar{\eta}_j (v^{(i)}, w^{(j)})\right|}{\sqrt{\left(\sum\limits_{i=1}^{m} |\xi_i|^2\right)\left(\sum\limits_{j=1}^{n} |\eta_j|^2\right)}} = \frac{\left|\sum\limits_{i,j} F_{ij} \xi_i \eta_j\right|}{\sqrt{\left(\sum\limits_{i=1}^{m} |\xi_i|^2\right)\left(\sum\limits_{j=1}^{n} |\eta_j|^2\right)}},$$

where F_{ij} are elements of the rectangular matrix F composed of the inner products $(v^{(i)}, w^{(j)})$ of the basis vectors $v^{(i)}$ and $w^{(j)}$ ($F = W^*V$, i.e., $F_{ij} = (v^{(i)}, w^{(j)})$).

As was shown in Chapter 3, it is possible to choose a unitary $m \times m$-matrix P and an $n \times n$-matrix Q ($P^*P = I_m$, $Q^*Q = I_n$) such that after the congruent transformation performed with the help of these matrices, the transformed matrix F becomes "diagonal," i.e., ($c_i \geq 0$):

$$C = Q^*FP = \begin{bmatrix} c_1 & & & \text{\Large 0} \\ & c_2 & & \\ & & \ddots & \\ \text{\Large 0} & & & c_n \end{bmatrix}$$

if $n = m$;

$$C = Q^*FP = \begin{bmatrix} c_1 & & \text{\Large 0} & & 0 \\ & c_2 & & & 0 \\ & & \ddots & & \vdots \\ \text{\Large 0} & & & c_n & 0 \end{bmatrix}$$

if $n < m$;

$$C = Q^*FP = \begin{bmatrix} c_1 & & & \text{\Large 0} \\ & c_2 & & \\ & & \ddots & \\ \text{\Large 0} & & & c_m \\ 0 & 0 & \dots & 0 \end{bmatrix}$$

if $n > m$. The transformations P and Q realize the passage to new orthonormal bases for R_1 and R_2. The new basis vectors form columns of the matrices $\widetilde{V} = VP$ and $\widetilde{W} = WQ$. Moreover, the fact that the matrix C with elements $c_{ij} = (\widetilde{v}^{(i)}, \widetilde{w}^{(j)})$ is "diagonal" means that every basis vector $\widetilde{v}^{(i)}$ in R_1 is orthogonal to all basis vectors $\widetilde{w}^{(j)}$, $j \neq i$, in R_2. The diagonal elements are expressed in terms of the inner products of the basis vectors: $c_i \equiv c_{ii} = (\widetilde{v}^{(i)}, \widetilde{w}^{(i)})$. Moreover, $c_i = |c_i| = |(\widetilde{v}^{(i)}, \widetilde{w}^{(i)})| \leq \|\widetilde{v}^{(i)}\| \, \|\widetilde{w}^{(i)}\| = 1$. Hence c_i are equal to the cosines of the angles between the basis vectors $\widetilde{v}^i$ in R_1 and the basis vectors $\widetilde{w}^i$ in R_2. The angle between two arbitrary vectors

$$\sum_{i=1}^{m} \alpha_i \widetilde{v}^{(i)} \in R_1, \quad \sum_{j=1}^{n} \beta_j \widetilde{w}^{(j)} \in R_2$$

can be found from the equality

$$\cos\varphi = \frac{\left| \sum\limits_{i=1}^{\min\{n,m\}} c_i \alpha_i \overline{\beta}_i \right|}{\sqrt{\left(\sum\limits_{i=1}^{m} |\alpha_i|^2\right) \left(\sum\limits_{j=1}^{n} |\beta|^2\right)}}$$

and, depending on α_i and β_j, can take any values between $\arccos[\max_i c_i]$ and $\arccos[\min_i c_i]$ if $m = n$ and between $\arccos[\max_i c_i]$ and $\pi/2$ if $m \neq n$. It is natural that the minimal angle $\arccos[\max_i c_i]$ is referred to as the *angle between the subspaces* R_1 *and* R_2. Hence the angle between any vector of R_1 and any vector of R_2 is not less than the angle between the subspaces R_1 and R_2.

Assume that $c_1 = \max_i c_i$ (otherwise, we renumber the basis vectors). Consider the Rayleigh quotient $\dfrac{(Au, u)}{(u, u)}$, where u runs over the two-dimensional subspace H_2 spanned by the basis vectors $\widetilde{v}^{(1)} \in R_1$ and $\widetilde{w}^{(1)} \in R_2$ such that the angle between $\widetilde{v}^{(1)}$ and $\widetilde{w}^{(1)}$ is minimal, i.e., is equal to the angle between the spaces R_1 and R_2. In other words, we consider vectors $u = \gamma\widetilde{v}^{(1)} + \delta\widetilde{w}^{(1)}$. Since R_1 and R_2 are invariant subspaces, we have

$$A\widetilde{v}^{(1)} = \sum_{i=1}^{m} \mu_i \widetilde{v}^{(i)}, \quad A\widetilde{w}^{(1)} = \sum_{j=1}^{n} \nu_j \widetilde{w}^{(j)}.$$

Using the relations $(\widetilde{v}^{(1)}, \widetilde{w}^{(j)}) = 0$, $j \geq 2$, and $(\widetilde{w}^{(1)}, \widetilde{v}^{(i)}) = 0$, $i \geq 2$, we find

$$\begin{aligned}(A\widetilde{v}^{(1)}, \widetilde{v}^{(1)}) = \mu_1, \quad (A\widetilde{w}^{(1)}, \widetilde{v}^{(1)}) = \nu_1 c_1,\\ (A\widetilde{v}^{(1)}, \widetilde{w}^{(1)}) = \mu_1 c_1, \quad (A\widetilde{w}^{(1)}, \widetilde{w}^{(1)}) = \nu_1.\end{aligned}$$

In other words, the set of the inner products (Au, u) for $u \in H_2$ does not change if we replace the operator A with the operator $\widehat{A}$ from H_2 to H_2 such that $\widehat{A}\widetilde{v}^{(1)} = \mu_1\widetilde{v}^{(1)}$ and $\widehat{A}\widetilde{w}^{(1)} = \nu_1\widetilde{w}^{(1)}$, i.e., $\widetilde{v}^{(1)}$ and $\widetilde{w}^{(1)}$ are eigenvectors of $\widehat{A}$ and

$$\frac{(Au, u)}{(u, u)} = \frac{(\widehat{A}u, u)}{(u, u)}$$

for $u \in H_2$. The Hausdorff sets of A and $\widehat{A}$ on H_2 coincide. We recall that the Hausdorff set $W(\widehat{A})$ of $\widehat{A}$ is an ellipse with foci at μ_1 and ν_1, and the angle φ between $\widetilde{v}^{(1)}$ and $\widetilde{w}^{(1)}$ can be found from the equality $\sin\varphi = |\mu_1 - \nu_1|/\widehat{L}$, where $\widehat{L}$ is the sum of the focal radii. Since $W(\widehat{A}) \subset W(A)$, the quantity $\widehat{L}$ does not exceed the smallest sum L of the distances from a boundary point of $W(A)$ to the foci μ_1 and ν_1. Therefore, $\sin\varphi \geq |\mu_1 - \nu_1|/L$. The points μ_1 and ν_1 lie in the Hausdorff sets $W(A_1)$ and $W(A_2)$ of the restrictions A_1 and A_2 of A to the subspaces R_1 and R_2, respectively. If the intersection of the Hausdorff sets $W(A_1)$ and $W(A_2)$ is empty and l is the distance between them, then $|\mu_1 - \nu_1| \geq l$. Consequently, $\sin\varphi \geq l/L$, which provides the required estimate for the angle between the invariant subspaces R_1 and R_2 in terms of the distance l between the Hausdorff sets $W(A_1)$ and $W(A_2)$ and the quantity L which characterizes how deep $W(A_1)$ and $W(A_2)$ are embedded into $W(A)$.

To conclude the section, we give an estimate for the angle between invariant subspaces R_1 and R_2 of a matrix A under the condition that the corresponding eigenvalues are grouped in clusters near two different boundary points of the Hausdorff set $W(A)$. Let R_1 correspond to the eigenvalues $\lambda_1(A), \lambda_2(A), \dots, \lambda_k(A)$ such that

$$|\lambda_j(A) - \lambda_0'| \leq \varepsilon', \quad \lambda_0' \in \partial W(A), \quad 1 \leq j \leq k,$$

and R_2 correspond to the eigenvalues $\lambda_{k+1}(A), \lambda_{k+2}(A), \dots, \lambda_{k+m}(A)$ such that

$$|\lambda_j(A) - \lambda_0''| \le \varepsilon'', \quad \lambda_0'' \in \partial W(A), \qquad k+1 \le j \le k+m.$$

As was shown above, the Hausdorff sets $W(A_1)$ and $W(A_2)$ of the restrictions A_1 and A_2 of the matrix A to the subspaces R_1 and R_2 lie in the disks of radii $\sqrt{2}k\varepsilon'$ and $\sqrt{2}m\varepsilon''$ centered at λ_0' and λ_0'', respectively. Therefore, the distance l between $W(A_1)$ and $W(A_2)$ is at least $|\lambda_0' - \lambda_0''| - \sqrt{2}k\varepsilon' - \sqrt{2}m\varepsilon''$, and the shortest path between $W(A_1)$ and $W(A_2)$ passing through a point of the boundary of $W(A)$ is shorter than $|\lambda_0' - \lambda_0''| + \sqrt{2}k\varepsilon' + \sqrt{2}m\varepsilon''$.

Assuming that $k\varepsilon' + m\varepsilon'' < |\lambda_0' - \lambda_0''|/\sqrt{2}$, for the angle φ between R_1 and R_2 we obtain the estimate

$$\sin\varphi \ge \frac{1 - \dfrac{1}{\sqrt{2}}\dfrac{(k\varepsilon' + m\varepsilon'')}{|\lambda_0' - \lambda_0''|}}{1 + \dfrac{1}{\sqrt{2}}\dfrac{(k\varepsilon' + m\varepsilon'')}{|\lambda_0' - \lambda_0''|}}.$$

It is clear that if ε' and ε'' are small compared to $|\lambda_0' - \lambda_0''|$, then the subspaces R_1 and R_2 are almost orthogonal.

11.4. Estimates for the resolvent and matrix exponential

An estimate for the resolvent in the case where λ does not belong to the Hausdorff set. Dependence on the location of the Hausdorff set in the complex plane. Sectorial operators and the Hausdorff sets.

Let a point λ in the complex plane not belong to the Hausdorff set $W(A)$ of a matrix A. Denote by $d = d(\lambda, W(A))$ the distance between λ and $W(A)$, i.e.,

$$d = \min_{x \ne 0} \left| \lambda - \frac{(Ax, x)}{(x, x)} \right|. \tag{11.4.1}$$

From (11.4.1) it follows that for any x ($\|x\| > 0$) we have

$$\left| \lambda - \frac{(Ax, x)}{(x, x)} \right| = \left| \frac{((\lambda I - A)x, x)}{\|x\|^2} \right| \ge d,$$
$$|((\lambda I - A)x, x)| \ge d\|x\|^2.$$

Consequently,

$$\|(\lambda I - A)x\| \, \|x\| \ge |((\lambda I - A)x, x)| \ge d\|x\|^2,$$
$$\|x\| \le \frac{\|(\lambda I - A)x\|}{d}.$$

Since λ lies outside $W(A)$ and, consequently, cannot coincide with any root of the characteristic polynomial $\det(\lambda I - A) = 0$, the matrix $\lambda I - A$ is invertible. Therefore, for any $z \ne 0$ we can construct a unique vector $x = (\lambda I - A)^{-1}z \ne 0$ such that

$$\|(\lambda I - A)^{-1}z\| \le \frac{\|z\|}{d}. \tag{11.4.2}$$

Since z is arbitrary, the following theorem holds.

THEOREM 11.4.1. *If a point λ in the complex plane lies outside the Hausdorff set $W(A)$ of a matrix A and the distance between λ and $W(A)$ is $d(\lambda, W(A))$, then*

$$\|(\lambda I - A)^{-1}\| \leq \frac{1}{d(\lambda, W(A))}.$$

From Theorem 11.4.1 we obtain the following corollaries.

COROLLARY 11.4.1. *If $|\lambda| > \|A\|$, then*

$$\|(\lambda I - A)^{-1}\| \leq \frac{1}{|\lambda| - \|A\|}.$$

COROLLARY 11.4.2. *If the Hausdorff set $W(A)$ of a matrix A lies in the left half-plane* ($\operatorname{Re}(Ax, x) \leq -\omega\|x\|^2$, $\omega \geq 0$) *and λ lies in the right half-plane* ($\operatorname{Re}\lambda > 0$)*, then the resolvent $(\lambda I - A)^{-1}$ satisfies the following estimate:*

$$\|(\lambda I - A)^{-1}\| \leq \frac{1}{\omega + \operatorname{Re}\lambda}.$$

It is obvious that the Hausdorff sets of the matrices A and A^* consisting of the values of quotients

$$\frac{(Au, u)}{(u, u)}, \quad \frac{(A^*u, u)}{(u, u)} = \frac{(u, Au)}{(u, u)} = \overline{\left[\frac{(Au, u)}{(u, u)}\right]}$$

can be obtained from one another by a symmetric reflection with respect to the real axis.

The Hausdorff set of the Hermitian matrix $\frac{1}{2}(A + A^*)$ formed by the real numbers

$$\frac{1}{2}\left\{\frac{(Au, u)}{(u, u)} + \overline{\left[\frac{(A^*u, u)}{(u, u)}\right]}\right\} = \operatorname{Re}\left\{\frac{(A^*u, u)}{(u, u)}\right\}$$

is the segment whose endpoints are the minimal and the maximal eigenvalues of this matrix. It is obvious that

$$\lambda_{\max}\left(\frac{1}{2}[A + A^*]\right) = \gamma = \max \operatorname{Re}\left\{\frac{(A^*u, u)}{(u, u)}\right\}. \tag{11.4.3}$$

Moreover, for any vector v

$$\left(\frac{1}{2}[A + A^*]v, v\right) \leq \gamma(v, v). \tag{11.4.4}$$

The matrix exponential e^{tA} (where t is real) is a matrix satisfying the matrix differential equation

$$\frac{d}{dt}e^{tA} = Ae^{tA} \tag{11.4.5}$$

(see §8.5), which allows us to compute and estimate the derivative of the quadratic form $(e^{tA}u, e^{tA}u) = \|e^{tA}u\|^2$:

$$\begin{aligned} \frac{d}{dt}(e^{tA}u, e^{tA}u) &= (Ae^{tA}u, e^{tA}u) + (e^{tA}u, Ae^{tA}u) \\ &= 2\left(\frac{1}{2}[A + A^*]e^{tA}u, e^{tA}u\right) \leq 2\gamma(e^{tA}u, e^{tA}u). \end{aligned} \tag{11.4.6}$$

In other words,

$$\begin{aligned}
&\frac{d}{dt}\|e^{tA}u\|^2 \le 2\gamma\|e^{tA}\|^2, \quad \frac{d}{dt}\|e^{tA}u\| \le \gamma\|e^{tA}u\|,\\
&\frac{d}{dt}[e^{-\gamma t}\|e^{tA}u\|] = e^{-\gamma t}\left[\frac{d}{dt}\|e^{tA}u\| - \gamma\|e^{tA}u\|\right] \le 0, \qquad (11.4.7)\\
&e^{-\gamma t}\|e^{tA}u\| \le e^{-\gamma\cdot 0}\|e^{0A}u\| = \|u\|.
\end{aligned}$$

The last inequality gives the following theorem.

THEOREM 11.4.2. *If the Hausdorff set $W(A)$ of a matrix A lies in the half-plane* $\operatorname{Re}\lambda \le \gamma$ $(\operatorname{Re}(Ax, x) \le \gamma\|x\|^2)$, *then* $\|e^{tA}\| \le e^{\gamma t}$ *for any real* γ.

We pass to the study of operators whose Hausdorff sets lie to the left of some line parallel to the imaginary axis and, in addition, are bounded from above and from below by rays that cut off an angle (sector) on the plane.

A linear operator given by an $N \times N$-matrix A is said to be *sectorial* if the Hausdorff set $W(A)$ lies in the half-plane $\operatorname{Re}\lambda \le \gamma$ and, at the same time, inside the angle formed by the rays

$$\cos\theta \operatorname{Re}(\lambda - \gamma) + \sin\theta \operatorname{Im}\lambda = 0,$$
$$\cos\theta \operatorname{Re}(\lambda - \gamma) - \sin\theta \operatorname{Im}\lambda = 0,$$

where $0 < \theta < \pi/2$.

In other words, an operator A is sectorial if there exist real numbers γ and θ $(0 < \theta < \pi/2)$ such that $\cos\theta \operatorname{Re}(\lambda - \gamma) + \sin\theta\,|\operatorname{Im}\lambda| \le 0$ for any point $\lambda \in W(A)$. By Theorem 11.4.2, a sectorial operator A satisfies the estimate $\|e^{tA}\| \le e^{\gamma t}$, which simplifies the study of the matrix exponential and the deduction of estimates from the representation of e^{tA} in the form of a contour integral.

We recall (see §8.5) that

$$e^{tA} = \frac{1}{2\pi i}\oint_\Gamma e^{\lambda t}(\lambda I - A)^{-1}\,d\lambda,$$

where the contour Γ surrounds all eigenvalues of A.

It is convenient to choose a contour as in Figure 11.4.1, where the shaded region represents the set $W(A)$ lying inside the sector bounded by the rays

$$(\alpha - \gamma)\cos\theta \pm \beta\sin\theta = 0 \quad (\alpha - \gamma < 0).$$

Let the contour consist of an arc

$$|\lambda - \gamma'| = \sqrt{(\alpha - \gamma')^2 + \beta^2} = R + \gamma', \quad R > \|A\|$$

and segments of length R lying on the rays that are parallel to the rays of the sector and start at the point $\alpha = \gamma' > \gamma$, $\beta = 0$. This contour (see Figure 11.4.1) surrounds the Hausdorff set $W(A)$ and, consequently, the spectrum of A. By Theorem 11.4.1, we have

$$\|e^{\lambda t}(\lambda I - A)^{-1}\| = e^{t\operatorname{Re}\lambda}\|(\lambda I - A)^{-1}\| \le \frac{e^{t\operatorname{Re}\lambda}}{(\gamma' - \gamma)\sin\theta}$$

along the lines that are at the distance $(\gamma' - \gamma)\sin\theta$ from the rays bounding the sector. On the arc we have

$$|e^{\lambda t}| = e^{t\operatorname{Re}\lambda} < e^{t(\gamma' - R\cos\theta)};$$

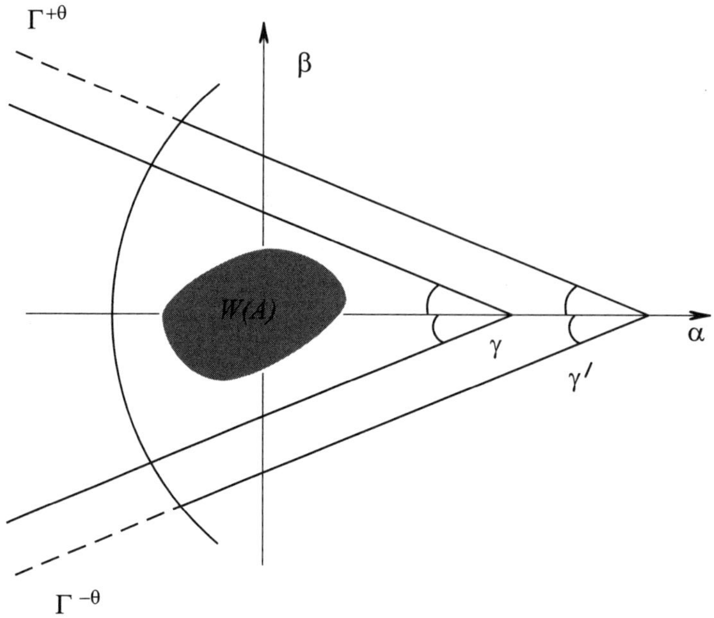

FIGURE 11.4.1

moreover,

$$\|(\lambda I - A)^{-1}\| \le \frac{1}{R - \|A\|}.$$

Therefore, on the arc Γ_R we have

$$\|e^{\lambda t}(\lambda I - A)^{-1}\| \le \frac{e^{-t(R\cos\theta - \gamma')}}{R - \|A\|}.$$

Consequently,

$$\left\| \frac{1}{2\pi i} \int\limits_{\Gamma_R} e^{\lambda t}(\lambda I - A)^{-1} d\lambda \right\| \le \frac{2R\theta}{2\pi} \frac{e^{-t(R\cos\theta - \gamma')}}{R - \|A\|}.$$

It is useful to consider the rays $\Gamma^{+\theta}$ and $\Gamma^{-\theta}$ that are extensions of the segments of the contour and estimate the integrals along them:

$$\begin{aligned}
&\left\| \frac{1}{2\pi i} \int\limits_{\Gamma^{\pm\theta}} e^{\lambda t}(\lambda I - A)^{-1} d\lambda \right\| \\
&\quad = \left\| \frac{1}{2\pi i} \int\limits_{-\infty}^{R} e^{-t[(\rho\cos\theta - \gamma') \pm i\rho\sin\theta]} \right. \\
&\qquad \left. \times \{[(\rho\cos\theta - \gamma') \pm i\rho\sin\theta]I - A\}^{-1}(\cos\theta \pm i\sin\theta)ds \right\| \\
&\quad \le \frac{1}{2\pi} \int\limits_{R}^{\infty} \frac{e^{-t(\rho\cos\theta - \gamma')}}{(\gamma' - \gamma)\sin\theta} d\rho = \frac{e^{-t(R\cos\theta - \gamma')}}{t\pi(\gamma' - \gamma)\sin 2\theta}.
\end{aligned}$$

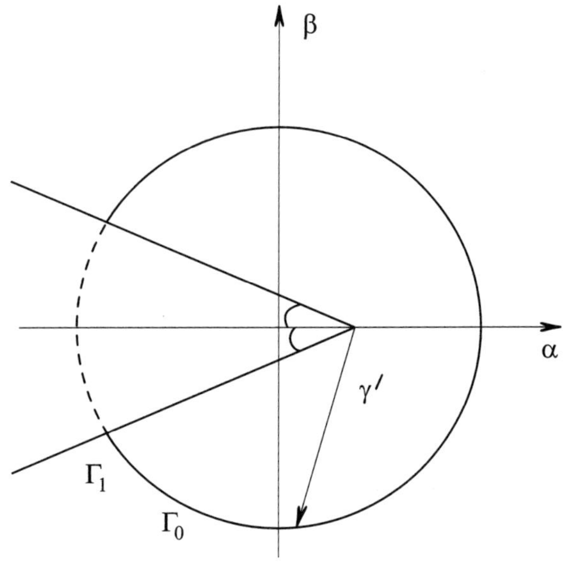

FIGURE 11.4.2

Since R was chosen arbitrarily, we have the integral representation

$$e^{tA} = \frac{1}{2\pi i} \int\limits_{\Gamma_>} e^{\lambda t}(\lambda I - A)^{-1} d\lambda, \quad t > 0,$$

where the open contour $\Gamma_>$ has infinite length and is bounded by two rays starting at the point $\lambda = \gamma'$ and having the angles $\pm\theta$ with the negative real half-axis.

If a circle $|\lambda - \gamma'| = R_0$ centered at the vertex of the sector $\Gamma_>$ does not pass through any point of the spectrum of A so that on this circle

$$\frac{\|(\lambda I - A)^{-1}\|}{|\lambda - \gamma'|} < M,$$

then, deforming the contour if necessary, we represent e^{tA} as the sum of two terms

$$e^{tA} = \frac{1}{2\pi i} \oint\limits_{\Gamma_0} e^{\lambda t}(\lambda I - A)^{-1} d\lambda + \frac{1}{2\pi i} \int\limits_{\Gamma_1} e^{\lambda t}(\lambda I - A)^{-1} d\lambda,$$

where Γ_0 is the circle $|\lambda - \gamma'| = R_0$ and Γ_1 consists of those parts of the rays forming $\Gamma_>$ that lie outside this circle and of the arc joining the endpoints of the rectilinear parts of the contour (see Figure 11.4.2).

We also assume that R_0 is so large that the arc of the circle $|\lambda - \gamma'| = R_0$ lying inside the domain bounded by the contour $\Gamma_>$ ($\lambda = \gamma' - R_0 e^{i\varphi}$, $-\theta \le \varphi \le \theta$) is located in the left half-plane. This assumption can be expressed as the inequality $R_0 \cos\theta - \gamma' > 0$. Hence we can estimate the decrease of the second term in the representation of e^{tA} as t increases, i.e.,

$$\left\|\frac{1}{2\pi i}\oint\limits_{\Gamma_1} e^{\lambda t}(\lambda I - A)^{-1}d\lambda\right\| \le 2\left\{\frac{1}{2\pi}\int\limits_{R_0}^{\infty}\frac{e^{-t(\rho\cos\theta-\gamma')}}{(\gamma'-\gamma)\sin\theta}ds\right\} + \frac{e^{-t(R_0\cos\theta-\gamma')}}{2\pi}M2\theta$$
$$\le \left[\frac{2}{t\pi(\gamma'-\gamma)\sin 2\theta} + \frac{\theta M}{\pi}\right]e^{-t(R_0\cos\theta-\gamma')}.$$

Denote by P the (commuting with A) projection ($P^2 = P$, $PA = AP$) onto the invariant subspace of A corresponding to the eigenvalues $\lambda_j(A)$ lying inside the set bounded by the contour Γ_0 (i.e., $|\lambda_j - \gamma'| < R_0$). The projection P can be computed with the help of the Riesz integral:

$$P = \frac{1}{2\pi i}\oint\limits_{\Gamma_0}(\lambda I - A)^{-1}d\lambda. \tag{11.4.8}$$

The integral along Γ_0 in the representation of e^{tA} can be written as follows:

$$\frac{1}{2\pi i}\oint\limits_{\Gamma_0} e^{\lambda t}(\lambda I - A)^{-1}d\lambda = e^{tA}P = Pe^{tA} = Pe^{tA}P = Pe^{tPA}P.$$

The behavior of this integral as $t \to \infty$ is completely determined by the eigenvalues surrounded by Γ_0.

In applications, the above scheme is convenient in the case of very large dimension and large norm of operators. It is often possible to indicate a contour Γ_0 (circle) surrounding only a few eigenvalues that are precisely the eigenvalues which must be taken into account in the study of the stability of the trivial solution to the system $\dot{x} = Ax$.

PROBLEM 11.4.1. In the case of stability, under the above conditions on the contour Γ_0, show the existence of a unique solution $H = H^* \ge 0$ to the matrix equations $HA + A^*H + P^*P = 0$, $H = HP = P^*H$, where P is the projection defined by (11.4.8). The solution is represented in the integral form

$$H = \int\limits_0^{\infty} P^* e^{tA^*} e^{tA} P dt,$$

and the norm $\|e^{tA}P\|$ can be estimated in terms of the parameter $\kappa = 2\|PA\|\,\|H\|$:

$$\|e^{tA}P\| \le \sqrt{\kappa}e^{-t\|PA\|/\kappa}\|P\|.$$

Hint. See §§9.2, 9.3, 10.1, and 10.5.

11.5. Sectorial operators

Sectorial operators. Sufficient conditions on a square matrix for the corresponding operator to be sectorial. Estimates for the matrix exponential of a sectorial operator and corollaries.

The construction of the Hausdorff set of a given operator is a rather complicated computational procedure. However, for some popular class of operators, which is important in applications, it is not hard to indicate convex polygons containing the Hausdorff sets. We consider sectorial operators whose Hausdorff sets lie inside an acute angle. We assume that the vertex is located on the real axis and the bisector of the angle is the negative x-axis. In other words, for a sectorial operator A the Rayleigh quotients $\dfrac{(Au,u)}{(u,u)} = \lambda \equiv \alpha + i\beta$, $u \neq 0$, satisfy the condition

$$|\beta| \leq (q_0 - \alpha)\tan\theta, \quad 0 < \theta < \pi/2, \ \alpha \leq q_0.$$

We give some sufficient conditions for an operator to be sectorial. Let

$$A = -M^*KM + L + H,$$

where the matrix K is symmetric and positive definite ($K = K^* \geq kI$, $k > 0$), the matrix L is skew-symmetric ($L = -L^*$), and $\|Lu\| \leq l\|Mu\|$, $\|H\| \leq h$, $l \geq 0$, $h \geq 0$. We show that the operator corresponding to the matrix A is sectorial.

For a sector containing the Hausdorff set of A we take the sector of angle 2θ, where $\tan\theta = l/(2\sqrt{kh})$, with the vertex $q_0 = 2h + 2h^{3/2}\sqrt{k}/l$. Indeed,

$$\frac{(Au,u)}{(u,u)} = \left\{-\frac{(KMu,Mu)}{(u,u)} + \frac{1}{2}\frac{([H+H^*]u,u)}{(u,u)} + \frac{1}{2}\frac{([L+L^*]u,u)}{(u,u)}\right\} + i\left\{\frac{1}{2i}\frac{([L-L^*]u,u)}{(u,u)} + \frac{1}{2i}\frac{([H-H^*]u,u)}{(u,u)}\right\}.$$

It is easy to see that both expressions in the braces are real, $((L+L^*)u,u) = 0$, and

$$-\frac{(KMu,Mu)}{(u,u)} + \frac{1}{2}\frac{([H+H^*]u,u)}{(u,u)} \leq -k\left(\frac{\|Mu\|}{\|u\|}\right)^2 + \|H\|,$$

$$\left|\frac{1}{2i}\frac{([L-L^*]u,u)}{(u,u)} + \frac{1}{2i}\frac{([H-H^*]u,u)}{(u,u)}\right| \leq \frac{\|Lu\|\,\|u\|}{\|u\|^2} + \|H\| \leq l\frac{\|Mu\|}{\|u\|} + \|H\|.$$

In other words,

$$\operatorname{Re}\frac{(Au,u)}{(u,u)} \leq -k\mu^2 + \|H\|, \qquad \left|\operatorname{Im}\frac{(Au,u)}{(u,u)}\right| \leq l\mu + \|H\|,$$

where $\mu = \dfrac{\|Mu\|}{\|u\|}$ is nonnegative. The first inequality shows that all numbers

$$\lambda = \alpha + i\beta = \frac{(Au,u)}{(u,u)}$$

lie in the half-plane $\alpha \leq \|H\| \leq h$. These inequalities also imply that

$$|\beta| \leq l\sqrt{\frac{\|H\|-\alpha}{k}} + \|H\| \leq l\sqrt{\frac{h-\alpha}{k}} + h = b(\alpha).$$

Since

$$b''(\alpha) = -\frac{l}{4\sqrt{k}}\frac{1}{(h-\alpha)^{3/2}} < 0$$

for $\alpha < h$, for all admissible α we have

$$|\beta| \leq b(\alpha) \leq b(0) + \alpha b'(0) = h + l\sqrt{h/k} - \frac{l}{2\sqrt{kh}}\alpha = \frac{l}{2\sqrt{kh}}\left[\frac{2\sqrt{k}h^{3/2}}{l} + 2h - \alpha\right].$$

Consequently, the operator A is sectorial.

In a finite-dimensional complex Euclidean space any operator can be regarded as sectorial. Indeed, the Hausdorff set $W(A)$ lies in the disk $|\alpha+i\beta| \leq \|A\|$ and the disk itself can be placed inside the sector $\alpha < \sqrt{2}\|A\|$, $\|\beta\| \leq (\sqrt{2}\|A\| - \alpha)\tan\frac{\pi}{4}$ with the vertex $q_0 = \sqrt{2}\|A\|$ and the angle $\pi/2$.

The notion of a sectorial operator appeared in the study of operators acting in infinite-dimensional Hilbert spaces. For a finite-dimensional case, this notion turns out to be convenient in the cases where the dimension is very large, q_0 is not too large, and the angle of the sector is less than π. Such a situation often appears if a finite-dimensional operator is an approximation of an infinite-dimensional sectorial operator. Perhaps, instead of the term "sectorial operator" we should use a more cumbersome but more definite term "q_0-θ-sectorial operator" with an indication of the vertex q_0 and the half-angle θ.

REMARK 11.5.1 (important). If the Hausdorff set $W(A)$ of an operator A lies in the sector

$$\alpha \leq q_0, \quad |\beta| \leq (q_0 - \alpha)\tan\theta,$$

then $W(A)$ lies in the larger sector

$$\alpha \leq q, \quad |\beta| \leq (q - \alpha)\tan\theta,$$

for any $q > q_0$. Moreover, the distance between a point of the initial sector and the rays $\beta = \pm(q-\alpha)\tan\theta$, $\alpha \leq q$, forming this larger sector is at least $(q - q_0)\sin\theta$, i.e., on these rays the resolvent $(\lambda I - A)^{-1}$ satisfies the estimate (see §11.4)

$$\|(\lambda I - A)^{-1}\| \leq \frac{1}{(q - q_0)\sin\theta}.$$

We also note that $\|e^{tA}\| \leq e^{q_0 t}$ for $t \geq 0$. This inequality was proved in §11.4 (see Theorem 11.4.2) in the case where the Hausdorff set lies in the half-plane $\operatorname{Re} t \leq q_0$.

Under some additional assumptions on a sectorial operator, more exact estimates for the matrix exponential can be derived.

Let

$$q > q_0, \quad \lambda = \alpha + i\beta = \frac{(Au, u)}{(u, u)}, \quad \alpha \leq q_0,$$
$$|\beta| \leq (q_0 - \alpha)\tan\theta, \quad 0 < \theta < \pi/2.$$

In other words, the sector contains the Hausdorff set of A. In addition, we assume that the estimate for the resolvent

$$\|(\lambda I - A)^{-1}\| \leq m/r$$

is valid on the circle $|\lambda - q| = r$ centered at a real point q. This circle realizes the dichotomy of the spectrum of A. The projection P_r ($AP_r = P_r A$) onto the linear span of all maximal invariant subspaces corresponding to the eigenvalues in the disk is computed by the formula

$$P_r = \frac{1}{2\pi i} \oint\limits_{|\lambda-q|=r} (\lambda I - A)^{-1} d\lambda = \frac{1}{2\pi i} \int\limits_0^{2\pi} \left[(q+re^{i\varphi})I - A\right]^{-1} d(q+re^{i\varphi})$$
$$= \frac{r}{2\pi} \int\limits_0^{2\pi} e^{i\varphi} \left[(q+re^{i\varphi})I - A\right]^{-1} d\varphi.$$

By assumption, $\|P_r\| \le m$. Since

$$P_r A = \frac{1}{2\pi i} \oint\limits_{|\lambda-q|=r} \lambda(\lambda I - A)^{-1} d\lambda,$$

we have $\|P_r A\| \le (r+|q|)m$. The matrix exponential e^{tA} can be represented as the sum of two terms:

$$e^{tA} = e^{tA} P_r + e^{tA}(I - P_r). \tag{11.5.1}$$

We note that

$$e^{tA} P_r = P_r e^{tA} P_r = P_r e^{tA},$$
$$e^{tA}(I - P_r) = (I - P_r) e^{tA} (I - P_r) = (I - P_r) e^{tA}.$$

Since the operator A is sectorial, in (11.5.1) we can estimate the second term represented as the contour integral

$$e^{tA}(I - P_r) = \frac{1}{2\pi i} \oint\limits_{\gamma} e^{\lambda t} (\lambda I - A)^{-1} d\lambda,$$

where the contour γ surrounds the eigenvalues $\lambda_j(A)$ lying outside the disk $|\lambda - q| < r$. We can assume that the contour contains some segments of the rays $\beta = \pm(q - \alpha \tan\theta)$ that bound the sector containing the Hausdorff set of A. We recall that $\|(\lambda I - A)^{-1}\| \le \dfrac{1}{[(q - q_0)\sin\theta]}$ for points $\lambda = \alpha + i\beta$ on these rays. Moreover, the contour contains the arcs $\lambda = q + re^{i\varphi}$ and $\lambda = q + Re^{i\varphi}$ $(\pi - \theta < \varphi < \pi + \theta)$ of the circles $|\lambda - q| = r$ and $|\lambda - q| = R$. For R we can take any sufficiently large number (e.g., $R > |q| + \|A\|$) so that the disk $|\lambda - q| < R$ contains the spectrum of A. The segments on the rays $\beta = \pm(q - \alpha)\tan\theta$ contain the endpoints of these arcs. Namely, the points

$$\begin{cases} \alpha = q - r\cos\theta, \\ \beta = r\sin\theta, \end{cases} \qquad \begin{cases} \alpha = q - R\cos\theta, \\ \beta = R\sin\theta \end{cases}$$

on the upper segment and the points

$$\begin{cases} \alpha = q - r\cos\theta, \\ \beta = -r\sin\theta, \end{cases} \qquad \begin{cases} \alpha = q - R\cos\theta, \\ \beta = -R\sin\theta \end{cases}$$

on the lower segment serve as endpoints. Under these assumptions, we can estimate the resolvent and all integrands in the representation of $e^{tA}(I - P_r)$.

On the upper segment we have $\lambda = \alpha + i\beta$, $\alpha = q - \rho\cos\theta$, $\beta = \rho\sin\theta$, $|d\lambda| = d\rho$ $(r \leq \rho \leq R)$. As is known,

$$\|(\lambda I - A)^{-1}\| \leq \frac{1}{(q - q_0)\sin\theta}.$$

Furthermore, $|e^{\lambda t}| = e^{t\alpha} = e^{t(q-\rho\cos\theta)}$. Consequently, along this ray for $t > 0$ we have

$$\begin{aligned}\left\| \frac{1}{2\pi i} \int_{q-re^{-i\theta}}^{q-Re^{-i\theta}} e^{t\lambda}(\lambda I - A)^{-1} d\lambda \right\| &\leq \frac{e^{tq}}{2\pi(q-q_0)\sin\theta} \int_r^R e^{-t\rho\cos\theta} d\rho \\ &\leq \frac{e^{qt}}{2\pi(q-q_0)\sin\theta} \int_r^\infty e^{-t\rho\cos\theta} d\rho \\ &= \frac{e^{(q-r\cos\theta)t}}{\pi t(q-q_0)\sin 2\theta}.\end{aligned} \tag{11.5.2}$$

Similarly, we can estimate the contour integral along the lower segment:

$$\left\| \frac{1}{2\pi i} \int_{q-Re^{i\theta}}^{q-re^{i\theta}} e^{t\lambda}(\lambda I - A)^{-1} d\lambda \right\| \leq \frac{e^{(q-r\cos\theta)t}}{\pi t(q-q_0)\sin 2\theta}. \tag{11.5.3}$$

On the arc $\lambda = Re^{i\varphi}$ $(\pi - \theta \leq \varphi \leq \pi + \theta)$ we have

$$\|(\lambda I - A)^{-1}\| \leq \frac{1}{R - |q| - \|A\|}, \qquad |e^{t\lambda}| \leq e^{(q-R\cos\theta)t}.$$

Therefore, along this arc,

$$\left\| \frac{1}{2\pi i} \int_{q-Re^{-i\theta}}^{q-Re^{i\theta}} e^{t\lambda}(\lambda I - A)^{-1} d\lambda \right\| \leq \frac{e^{(q-R\cos\theta)t}}{R - |q| - \|A\|} \frac{2R}{2\pi}. \tag{11.5.4}$$

On the arc $\lambda = re^{i\varphi}$ $(\pi - \theta \leq \varphi \leq \pi + \theta)$ we have $\|(\lambda I - A)^{-1}\| \leq m/r$ and $|e^{t\lambda}| \leq e^{(q-r\cos\theta)t}$. Therefore, the contour integral along this arc is estimated as follows:

$$\left\| \frac{1}{2\pi i} \int_{q-re^{i\theta}}^{q-re^{-i\theta}} e^{t\lambda}(\lambda I - A)^{-1} d\lambda \right\| \leq \frac{\theta m}{\pi} e^{(q-r\cos\theta)t}. \tag{11.5.5}$$

Combining the estimates (11.5.2)–(11.5.5), we obtain

$$\begin{aligned}\|e^{tA}(I - P_r)\| &= \left\| \frac{1}{2\pi i} \oint_\gamma e^{\lambda t}(\lambda I - A)^{-1} d\lambda \right\| \\ &\leq \frac{2e^{(q-r\cos\theta)t}}{\pi t(q-q_0)\sin 2\theta} + \frac{\theta m}{\pi} e^{(q-r\sin\theta)t} + \frac{\theta R e^{(q-R\cos\theta)t}}{\pi(R - |q| - \|A\|)}.\end{aligned}$$

In this estimate we can choose R as large as desired (we need only the inequality $R > |q| + \|A\|$). Hence for $t > 0$ we can drop the last term. Thus,

$$\|e^{tA}(I - P_r)\| \leq \left(\frac{2}{\pi t(q - q_0)\sin 2\theta} + \frac{\theta m}{\pi}\right) e^{(q - r\cos\theta)t},$$

$$\|e^{tA}\| \leq \|e^{tA}P_r\| + \left(\frac{2}{\pi t(q - q_0)\sin 2\theta} + \frac{\theta m}{\pi}\right) e^{(q - r\cos\theta)t}.$$

From the estimate $\|e^{tA}\| \leq e^{q_0 t}$, for $t > 0$ we find

$$\|e^{tA}\| \leq \min\left\{e^{q_0 t}, \|e^{tA}P_r\| + \left(\frac{2}{\pi t(q - q_0)\sin 2\theta} + \frac{\theta m}{\pi}\right) e^{(q - r\cos\theta)t}\right\}. \tag{11.5.6}$$

The inequality (11.5.6) is useful in the analysis of the Lyapunov stability of equations of the form $\dot{x} = Ax$ that describe processes with very many degrees of freedom (e.g., models of systems with an infinite number of degrees of freedom constructed by the finite element method). In addition, the dimension of the vector x, the number of columns and rows of A, and the norm $\|A\|$ can be very large. However, if A is a sectorial operator and if it is possible to justify the existence of a circle $|\lambda - q| = r$ such that the norm $\|(\lambda I - A)^{-1}\|$ is not too large and $q - r\cos\theta < 0$ on this circle, then, in view of the above estimate, the study of stability is reduced to the study of the norm $\|e^{tA}P_r\|$, i.e., to the following question: Do the solutions in some subspace of rather small dimension increase or decrease as $t \to \infty$? To analyse this question, quadratic Lyapunov functions (see §§9.3, 10.1, and 10.5) can be used. To use this tool, it is essential not only that AP_r has small rank but also that the norm $\|AP_r\|$ is not too large. We recall the estimate $\|P_rA\| = \|AP_r\| \leq (r + |q|)m$ obtained above.

Part 3

Application of Spectral Analysis. The Most Important Algorithms

CHAPTER 12

Matrix Operators as Models of Differential Operators

12.1. A typical example of a sectorial operator

Finite-dimensional approximations to a second order differential operator by sectorial operators. The weak approximation method. Comparison with the Galerkin method. The quantitative characteristics of approximation operators and some estimates.

In applications, finite-dimensional sectorial operators usually appear as approximations to second order differential operators. The differential operators are assumed to act in an infinite-dimensional function space. To define an approximation, it is necessary to introduce a finite-dimensional function space that will serve as a model of the infinite-dimensional function space. After that it is necessary to simulate the action of the differential operators. For example, for a model space we can take the linear span of a finite number of elements of the original infinite-dimensional function space. However, this approach leads to some problems.

Considering simple but typical examples, we will explain how to overcome these difficulties. In a connected three-dimensional bounded domain D, we consider the differential operator

$$Lu = \frac{\partial}{\partial x}\left(K\frac{\partial u}{\partial x}\right)+\frac{\partial}{\partial y}\left(K\frac{\partial u}{\partial y}\right)+\frac{\partial}{\partial z}\left(K\frac{\partial u}{\partial z}\right)+a\frac{\partial u}{\partial x}+b\frac{\partial u}{\partial y}+c\frac{\partial u}{\partial z},$$

which is applied to sufficiently smooth functions $u(x, y, z)$ vanishing on the boundary ∂D of D. The coefficients $K(x, y, z)$, $a(x, y, z)$, $b(x, y, z)$, and $c(x, y, z)$ are also assumed to be sufficiently smooth, and the real function $K(x, y, z)$ is assumed to be strictly positive and bounded, i.e.,

$$\kappa_1 \geq K(x, y, z) \geq \kappa_0 > 0.$$

Sometimes, it is convenient to represent L as the sum $L = L^{[2]} + L^{[1]}$, where

$$L^{[2]}u = \frac{\partial}{\partial x}\left(K\frac{\partial u}{\partial x}\right)+\frac{\partial}{\partial y}\left(K\frac{\partial u}{\partial y}\right)+\frac{\partial}{\partial z}\left(K\frac{\partial u}{\partial z}\right),$$
$$L^{[1]}u = a\frac{\partial u}{\partial x}+b\frac{\partial u}{\partial y}+c\frac{\partial u}{\partial z}.$$

Let the function space be equipped with the inner product

$$(f, g) = \iiint\limits_D f(x, y, z)\cdot\overline{g}(x, y, z)\, dx\, dy\, dz$$

and the corresponding norm $\|f\| = \sqrt{(f,f)}$. With the help of the inner product we construct the bilinear forms

$$(Lu,v) = (L^{[2]}u,v) + (L^{[1]}u,v), \quad (L^{[2]}u,v), \quad (L^{[1]}u,v)$$

which play an important role later in this volume. Using the equalities

$$\begin{aligned}(L^{[2]}u,v) &= \iiint\limits_D \left[\frac{\partial}{\partial x}\left(K\frac{\partial u}{\partial x}\right)+\frac{\partial}{\partial y}\left(K\frac{\partial u}{\partial y}\right)+\frac{\partial}{\partial z}\left(K\frac{\partial u}{\partial z}\right)\right]\overline{v}\,dx\,dy\,dz \\ &= -\iiint\limits_D K\left(\frac{\partial u}{\partial x}\frac{\partial \overline{v}}{\partial x}+\frac{\partial u}{\partial y}\frac{\partial \overline{v}}{\partial y}+\frac{\partial u}{\partial z}\frac{\partial \overline{v}}{\partial z}\right)dx\,dy\,dz \\ &\quad + \iint\limits_{\partial D} \overline{v}K\left(\frac{\partial u}{\partial x}\,dy\,dz + \frac{\partial u}{\partial y}\,dz\,dx + \frac{\partial u}{\partial z}\right)dx\,dy\end{aligned}$$

and the boundary condition $v\big|_{\partial D} = 0$, we establish that

$$\begin{aligned}(L^{[2]}u,v) &= -\iiint\limits_D K(x,y,z)(u_x\overline{v}_x + u_y\overline{v}_y + u_z\overline{v}_z)\,dx\,dy\,dz, \\ (L^{[2]}u,u) &\le 0.\end{aligned} \tag{12.1.1}$$

The bilinear form $(-L^{[2]}u,v)$ satisfies all the axioms of the inner product. Therefore,

$$|(L^{[2]}u,v)| \le \sqrt{(-L^{[2]}u,u)(-L^{[2]}v,v)}.$$

Moreover,

$$\begin{aligned}|(L^{[1]}u,v)| &= \left|\iiint\limits_D \overline{v}(au_x + bu_y + cu_z)\,dx\,dy\,dz\right| \\ &\le \left|\iiint\limits_D \frac{1}{\sqrt{K}}\overline{v}(\sqrt{K}au_x + \sqrt{K}bu_y + \sqrt{K}cu_z)\,dx\,dy\,dz\right| \\ &\le \sqrt{(v,v)}\frac{m}{\sqrt{\kappa_0}}\sqrt{\iiint\limits_D K(|u_x|^2 + |u_y|^2 + |u_z|^2)\,dx\,dy\,dz} \\ &= \frac{m}{\sqrt{\kappa_0}}\sqrt{(-L^{[2]}u,u)}\cdot\|v\|,\end{aligned} \tag{12.1.2}$$

where

$$m = \max_D \sqrt{|a|^2 + |b|^2 + |c|^2}. \tag{12.1.3}$$

The above estimates will be used in order to determine a domain in the complex plane in which the Rayleigh quotients $\dfrac{(Lu,u)}{(u,u)}$ lie. Introduce the notation

$$\mu = \sqrt{(-L^{[2]}u,u)/(u,u)} \quad (\mu > 0).$$

From (12.1.1)–(12.1.3) we obtain

$$\operatorname{Re}\frac{(Lu,u)}{(u,u)} \le -\mu^2 + \frac{m}{\sqrt{\kappa_0}}\mu, \qquad \left|\operatorname{Im}\frac{(Lu,u)}{(u,u)}\right| \le \frac{m}{\sqrt{\kappa_0}}\mu.$$

Consequently,

$$\frac{\left|\operatorname{Im}\dfrac{(Lu,u)}{(u,u)}\right|}{\left(\dfrac{m}{\sqrt{\kappa_0}}\right)^2-\operatorname{Re}\dfrac{(Lu,u)}{(u,u)}}\le\frac{\dfrac{m}{\sqrt{\kappa_0}}\mu}{\left(\dfrac{m}{\sqrt{\kappa_0}}\right)^2-\dfrac{m}{\sqrt{\kappa_0}}\mu+\mu^2}.$$

The ratio on the right-hand side of this inequality vanishes for $\mu=0$. For positive μ it attains the maximal value 1 at $\mu=m/\sqrt{\kappa_0}$. Indeed, the derivative

$$\frac{d}{d\mu}\frac{\dfrac{m}{\sqrt{\kappa_0}}\mu}{\left(\dfrac{m}{\sqrt{\kappa_0}}\right)^2-\dfrac{m}{\sqrt{\kappa_0}}\mu+\mu^2}=\frac{\dfrac{m}{\sqrt{\kappa_0}}\left[\dfrac{m^2}{\kappa_0}-\mu^2\right]}{\left[\left(\dfrac{m}{\sqrt{\kappa_0}}\right)^2-\dfrac{m}{\sqrt{\kappa_0}}\mu+\mu^2\right]^2}$$

vanishes only for $\mu=m/\sqrt{\kappa_0}$. We obtain the inequality

$$\left|\operatorname{Im}\frac{(Lu,u)}{(u,u)}\right|+\operatorname{Re}\frac{(Lu,u)}{(u,u)}\le\frac{m^2}{\kappa_0},\tag{12.1.4}$$

which means that for any $u(x,y,z)$ the complex number $\dfrac{(Lu,u)}{(u,u)}=\lambda=\alpha+i\beta$ lies in the sector formed by the lines $\alpha\pm\beta=m^2/\kappa_0$ forming the right angle, with the vertex at $\lambda=m^2/\kappa_0$, and such that the real half-axis $\beta=0$, $\alpha\le m^2/\kappa_0$ is the bisector.

To obtain a finite-dimensional approximation of L and the corresponding approximations of $L^{[2]}$ and $L^{[1]}$, it suffices to take a finite-dimensional subspace of functions that are regular in D and satisfy the boundary condition $u\big|_{\partial D}=0$. For example, we can take the linear span of the eigenvectors $u^{(j)}$ of the Laplace operator

$$\frac{\partial^2u^{(j)}}{\partial x^2}+\frac{\partial^2u^{(j)}}{\partial y^2}+\frac{\partial^2u^{(j)}}{\partial z^2}=\lambda^{(j)}u^{(j)},\quad u^{(j)}\big|_{\partial D}=0,$$

if $u^{(j)}$ can be computed in D. We emphasize the following problem. Assume that the coefficients a, b, and c are nonzero or that $K(x,y,z)$ is not constant. In this case, applying the operator L to $u^{(j)}$, we obtain the functions $Lu^{(j)}$ that do not belong to our finite-dimensional space. As a rule, the operator L transforms the original finite-dimensional space to some other finite-dimensional space. To obtain the approximation $\widetilde{L}$ of L, it is necessary to replace the vectors $Lu^{(j)}$ by their projections onto the original space.

Usually, the operator $\widetilde{L}$, which transforms the basis functions $u^{(1)},u^{(2)},\dots$ of the original space to their linear combinations

$$\widetilde{L}u^{(i)}=\sum_j l_{ij}u^{(j)},\tag{12.1.5}$$

is taken so that

$$(\widetilde{L}u^{(i)},u^{(j)})=(Lu^{(i)},u^{(j)})\tag{12.1.6}$$

for all i and j. Furthermore,

$$(\widetilde{L}u,u)=\sum_{i,k}\xi_i\bar{\xi}_k(\widetilde{L}u^{(i)},u^{(k)})=(Lu,u)$$

for any vector $u=\sum\limits_i\xi_iu^{(i)}$.

The Hausdorff set of $\widetilde{L}$ is a subset of the Hausdorff set of L; namely, the set of $\lambda = \dfrac{(\widetilde{L}u, u)}{(u,u)} = \dfrac{(Lu,u)}{(u,u)}$, where $u(x,y,z)$ runs over the model space. In the above example, the operator $\widetilde{L}$ constructed in accordance with the condition (12.1.6) is sectorial and satisfies (12.1.4).

We show that if the functions $u^{(j)}$ are linearly independent, i.e., if the Gram matrix

$$G = \begin{bmatrix} g_{11} & g_{12} & \cdots \\ g_{21} & g_{22} & \cdots \\ \cdots & \cdots & \cdots \\ \cdots & \cdots & \cdots \end{bmatrix}$$

with elements $g_{ik} = \overline{g}_{ki} = (u^{(i)}, u^{(k)})$ is positive definite, then the coefficients l_{ij} in (12.1.5) are uniquely determined from (12.1.6). Indeed, because of the equalities

$$(Lu^{(i)}, u^{(j)}) = (\widetilde{L}u^{(i)}, u^{(j)}) = \Big(\sum_k l_{ik}u^{(k)}, u^{(j)}\Big) = \sum_k l_{ik}(u^{(k)}, u^{(j)})$$
$$= \sum_k l_{ik}g_{kj} = \sum_k \overline{g}_{jk}l_{ik},$$

the l_{ik}'s satisfy the equations

$$\sum_k \overline{g}_{jk}l_{ik} = (Lu^{(i)}, u^{(j)}),$$

where the matrix $\overline{G}$ with elements $\overline{g}_{jk}$ is nonsingular. Therefore, these equations are uniquely solvable with respect to l_{ik}. Similarly, from the equalities

$$(\widetilde{L}^{[2]}u^{(i)}, u^{(j)}) = (L^{[2]}u^{(i)}, u^{(j)}), \quad (\widetilde{L}^{[1]}u^{(i)}, u^{(j)}) = (L^{[1]}u^{(i)}, u^{(j)})$$

we find the approximations $\widetilde{L}^{[2]}$ and $\widetilde{L}^{[1]}$ such that

$$\widetilde{L}^{[2]}u^{(i)} = \sum_k l_{ik}{}^{[2]}u^{(k)}, \quad \widetilde{L}^{[1]}u^{(i)} = \sum_k l_{ik}{}^{[1]}u^{(k)}$$

of $L^{[2]}$ and $L^{[1]}$. We emphasize that in the construction of the approximations $\widetilde{L}$ and $\widetilde{L}^{[2]}$ of the operators L and $L^{[2]}$ we can find $(Lu^{(i)}, u^{(j)})$ and $(L^{[2]}u^{(i)}, u^{(j)})$ without computing the second order derivatives of the basis functions although they appear in the definitions of the operators. Indeed, it suffices to use the equalities (12.1.7)

$$(L^{[2]}u^{(i)}, u^{(j)}) = -\iiint_D K\left(\frac{\partial u^{(i)}}{\partial x}\frac{\partial \overline{u}^{(j)}}{\partial x} + \frac{\partial u^{(i)}}{\partial y}\frac{\partial \overline{u}^{(j)}}{\partial y} + \frac{\partial u^{(i)}}{\partial z}\frac{\partial \overline{u}^{(j)}}{\partial z}\right) dx\,dy\,dz,$$

$$(Lu^{(i)}, u^{(j)}) = (L^{[2]}u^{(i)}, u^{(j)}) + \iiint_D \overline{u}^{(j)}(a{u_x}^{(i)} + b{u_y}^{(i)} + c{u_z}^{(i)})\,dx\,dy\,dz.$$

The computation of elements of a matrix that approximates the second order differential operator is based on formula (12.1.7); moreover, only the first order derivatives of the basis functions are used. This procedure is called the *weak approximation method.* In fact, the method realizes Galerkin's generalization of the Rietz method for self-adjoint problems in calculus of variations.

In the most general form of the Galerkin method the space of functions u and the space of images Lu do not necessarily coincide. In the original space and in the space of images, we choose different bases $u^{(i)}$, $i = 1, 2, \dots, N_0$, and $v^{(j)}$, $j = 1, 2, \dots, N_1$, respectively. (As a rule, N_0 and N_1 are different.) In accordance with the Galerkin method, the action of the operator L on the linear combination $u = \sum_{i=1}^{N_0} \xi_i u^{(i)}$ is modeled by the operator

$$\widetilde{L}u = \sum_{i=1}^{N_0} \xi_i \left[\sum_{j=1}^{N_1} l_{ij} v^{(j)} \right], \tag{12.1.8}$$

where l_{ij} are found from the conditions

$$(Lu^{(i)}, v^{(j)}) = \sum_k l_{ik} (v^{(k)}, v^{(j)}). \tag{12.1.9}$$

Moreover, in realizations of the Galerkin method, test functions $v^{(j)}$ are often used. Such test functions do not lie in the space of images of u. For example, if L is a second order differential operator and among $u^{(i)}$ there are functions having piecewise continuous first order derivatives, then $Lu^{(i)}$ are not defined. However, this fact does not prevent us from using (12.1.8) and (12.1.9) if we apply a method based on (12.1.7) (if $u^{(i)}$ and $v^{(i)}$ coincide) of computing $(Lu^{(i)}, v^{(j)})$.

We will not check if $\widetilde{L}$ is an approximation of L. The detailed study of this question and an accurate exact formulation of the notion of "weak approximation" are discussed in mathematical physics, the theory of differential equations, functional analysis, and applied mathematics. Our goal is to explain how to obtain quantitative characteristics for typical model operators $\widetilde{L}$ in order to investigate spectral properties and describe the location of spots of the ε-spectra of operators. We discuss how to compute the singular values of the operator $\widetilde{L}$ or, if it is difficult, how to estimate the singular values (see §§12.2–12.4).

Denote the inner product in the space of functions $u(x, y, z)$ by $(\cdot, \cdot)$ and in the space of images $\widetilde{L}$ by $[\cdot, \cdot]$. Consider the positive definite Hermitian Gram matrices of orders N_0 and N_1

$$\overset{0}{G} = \begin{bmatrix} \overset{0}{g}_{11} & \overset{0}{g}_{12} & \cdots \\ \overset{0}{g}_{21} & \overset{0}{g}_{22} & \cdots \\ \cdots & \cdots & \cdots \\ \cdots & \cdots & \cdots \end{bmatrix}, \quad \overset{1}{G} = \begin{bmatrix} \overset{1}{g}_{11} & \overset{1}{g}_{12} & \cdots \\ \overset{1}{g}_{21} & \overset{1}{g}_{22} & \cdots \\ \cdots & \cdots & \cdots \\ \cdots & \cdots & \cdots \end{bmatrix}$$

with elements $\overset{0}{g}_{ij} = (u^{(i)}, u^{(j)})$ and $\overset{1}{g}_{ij} = (v^{(i)}, v^{(j)})$, respectively. If $u^{(1)}, u^{(2)}, \dots, u^{(N_0)}$ form an orthonormal basis, then $\overset{0}{G} = I_{N_0}$ is the identity matrix and if $u^{(1)}, u^{(2)}, \dots, u^{(N_0)}$ is close to an orthonormal basis, then $\overset{0}{G} = I + E$, where the norm of $E = E^*$ is sufficiently small ($\|E\| \le e < 1$). Denote by H the rectangular matrix with elements $h_{ij} = [Lu^{(i)}, v^{(j)}]$. The matrix H has N_0 rows and N_1 columns. The Gram matrices can be simplified using unitary transformations U and V ($UU^* = I_{N_0}$, $VV^* = I_{N_1}$) that reduce the original matrices to the diagonal canonical form

$$U\overset{0}{G}U^* = \begin{bmatrix} \overset{0}{g}_1 & & & 0 \\ & \overset{0}{g}_2 & & \\ & & \ddots & \\ 0 & & & \overset{0}{g}_{N_0} \end{bmatrix}, \quad V\overset{1}{G}V^* = \begin{bmatrix} \overset{1}{g}_1 & & & 0 \\ & \overset{1}{g}_2 & & \\ & & \ddots & \\ 0 & & & \overset{1}{g}_{N_1} \end{bmatrix}$$

with real positive diagonal elements so that

$$\|\overset{0}{G}\| = \max_i \overset{0}{g}_i, \quad \|\overset{0}{G}{}^{-1}\| = \max_i \overset{0}{g}_i{}^{-1}, \quad \|\overset{1}{G}\| = \max_i \overset{1}{g}_i, \quad \|\overset{1}{G}{}^{-1}\| = \max_i \overset{1}{g}_i{}^{-1}.$$

In particular, if $\overset{0}{G} = I_{N_0} + E$ and $\|E\| \le e < 1$, then

$$\|\overset{0}{G}\| \le 1 + e, \quad \|\overset{0}{G}{}^{-1}\| \le \frac{1}{1-e}.$$

Introduce the square nonsingular matrices

$$\overset{0}{T} = \begin{bmatrix} \overset{0}{g}_1{}^{-1/2} & & & 0 \\ & \overset{0}{g}_2{}^{-1/2} & & \\ & & \ddots & \\ 0 & & & \overset{0}{g}_{N_0}{}^{-1/2} \end{bmatrix} U = \begin{bmatrix} \overset{0}{t}_{11} & \overset{0}{t}_{12} & \dots \\ \overset{0}{t}_{21} & \overset{0}{t}_{22} & \dots \\ \dots & \dots & \dots \\ \dots & \dots & \dots \end{bmatrix},$$

$$\overset{1}{T} = \begin{bmatrix} \overset{1}{g}_1{}^{-1/2} & & & 0 \\ & \overset{1}{g}_2{}^{-1/2} & & \\ & & \ddots & \\ 0 & & & \overset{1}{g}_{N_1}{}^{-1/2} \end{bmatrix} V = \begin{bmatrix} \overset{1}{t}_{11} & \overset{1}{t}_{12} & \dots \\ \overset{1}{t}_{21} & \overset{1}{t}_{22} & \dots \\ \dots & \dots & \dots \\ \dots & \dots & \dots \end{bmatrix}$$

of orders N_0 and N_1, respectively, such that $\overset{0}{G} = \overset{0}{T}{}^{-1}\overset{0}{T}{}^{-*}$ and $\overset{1}{G} = \overset{1}{T}{}^{-1}\overset{1}{T}{}^{-*}$. It is convenient to pass to the orthonormal bases

$$\widehat{u}^{(i)} = \sum_{k=1}^{N_0} \overset{0}{t}_{ik} u^{(k)}, \quad \widehat{v}^{(j)} = \sum_{k=1}^{N_1} \overset{1}{t}_{jk} v^{(k)}.$$

It is easy to verify that

$$(\widehat{u}^{(i)}, \widehat{u}^{(j)}) = \delta_{ij}, \quad [\widehat{v}^{(i)}, \widehat{v}^{(j)}] = \delta_{ij}, \quad \widehat{H} = [\widetilde{L}\widehat{u}^{(i)}, \widehat{v}^{(j)}] = \overset{0}{T}H\overset{1}{T}{}^*.$$

Applying $\widetilde{L}$ to the vector $u = \sum_{i=1}^{N_0} \xi_i \widehat{u}^{(i)}$, we obtain the vector

$$v = \sum_{i=1}^{N_0} \xi_i \left[\sum_{j=1}^{N_1} [\widetilde{L}\widehat{u}^{(i)}, \widehat{v}^{(j)}] \cdot \widehat{v}^{(j)} \right] = \sum_{i=1}^{N_0} \sum_{j=1}^{N_1} \xi_i \widehat{H}_{ij} \widehat{v}^{(j)} = \sum_{j=1}^{N_1} \eta_j \widehat{v}^{(j)}$$

such that

$$\begin{pmatrix} \eta_1 \\ \eta_2 \\ \vdots \\ \eta_{N_1} \end{pmatrix} = \widehat{H}^T \begin{pmatrix} \xi_1 \\ \xi_2 \\ \vdots \\ \xi_{N_0} \end{pmatrix} = (\overset{0}{T}H\overset{1}{T}{}^*)^T \begin{pmatrix} \xi_1 \\ \xi_2 \\ \vdots \\ \xi_{N_0} \end{pmatrix}.$$

It is natural to identify the singular values of the transposed matrix $\widehat{H}^T$ (they coincide with the singular values of $\widehat{H}$) with the singular values of the operator $\widetilde{L}$, i.e., $\sigma_j(\widetilde{L}) \equiv \sigma_j(\widehat{H}) = \sigma_j(\overset{0}{T}H\overset{1}{T}{}^*)$ $(\sigma_j \leq \sigma_{j+1})$. Such $\sigma_j(\widetilde{L})$ are independent of the choice of bases in the finite-dimensional spaces of test functions u, v. For a given operator L they are determined by the spaces themselves and the inner products $(\cdot,\cdot)$ and $[\cdot,\cdot]$. We left this assertion to the reader as an exercise.

The above arguments make clear the following relation:

$$\frac{\sigma_j(H)}{\sqrt{\|\overset{0}{G}\|\,\|\overset{1}{G}\|}} = \sigma_1(\overset{0}{T})\sigma_1(\overset{1}{T})\sigma_j(H) \leq \sigma_j(\widehat{H}) = \sigma_j(\widetilde{L}) = \sigma_j(\widehat{H})$$

$$\leq \sigma_{N_0}(\overset{0}{T})\sigma_{N_1}(\overset{1}{T})\sigma_j(H) = \sigma_j(H)\sqrt{\|\overset{0}{G}{}^{-1}\|\,\|\overset{1}{G}{}^{-1}\|}.$$

In particular, if $\overset{0}{G} = I_{N_0} + E$, $\|E\| \leq e < 1$, and $\overset{1}{G} = I_{N_1}$, then

$$\frac{\sigma_j(H)}{\sqrt{1+e}} \leq \sigma_j(\widetilde{L}) \leq \frac{\sigma_j(H)}{\sqrt{1-e}}. \tag{12.1.10}$$

12.2. Finite-dimensional models of first order operators

Finite-dimensional approximations of a first order differential operator. A finite-dimensional approximation of the gradient. Properties of the matrix associated with the operator approximating the gradient. Estimates for singular values of this matrix.

Let $u(x,y,z)$ be a sufficiently smooth function defined in the closed cube $0 \leq x \leq Q$, $0 \leq y \leq Q$, $0 \leq z \leq Q$ and vanishing on the boundary of the cube. The vector of the derivative analog $\widehat{d}$ (gradient) maps u to three scalar functions u_x, u_y, and u_z which can be regarded as three components of the vector-valued function

$$\widehat{d}u(x,y,z) = \begin{pmatrix} u_x(x,y,z) \\ u_y(x,y,z) \\ u_z(x,y,z) \end{pmatrix}. \tag{12.2.1}$$

We describe how to construct a finite-dimensional approximation $\widetilde{d}$ of the operator $\widehat{d}$.

First, in the space of scalar functions u and in the space of vector-valued functions f with components f_1, f_2, and f_3 we introduce the inner products

$$\begin{aligned} (u,w) &= \int_0^Q\int_0^Q\int_0^Q u(x,y,z)\overline{w}(x,y,z)\,dx\,dy\,dz, \\ [f,g] &= \int_0^Q\int_0^Q\int_0^Q (f_1\overline{g}_1 + f_2\overline{g}_2 + f_3\overline{g}_3)\,dx\,dy\,dz. \end{aligned} \tag{12.2.2}$$

After that we choose finite sets of basis elements. In the space of scalar functions we suggest to use the basis elements $u^{(i)}$ of the form

$$u_{klm} = (2/Q)^{3/2}\sin(k\pi x/Q)\sin(l\pi y/Q)\sin(m\pi z/Q). \tag{12.2.3}$$

In the space of vector-valued functions the number of basis elements $v^{(i)}$ is three times larger:

$$v'_{klm} = \begin{bmatrix} \left(\frac{2}{Q}\right)^{3/2} \cos\frac{k\pi x}{Q} \sin\frac{l\pi y}{Q} \sin\frac{m\pi z}{Q} \\ 0 \\ 0 \end{bmatrix},$$

$$v''_{klm} = \begin{bmatrix} 0 \\ \left(\frac{2}{Q}\right)^{3/2} \sin\frac{k\pi x}{Q} \cos\frac{l\pi y}{Q} \sin\frac{m\pi z}{Q} \\ 0 \end{bmatrix},$$

$$v'''_{klm} = \begin{bmatrix} 0 \\ 0 \\ \left(\frac{2}{Q}\right)^{3/2} \sin\frac{k\pi x}{Q} \sin\frac{l\pi y}{Q} \cos\frac{m\pi z}{Q} \end{bmatrix}.$$

The integer-valued parameters k, l, and m must satisfy the condition $k, l, m \geq 1$, $k^2 + l^2 + m^2 \leq r^2$ with some $r \geq 1$. For a finite-dimensional space of functions on which the model operator $\widetilde{d}$ acts we take the linear span R_0 of a finite set of elements $u_{klm}(x, y, z)$. The action of $\widetilde{d}$ on the basis elements $u_{klm}(x, y, z)$ is defined by the formula

$$\widetilde{d}u_{klm} = \widehat{d}u_{klm} = \frac{k\pi}{Q}v'_{klm} + \frac{l\pi}{Q}v''_{klm} + \frac{m\pi}{Q}v'''_{klm}.$$

The linear operator is extended to the entire linear span R_0 of u_{klm} in a natural way. It is obvious that the image $\widetilde{d}u = \widehat{d}u$ of $u \in R_0$ under the action of the operator $\widetilde{d}$ belongs to the linear span of the vector basis elements v'_{klm}, v''_{klm}, and v'''_{klm}. We regard this linear span as a finite-dimensional linear space R_1, so that $\widetilde{d}R_0 \subset R_1$.

It is easy to check that the chosen bases for R_0 and R_1 are orthonormal and that the dimension of the space R_1 is three times more than the dimension of the space R_0. The dimension of the space R_0 is equal to the number of the points with positive integer-valued coordinates k, l, and m such that $k^2 + l^2 + m^2 \leq r^2$ (no basis function u_{klm} is associated to the point $(0, 0, 0)$).

The matrix describing the action of the operator $\widetilde{d}$ in the chosen bases consists of columns such that only three elements of each column differ from zero and are equal to $k\pi/Q$, $l\pi/Q$, $m\pi/Q$; moreover, in the remaining columns, all elements of rows containing these elements must be zero. Taking into account the structure of this matrix, we see that the singular values are all possible values of the expression $\pi\sqrt{(k^2 + l^2 + m^2)}/Q$ that do not exceed $\pi r/Q$.

We note that among the singular values there are families of equal numbers because the same value $\sqrt{k^2 + l^2 + m^2}$ can be obtained for different integer-valued parameters k, l, and m. For example, in the ball $k^2 + l^2 + m^2 \leq 50$ there are 6 different points with positive integer-valued coordinates k, l, m at the same distance $\sqrt{k^2 + l^2 + m^2} = \sqrt{50}$ from the center. We indicate all possible coordinates k, l, m of such points:

$$(3, 4, 5), (3, 5, 4), (5, 3, 4), (5, 4, 3), (4, 3, 5), (4, 5, 3).$$

In other words, the number $\sigma = \pi\sqrt{50}/Q$ can occur 6 times in the sequence of the singular values of the finite-dimensional approximation $\widetilde{d}$ of the operator $\widehat{d}$ if the dimension of the space R_0 is large.

For the operator $\widetilde{d}$ the number $j \le s$ of the singular values $\sigma_j(\widetilde{d})$ coincides with the number of all admissible triples (k,l,m) such that $\pi\sqrt{(k^2+l^2+m^2)}/Q \le s$, i.e., j coincides with the number of triples (k,l,m) such that

$$\sqrt{k^2+l^2+m^2} \le \frac{Qs}{\pi} = r.$$

We now estimate j for a given s. With an integer-valued triple (k,l,m), $k \ge 1$, $l \ge 1$, $m \ge 1$, we associate the unit cube $k \le x \le k+1$, $l \le y \le l+1$, $1 \le z \le l+1$ such that (k,l,m) is a vertex of the cube. The sum of volumes of all such cubes lying in the ball

$$x^2+y^2+z^2 \le (r+\sqrt{3})^2 = \left(\frac{Qs}{\pi}+\sqrt{3}\right)^2$$

is not less than the number of integer-valued k, l, m such that $\sqrt{k^2+l^2+m^2} \le r$. Since the volume of a ball of radius $r+\sqrt{3}$ is $\frac{4\pi}{3}(r+\sqrt{3})^2$ and the cubes under consideration are located in the octant $x \ge 0$, $y \ge 0$, $z \ge 0$, we conclude that the sum of volumes of all cubes is not larger than $\frac{\pi}{6}(r+\sqrt{3})^3$. Therefore,

$$j \le \frac{\pi}{6}(r+\sqrt{3})^3 = \frac{\pi}{6}\left(\frac{Q\sigma_j}{\pi}+\sqrt{3}\right)^3$$

or

$$\sigma_j \ge \frac{\pi}{Q}\left(\left(\frac{6j}{\pi}\right)^{1/3} - \sqrt{3}\right).$$

The smallest singular value $\sigma_1(\widetilde{d})$ corresponds to one of the triples with the smallest value $k^2+l^2+m^2 = 3$. It is obvious that $\sigma_1(\widetilde{d}) = \sqrt{3}\pi/Q$.

We have shown that for the finite-dimensional approximation $\widetilde{d}$ of the gradient $\widehat{d}$ the singular values $\sigma_j(\widetilde{d})$ satisfy the estimate

$$\sigma_j(\widetilde{d}) \ge \max\left\{\sqrt{3}\pi/Q, \frac{\pi}{Q}((6j/\pi)^{1/3} - \sqrt{3})\right\}. \tag{12.2.4}$$

The number N_0 of the singular values $\sigma_j(\widetilde{d})$ coincides with the dimension $N_0(r)$ of the space R_0 in which the operator $\widetilde{d}$ acts.

The estimate (12.2.4) implies, for any $p > 3$, the inequality

$$\sum_{j=1}^{i=N_0} \sigma_j^{-p}(\widetilde{d}) \le \left(\frac{Q}{\pi}\right)^p \varphi(p),$$

where the right-hand side is independent of the number of summands, i.e., it is independent of the dimension N_0 of the finite-dimensional space in which the model operator $\widetilde{d}$ acts.

Later in this volume, we will establish similar estimates for the sums of powers of singular values of other finite-dimensional approximations of differential operators. In particular, we will consider a model sectorial operator $\widetilde{L}$ (see §12.1).

The operator $\widehat{d}$ transforms a scalar function u to a vector-valued function with components u_x, u_y, u_z, whereas the differential operator

$$L^{[1]}u = a\frac{\partial u}{\partial x} + b\frac{\partial u}{\partial y} + c\frac{\partial u}{\partial z}$$

transforms a scalar function u to a scalar function. It is obvious that at any point (x, y, z),

$$|\{L^{[1]}u\}_{x,y,z}| \le \max_{x,y,z} \sqrt{a^2+b^2+c^2}\sqrt{u_x^2+u_y^2+u_z^2}.$$

Consequently,

$$\int_0^L\int_0^L\int_0^L |L^{[1]}u|^2\,dx\,dy\,dz \le m^2[\widehat{d}u, \widehat{d}u],$$

where $m = \max\{\sqrt{a^2+b^2+c^2}\}$ (see (12.1.3)).

Note that for the finite-dimensional approximation $\widetilde{d}$ of the operator $\widehat{d}$ the choice of the special basis $v^{(j)}$ for the space R_1 yields the equality $\widetilde{d}u = \widehat{d}u$ for all $u \in R_0$, i.e., for all u represented as $u = \sum_i \xi_i u^{(i)}$. We have $[\widetilde{d}u, \widetilde{d}u] = [\widehat{d}u, \widehat{d}u]$.

Constructing a finite-dimensional approximation $\widetilde{L}^{[1]}$ of $L^{[1]}$, we can choose R_1 so that it coincides with R_0. In this case, it is natural to choose the basis $v^{(j)}$ coinciding with $u^{(j)}$ (the orthonormal functions u_{klm} defined in (12.2.3)). A function $u \in R_0$ is transformed to the function

$$\widetilde{L}^{[1]}u = \sum_{j=1}^{j=N_0} (L^{[1]}u, u^{(j)})u^{(j)},$$

which is a partial sum of the Fourier series of $L^{[1]}u$ with respect to the orthonormal functions $u^{(j)}$. Moreover,

$$\begin{aligned}(\widetilde{L}^{[1]}u, \widetilde{L}^{[1]}u) &= \int_0^L\int_0^L\int_0^L |\widetilde{L}^{[1]}u|^2\,dx\,dy\,dz \le m^2[\widehat{d}u, \widehat{d}u] \le \frac{m^2}{\kappa_0}[K\widehat{d}u, \widehat{d}u] \\ &= \frac{m^2}{\kappa_0}\int_0^L\int_0^L\int_0^L K(u_x\overline{u}_x + u_y\overline{u}_y + u_z\overline{u}_z)\,dx\,dy\,dz \\ &= -\frac{m^2}{\kappa_0}(L^{[2]}u, u) = -\frac{m^2}{\kappa_0}(\widetilde{L}^{[2]}u, u).\end{aligned}$$

We recall (see §12.1) that

$$L^{[2]}u = \frac{\partial}{\partial x}\left(K\frac{\partial u}{\partial x}\right) + \frac{\partial}{\partial y}\left(K\frac{\partial u}{\partial y}\right) + \frac{\partial}{\partial z}\left(K\frac{\partial u}{\partial z}\right),$$

$$\kappa_0 \le K \le \kappa_1, \quad (L^{[2]}u, u) \le 0,$$

where $\widetilde{L}^{[2]}$ is the finite-dimensional approximation of $L^{[2]}$ constructed by the weak approximation method, so that $\widetilde{L}^{[2]}u \in R_0$ for $u \in R_0$. By the Cauchy–Schwarz–Bunyakovskii inequality, for any functions $u \in R_0$ and $v \in R_0$ we have

$$|(\widetilde{L}^{[1]}u, v)| = |(L^{[1]}u, v)| \le \sqrt{(L^{[1]}u, L^{[1]}u)(v, v)} \le \frac{m}{\sqrt{\kappa_0}}\sqrt{-(\widetilde{L}^{[2]}u, u)}$$

or, equivalently,

$$\|\widetilde{L}^{[1]}u\|_{R_0} \leq \frac{m}{\sqrt{\kappa_0}}\sqrt{-(\widetilde{L}^{[2]}u, u)}. \tag{12.2.5}$$

We will use these inequalities in §12.3 where we study the singular values of the finite-dimensional approximation $\widetilde{L}$ of the second order differential operator $L = L^{[2]} + L^{[1]}$.

12.3. Finite-dimensional approximations of second order operators

A finite-dimensional approximation of the Laplace operator and its properties. Comparison with the weak approximation method. Estimates for singular values of finite-dimensional approximations.

We begin with the simplest elliptic second order operator. Namely, consider the Laplace operator

$$\Delta = \frac{\partial^2}{\partial x^2} + \frac{\partial^2}{\partial y^2} + \frac{\partial^2}{\partial z^2}$$

which transforms a function $u(x, y, z)$ defined in the cube $0 \leq x \leq Q$, $0 \leq y \leq Q$, $0 \leq z \leq Q$ to the function

$$\Delta u = \frac{\partial^2 u}{\partial x^2} + \frac{\partial^2 u}{\partial y^2} + \frac{\partial^2 u}{\partial z^2}.$$

According to §12.1, we start the construction of an approximation of the Laplace operator with the choice of a finite-dimensional space R_0 in which the model operator acts. For R_0 we take the linear span of the basis functions

$$u_{klm} = \left(\frac{2}{Q}\right)^{3/2} \sin\frac{k\pi x}{Q} \sin\frac{l\pi y}{Q} \sin\frac{m\pi z}{Q} \tag{12.3.1}$$

such that $3 \leq k^2 + l^2 + m^2 \leq r_0^2$ for some r_0. This basis is orthonormal and consists of the eigenfunctions of the Laplace operator

$$\Delta u_{klm} = -\frac{\pi^2}{Q^2}(k^2 + l^2 + m^2)u_{klm}. \tag{12.3.2}$$

Introduce the notation

$$D(r_0) = \{(k, l, m) \mid k \geq 1, l \geq 1, m \geq 1, k^2 + l^2 + m^2 \leq r_0^2\}.$$

The image of a function $u(x, y, z) \in R_0$ of the form

$$u = \sum_{(k,l,m)\in D(r_0)} \xi_{klm} u_{klm} (\equiv \sum \xi_j u^{(j)})$$

under the action of the operator Δ will be approximated by the function

$$\widetilde{\Delta} u = \Delta u = -\sum_{(k,l,m)\in D(r_0)} \frac{\pi^2}{Q^2}(k^2 + l^2 + m^2)\xi_{klm} u_{klm}. \tag{12.3.3}$$

Actually, this result is not an approximation but an exact result. A vector of R_0 with components ξ_{klm} goes to the vector of the same space R_0 with components $\eta_{klm} = -\frac{\pi^2}{Q^2}(k^2 + l^2 + m^2)\xi_{klm}$. Thus, in the chosen orthonormal basis the operator Δ is represented as the diagonal matrix with diagonal elements $-\frac{\pi^2}{Q^2}(k^2 + l^2 + m^2)$.

These diagonal elements taken with the opposite signs are singular values of the finite-dimensional approximation $\widetilde{\Delta}$ of Δ. If $\sigma_1 \le \sigma_2 \le \cdots \le \sigma_{N_0}$ and $\sigma_j = \frac{\pi^2}{Q^2}(k_j^2 + l_j^2 + m_j^2)$, then in the set $D(Q\sqrt{\sigma_j}/\pi)$ there are at least j different points (k, l, m). To such a point we associate the unit cube $k \le x \le k+1$, $l \le y \le l+1$, $m \le z \le l+1$ with diagonal $\sqrt{3}$. We note that these cubes lie in the octant $x \ge 0$, $y \ge 0$, $z \ge 0$ in the ball

$$x^2 + y^2 + z^2 \le \left(\frac{Q\sqrt{\sigma_j}}{\pi} + \sqrt{3}\right)^2,$$

i.e., the volume of the part of the octant lying in the ball cannot be less than $j+1$:

$$\frac{1}{8}\frac{4\pi}{3}\left(\frac{Q\sqrt{\sigma_j}}{\pi} + \sqrt{3}\right)^3 \ge j+1.$$

Moreover, $\sigma_j \ge \sigma_1 = 3\pi^2/Q^2$. Therefore,

$$\sigma_j(\widetilde{\Delta}) \ge \frac{\pi^2}{Q^2}[\max\{((6(j+1))/\pi)^{1/3} - \sqrt{3}, \sqrt{3}\}]^2. \tag{12.3.4}$$

We see that $\sigma_j^{-1}(\widetilde{\Delta}) = O(j^{-2/3})$. Consequently, for $p > 3/2$ we have

$$\sum_{j=1}^{N_0(r_0)} \sigma_j(\widetilde{\Delta})^{-p} \le \frac{Q^{2p}}{\pi^{2p}}\theta(p),$$

where $\theta(p)$ depends only on the exponent p but is independent of r_0, i.e., it is independent of the dimension N_0 of the model space R_0.

If a finite-dimensional approximation of the Laplace operator is constructed using the weak approximation method as in §12.1, then we obtain the same diagonal matrix which, acting on the vector with components ξ_{klm}, transforms the function

$$u = \sum_{(k,l,m)\in D(r_0)} \xi_{klm} u_{klm}$$

to the vector with components $\eta_{klm} = -\frac{\pi^2}{Q^2}(k^2 + l^2 + m^2)\xi_{klm}$ corresponding to the function

$$\Delta u = \sum_{(k,l,m)\in D(r_0)} \eta_{klm} u_{klm}.$$

Indeed, taking the space R_1 to be equal to R_0 and using (12.1.8) and (12.1.9), where $v^{(j)} \equiv u^{(j)}$ runs over the basis functions u_{klm}, we must approximate the operator Δ by the operator $\widetilde{\Delta}$ so that

$$\widetilde{\Delta} u \equiv \widetilde{\Delta}\left(\sum_{(k,l,m)\in D(r_0)} \xi_{klm} u_{klm}\right) = \sum_{k,l,m} \xi_{klm}\left[\sum_{k',l',m'} l_{klm}^{k'l'm'} u_{k'l'm'}\right],$$

$$\begin{aligned} l_{klm}^{k'l'm'} &= (\Delta u_{klm}, u_{k',l',m'}) \\ &= \begin{cases} -\frac{\pi^2}{Q^2}(k^2 + l^2 + m^2), & k = k', l = l', m = m', \\ 0, & \text{if at least one of the equalities} \\ & k = k',\ l = l',\ m = m' \text{ fails.} \end{cases} \end{aligned} \tag{12.3.5}$$

We have proved that the rule (12.3.5) based on the weak approximation method coincides with the natural construction (12.3.3).

Recall (see §12.1) that for sufficiently smooth functions $u(x, y, z)$ and $v(x, y, z)$ vanishing on the boundary of the cube $0 \le x \le Q$, $0 \le y \le Q$, $0 \le z \le Q$ we have

$$\begin{aligned}(\Delta u, v) &= \int_0^Q\int_0^Q\int_0^Q \left(\frac{\partial^2 u}{\partial x^2} + \frac{\partial^2 u}{\partial y^2} + \frac{\partial^2 u}{\partial z^2}\right)\overline{v} dx\, dy\, dz \\ &= -\int_0^Q\int_0^Q\int_0^Q \left(\frac{\partial u}{\partial x}\frac{\partial \overline{v}}{\partial x} + \frac{\partial u}{\partial y}\frac{\partial \overline{v}}{\partial y} + \frac{\partial u}{\partial z}\frac{\partial \overline{v}}{\partial z}\right) dx\, dy\, dz = -[\widehat{d}u, \widehat{d}v],\end{aligned} \tag{12.3.6}$$

where $\widehat{d}$ is the gradient (see (12.2.1)) and $[\cdot,\cdot]$ denotes the inner product (see (12.2.2)). Formula (12.3.6) shows that the inner products $(\Delta u^{(i)}, u^{(j)}) = (\widetilde{\Delta} u^{(i)}, u^{(j)})$ can be found without computing the second order derivatives of $u^{(i)}$ and $u^{(j)}$.

Now we consider an approximation of the operator $L^{[2]}$ which is slightly more complicated than the Laplace operator Δ:

$$L^{[2]}u = \frac{\partial}{\partial x}\left(K\frac{\partial u}{\partial x}\right) + \frac{\partial}{\partial y}\left(K\frac{\partial u}{\partial y}\right) + \frac{\partial}{\partial z}\left(K\frac{\partial u}{\partial z}\right),$$

$$0 < \kappa_0 \le K = K(x, y, z) \le \kappa_1.$$

This operator acts on the same space of sufficiently smooth functions $u(x, y, z)$ defined in the cube $0 \le x \le Q$, $0 \le y \le Q$, $0 \le z \le Q$ and vanishing on the boundary. As was mentioned,

$$\begin{gathered}(L^{[2]}u, v) = -(K\widehat{d}u, \widehat{d}v), \\ -\kappa_0(\Delta u, u) \le (-L^{[2]}u, u) \le \kappa_1(\widehat{d}u, \widehat{d}u) = -\kappa_1(\Delta u, u).\end{gathered} \tag{12.3.7}$$

To construct a finite-dimensional approximation $\widetilde{L}^{[2]}$ of the operator $L^{[2]}$, we choose $R_0 = R_1$ as in the case of the Laplace operator. The basis $u^{(j)}$ for R_0 is the same. The inequalities (12.3.7) lead to the relation

$$\kappa_0\sigma_j(\widetilde{\Delta}) \le \sigma_j(-\widetilde{L}^{[2]}) = \lambda_j(-\widetilde{L}^{[2]}) \le \kappa_1\sigma_j(\widetilde{\Delta})$$

and for any $d_0 > 0$,

$$\begin{gathered}d_0 + \kappa_0\sigma_j(\widetilde{\Delta}) \le \lambda_j(d_0 I_{N_0} - \widetilde{L}^{[2]}) = \sigma_j(d_0 I_{N_0} - \widetilde{L}^{[2]}) \le d_0 + \kappa_1\sigma_j(\widetilde{\Delta}), \\ d_0 \le d_0 + \kappa_0\sigma_1(\widetilde{\Delta}) \le \sigma_1(d_0 I_{N_0} - \widetilde{L}^{[2]}), \\ \|(d_0 I_{N_0} - \widetilde{L}^{[2]})^{-1}\| \le 1/d_0,\end{gathered} \tag{12.3.8}$$

where $\lambda_j(-\widetilde{L}^{[2]})$ and $\lambda_j(d_0 I_{N_0} - \widetilde{L}^{[2]})$ denote the jth eigenvalues which are real and positive.

Since $\sigma_j^{-1}(\widetilde{\Delta}) = O(j^{-2/3})$ (see (12.3.4)), for $\sigma_j(d_0 I_{N_0} - \widetilde{L}^{[2]})$ we have the similar relation

$$\sigma_j^{-1}(d_0 I_{N_0} - \widetilde{L}^{[2]}) = O(j^{-2/3}).$$

Recall that $\widetilde{L}^{[1]}$ is a finite-dimensional approximation of the first order differential operator

$$L^{[1]} = a\frac{\partial u}{\partial x} + b\frac{\partial u}{\partial y} + c\frac{\partial u}{\partial z}.$$

As was shown in §12.1, the second order operator $L = L^{[2]} + L^{[1]}$ is sectorial. Therefore,

$$\left|\mathrm{Im}\left(\frac{(Lu,u)}{(u,u)}\right)\right| + \mathrm{Re}\frac{(Lu,u)}{(u,u)} \le \frac{m^2}{\kappa_0}. \tag{12.3.9}$$

The inequality (12.3.9) remains valid for any finite-dimensional approximation $\widetilde{L} = \widetilde{L}^{[2]} + \widetilde{L}^{[1]}$ of the operator L. By (12.3.9), for $d_0 > m^2/\kappa_0$ the operator $d_0 I_{N_0} - \widetilde{L}$ is invertible, hence nondegenerate, i.e.,

$$\|(d_0 I_{N_0} - \widetilde{L})^{-1}\|_{R_0} \le \gamma_0(d_0) = \left(d_0 - \frac{m^2}{\kappa_0}\right)^{-1}, \quad d_0 > \frac{m^2}{\kappa_0}. \tag{12.3.10}$$

In §12.2 we proved the inequality (12.2.5). Now, it is convenient to rewrite this inequality as follows:

$$\|\widetilde{L}^{[1]}u\|_{R_0} \le \frac{m}{\sqrt{\kappa_0}}\sqrt{([d_0 I_{N_0} - \widetilde{L}^{[2]}]u, u)}.$$

Replacing u by $[d_0 I_{N_0} - \widetilde{L}^{[2]}]^{-1}u$ and using (12.3.8) and the fact that the operator $[d_0 I_{N_0} - \widetilde{L}^{[2]}]^{-1}$ is self-adjoint, we obtain

$$\begin{aligned} &\|\widetilde{L}^{[1]}(d_0 I_{N_0} - \widetilde{L}^{[2]})^{-1}\| \le \frac{m}{\sqrt{\kappa_0 d_0}}, \\ &\|[d_0 I_{N_0} - \widetilde{L}][d_0 I_{N_0} - \widetilde{L}^{[2]}]^{-1}u\| = \|u - \widetilde{L}^{[1]}(d_0 I_{N_0} - \widetilde{L}^{[2]})^{-1}u\| \\ &\qquad \le 1 + \frac{m}{\sqrt{\kappa_0 d_0}}. \end{aligned} \tag{12.3.11}$$

It is obvious that

$$\left(1 - \frac{m}{\kappa_0 d_0}\right)\|u\| \le \|u - \widetilde{L}^{[1]}(d_0 I_{N_0} - \widetilde{L}^{[2]})^{-1}u\| \le \left(1 + \frac{m}{\kappa_0 d_0}\right)\|u\|.$$

We can choose d_0 such that

$$\frac{\|u\|}{2} \le \|[I_{N_0} - \widetilde{L}^{[1]}(d_0 I_{N_0} - \widetilde{L}^{[2]})^{-1}]u\| \le \frac{3\|u\|}{2}, \tag{12.3.12}$$

which implies that for this d_0 the ratio

$$\frac{\sigma_j(d_0 I_{N_0} - \widetilde{L})}{\sigma_j(d_0 I_{N_0} - \widetilde{L}^{[2]})}$$

of the singular values of the operators

$$\begin{gathered} d_0 I_{N_0} - \widetilde{L}^{[2]}, \\ d_0 I_{N_0} - \widetilde{L} = [I_{N_0} - \widetilde{L}^{[1]}(d_0 - \widetilde{L}^{[2]})^{-1}](d_0 - \widetilde{L}^{[2]}) \end{gathered}$$

is located between 1/2 and 3/2. These constants are independent of j and the dimension N_0 of the space R_0. Therefore,

$$\begin{aligned}\sigma_j^{-1}([d_0 I_{N_0} - \widetilde{L}]) &= O(j^{-2/3}),\\ \sigma_j([d_0 I_{N_0} - \widetilde{L}]) &= O(j^{2/3}),\\ \sigma_j(\widetilde{L}) &= O(j^{2/3}).\end{aligned}$$

Thus, finite-dimensional approximations of the operator L are sectorial operators (see §12.1); moreover, for any $p > 3/2$ there exists a constant Φ depending only on d_0, p, and the coefficients of L such that

$$\sum_{j=1}^{N_0} \sigma_j^p([d_0 I_{N_0} - \widetilde{L}]^{-1}) \leq \Phi. \tag{12.3.13}$$

It is essential that the constant Φ is independent of the dimension N_0 of the finite-dimensional space R_0 in which $\widetilde{L}$ acts.

Our proof was based on the choice of the special basis for the space R_0, namely,

$$u_{klm} = \left(\frac{2}{Q}\right)^{3/2} \sin\frac{k\pi x}{Q} \sin\frac{l\pi y}{Q} \sin\frac{m\pi z}{Q}.$$

In §12.4 we will show that such a choice of a basis (as well as the choice of the domain D in the form of a cube) is not crucial for the estimate (12.3.13).

12.4. The finite element method

The construction of finite-dimensional approximations of a second order differential operator defined in a bounded polyhedron. Finite elements. The finite element method. The boundedness of the spectral condition parameter in the approximation by the finite element method.

We return to finite-dimensional approximations $\widetilde{L}$ of the differential operator

$$Lu = \frac{\partial}{\partial x}\left(K\frac{\partial u}{\partial x}\right) + \frac{\partial}{\partial y}\left(K\frac{\partial u}{\partial y}\right) + \frac{\partial}{\partial z}\left(K\frac{\partial u}{\partial z}\right) + a\frac{\partial u}{\partial x} + b\frac{\partial u}{\partial y} + c\frac{\partial u}{\partial z}, \tag{12.4.1}$$

but under the assumption that the domain D is not necessarily a cube. We assume that D (or some approximation of D) is a bounded polyhedron lying in the cube $0 \leq x \leq Q$, $0 \leq y \leq Q$, $0 \leq z \leq Q$. We make the so-called simplicial partition of the polyhedron, i.e., we consider it to be the union of sufficiently small tetrahedra (simplices) satisfying certain conditions. Namely, we require that two different tetrahedra are either disjoint, or their intersection is a single point (a common vertex), or a segment (a common edge), or a single plane triangle (a common face). The operator L is applied to sufficiently smooth functions defined in D and vanishing on the boundary of D.

To construct a finite-dimensional approximation $\widetilde{L}$ of the operator L, for the space R_0 on which $\widetilde{L}$ acts we take the space of functions that are defined and continuous in D, vanish on the boundary of D, and are linear functions in every tetrahedron of the simplicial partition. Such a function is uniquely determined by its values at the interior vertices of the partition, i.e., at those vertices of the tetrahedra which do not belong to the boundary of the polyhedron. The dimension N_0 of the space R_0 coincides with the number of the interior vertices of the partition. Any function from the space R_0 can be represented as a linear combination of special basis functions u that vanish at all vertices of the partition, except for one vertex A at which $u = 1$. Such a function differs from zero only at the points of those tetrahedra that have A as one of the vertices. The basis functions are called *finite elements* and the approximation $\widetilde{L}$ of the operator L with the space R_0 is called the *approximation by the finite element method.* In spite of the fact that functions of R_0 have no second derivatives, the inner products (Lu, v) written in the form

$$(12.4.2)\quad (Lu, v) = \iiint_D [\overline{v}(au_x + bu_y + cu_z) - K(u_x\overline{v}_x + u_y\overline{v}_y + u_z\overline{v}_z)]\, dx\, dy\, dz,$$

can be easily computed for $u, v \in R_0$. We emphasized this fact in §12.1 when the weak approximation method was mentioned for the first time.

For any choice of a basis in R_0 the model operator $\widetilde{L}$, being a formal approximation of the operator L, is the same for any choice of the construction method based on the procedure from §12.1, although the corresponding matrices depend on the choice of the basis. To construct $\widetilde{L}$, we use the basis composed of orthonormal linear combinations of finite elements $u^{(1)}, u^{(2)}, \dots, u^{(N_0)}$ so that

$$(u^{(i)}, u^{(j)}) = \begin{cases} 0 & \text{for } i \neq j, \\ 1 & \text{for } i = j. \end{cases}$$

Denote by H the square matrix with elements $h_{ij} = (Lu^{(i)}, u^{(j)})$. Following the procedure from §12.1, we set

$$\widetilde{L}u^{(i)} = \sum_{j=1}^{N_i} h_{ij}u^{(j)}$$

so that the action of the operator $\widetilde{L}$ on any function

$$u = \sum_{j=1}^{N_0} \xi_j u^{(i)} \in R_0,$$

i.e., the computation of $\widetilde{L}u = v = \sum_{j=1}^{N_0} \eta_j u^{(j)}$, is reduced to the following rules:

$$(12.4.3)\qquad \widetilde{L}u = \sum_{j=1}^{N_0}\left(\sum_{i=1}^{N_0} \xi_i h_{ij} u^{(j)}\right), \qquad \eta_j = \sum_{i=1}^{N_0} h_{ij}\xi_i.$$

Denoting by ξ and η the vectors with components ξ_i and η_j, we rewrite the last equality in (12.4.3) in the symbolic form as follows: $\eta = H^T\xi$, where H^T is the transposed matrix to H.

For the singular values of the operator $\widetilde{L}$ it is natural to take the singular values of H^T (which coincide with the singular values $\sigma_j(H)$):

$$\sigma_j(\widetilde{L}) = \sigma_j(H).$$

The constructed operator $\widetilde{L}$ is sectorial. Indeed, we can repeat all arguments from §12.1 because no assumption on the basis elements $u^{(i)}$ was imposed there except for the assumption that $(Lu^{(i)}, u^{(j)})$ can be found by formula (12.4.2). However, the estimates for singular values of the operator $d_0 I_{N_0} - \widetilde{L}$ (see §12.3) can fail for $\widetilde{L}$ because the basis elements

$$u_{klm} = \left(\frac{2}{Q}\right)^{3/2} \sin\frac{k\pi x}{Q} \sin\frac{l\pi y}{Q} \sin\frac{m\pi z}{Q}$$

were essentially used in the derivation of these estimates. However, we can claim that the operators $q_0 I - L$ exist and for every $p > 3/2$ there is a constant Φ such that

$$\sum_{j=1}^{N_0} \sigma_j^p([q_0 I_{N_0} - \widetilde{L}]^{-1}) = \sum_{j=1}^{N_0} \sigma_j^{-p}(q_0 I_{N_0} - \widetilde{L}) < \Phi$$

for any approximation $\widetilde{L}$ obtained by the finite element method. This estimate is independent of the dimension N_0 of the space R_0, i.e., of the number of tetrahedra in the simplicial partition and their size.

We extend the functions $u^{(j)}$ to the entire cube $0 \le x \le Q$, $0 \le y \le Q$, $0 \le z \le Q$ which contains the domain D so that $u^{(j)}$ vanish outside D. It is obvious that the inner products $(u^{(i)}, u^{(j)})$ and $(Lu^{(i)}, u^{(j)})$ in the construction of $\widetilde{L}$ do not change if we integrate over the entire cube.

A theorem in analysis says that if u is a continuous function such that its first order derivatives are bounded and piecewise continuous, have a finite number of sufficiently smooth curves of discontinuity, and vanish on the boundary of the cube, then u can be approximated by partial Fourier series:

$$\widehat{u} = \sum_{(k,l,m)\in D(r_0)} \xi_{klm} \left(\frac{2}{Q}\right)^{3/2} \sin\frac{k\pi x}{Q} \sin\frac{l\pi y}{Q} \sin\frac{m\pi z}{Q}. \tag{12.4.4}$$

Moreover, for any small positive ε_0 and ε_1 we can choose a number of terms of the series and the coefficients ξ_{klm} so that the following inequalities hold:

$$\begin{gathered} \int_0^Q\int_0^Q\int_0^Q |u(x,y,z) - \widehat{u}(x,y,z)|^2\, dx\, dy\, dz < \varepsilon_0, \\ \int_0^Q\int_0^Q\int_0^Q (|u_x - \widehat{u}_x|^2 + |u_y - \widehat{u}_y|^2 + |u_z - \widehat{u}_z|^2)\, dx\, dydz < \varepsilon_1. \end{gathered} \tag{12.4.5}$$

The assumption that the first derivatives are piecewise continuous and bounded can be substantially weakened. In fact, a similar approximation theorem is valid for any function from the Sobolev space $\overset{0}{W}{}_2^1$, i.e., the space of square integrable functions that have the generalized first order derivatives and vanish on the boundary of the cube. For our purpose it suffices to know that the approximation (12.4.5) with any ε_0 and ε_1 exists for the basis functions $u^{(j)}$ from R_0. Fix a space R_0 of dimension N_0. By the approximation theorem, there exists r_0 such that for all $u^{(1)}, u^{(2)}, \ldots, u^{(N_0)}$ we can find functions $\widehat{u}^{(j)}$ of the form (12.4.4) such that for any given e_0, e_1 $(0 < e_0 < 1,\ 0 < e_1 < 1)$ the matrices

$$\widehat{G} = \begin{bmatrix} \widehat{g}_{11} & \widehat{g}_{12} & \cdots \\ \widehat{g}_{21} & \widehat{g}_{22} & \cdots \\ \cdots & \cdots & \cdots \\ \cdots & \cdots & \cdots \end{bmatrix}, \qquad G = I_{N_0} = \begin{bmatrix} g_{11} & g_{12} & \cdots \\ g_{21} & g_{22} & \cdots \\ \cdots & \cdots & \cdots \\ \cdots & \cdots & \cdots \end{bmatrix},$$

$$\widehat{H} = \begin{bmatrix} \widehat{h}_{11} & \widehat{h}_{12} & \cdots \\ \widehat{h}_{21} & \widehat{h}_{22} & \cdots \\ \cdots & \cdots & \cdots \\ \cdots & \cdots & \cdots \end{bmatrix}, \qquad H = \begin{bmatrix} h_{11} & h_{12} & \cdots \\ h_{21} & h_{22} & \cdots \\ \cdots & \cdots & \cdots \\ \cdots & \cdots & \cdots \end{bmatrix}$$

with elements

$$\widehat{g}_{ij} = (\widehat{u}^{(i)}, \widehat{u}^{(j)}), \quad g_{ij} = (u^{(i)}, u^{(j)}) = \delta_{ij},$$
$$\widehat{h}_{ij} = (L\widehat{u}^{(i)}, \widehat{u}^{(j)}), \quad h_{ij} = (Lu^{(i)}, u^{(j)})$$

satisfy the following inequalities:

$$\|\widehat{G} - I_{N_0}\| \le e_0, \quad \|\widehat{H} - H\| \le e_1. \tag{12.4.6}$$

Using the matrix H and taking into account that $u^{(j)}$ form an orthonormal basis for R_0, i.e., $G = I_{N_0}$, we construct the approximation $\widetilde{L}$ of L. The singular values $\sigma_j(\widetilde{L})$ of $\widetilde{L}$ coincide with the singular values of the matrix H $(\sigma_j(\widetilde{L}) = \sigma_j(H))$.

Using the matrices $\widehat{G}$ and $\widehat{H}$, we construct another approximation $\widehat{L}$ of the operator L acting in the space R_0 of the same dimension N_0. The space $\widehat{R}_0$ is the linear span of the vectors $\widehat{u}^{(1)}, \widehat{u}^{(2)}, \dots, \widehat{u}^{(N_0)}$. For sufficiently small e_0 the vectors $\widehat{u}^{(1)}, \widehat{u}^{(2)}, \dots, \widehat{u}^{(N_0)}$ are linearly independent because the Gram matrix is positive definite in view of the first inequality in (12.4.6).

Starting with the matrices $\widehat{G}$ and $\widehat{H}$, we can find the singular values of $\widehat{L}$. Indeed, it suffices to use the estimate

$$\frac{\sigma_j(\widehat{H})}{\sqrt{1+e_0}} \le \sigma_j(\widehat{L}) \le \frac{\sigma_j(\widehat{H})}{\sqrt{1-e_0}} \tag{12.4.7}$$

(see (12.1.10), we emphasize that the notations are different). The second inequality in (12.4.6) leads to the relations

$$\begin{gathered} |\sigma_j(H) - \sigma_j(\widehat{H})| < e_1, \\ \sigma_j(\widehat{H}) - e_0 \le \sigma_j(\widetilde{L}) = \sigma_j(H) \le \sigma_j(\widehat{H}) + e_0. \end{gathered} \tag{12.4.8}$$

From (12.4.7) and (12.4.8) we obtain

$$\sqrt{1+e_0}\sigma_j(\widehat{L}) + e_1 \ge \sigma_j(\widetilde{L}) \ge \sqrt{1-e_0}\sigma_j(\widehat{L}) - e_1. \tag{12.4.9}$$

We will show that the singular values $\sigma_j(\widehat{L})$ satisfy the relation

$$\sigma_j(\widehat{L}) = O(j^{2/3}), \tag{12.4.10}$$

where the constant in the estimate $O(j^{2/3}) \le \text{const } j^{2/3}$ depends only on the coefficients of the operator L and the size of the cube. By (12.4.9), the relation similar to (12.4.10) holds for $\widetilde{L}$ and $(q_0 I - \widetilde{L})$:

$$\sigma_j[(q_0 I - \widetilde{L})] = O(j^{2/3}).$$

To prove (12.4.10), we note that $\widehat{R}_0$ is a subspace of the linear space $\widehat{\widehat{R}}_0$ of larger but finite dimension formed by all sums of functions u_{klm} of the form (12.4.4) for sufficiently large r_0. It is necessary that the functions $\widehat{u}^{(1)}, \widehat{u}^{(2)}, \ldots, \widehat{u}^{(N_0)}$ be present among such sums. The operator $\widehat{L}$ can be extended to the entire space $\widehat{R}_0$, and the extension coincides with the operator $\widetilde{L}$ (see §12.3). By the alternation theorem (see §5.4), the singular values with the same numbers can only decrease. Therefore, the assertion that $\sigma_j = O(j^{2/3})$ in the extended space implies the same assertion for $\sigma_j(\widehat{L})$.

This completes our discussion of the fact that the approximation $\widetilde{L}$ of the second order differential operator (12.4.1) constructed by the finite element method, as well as the approximation from §12.3, is a sectorial operator and, in addition, satisfies the estimates

$$\sigma_j[(q_0 I - \widetilde{L})] = O(j^{2/3}),$$

$$\sum_{j=1}^{N_0} \sigma_j^p[(q_0 I - \widetilde{L})^{-1}] \leq \Phi \quad (p > 3/2)$$

with constants independent of the dimension N_0.

We recall the definition of the spectral condition number (see §7.8):

$$\nu^{(p)}(C) = \frac{1}{\sigma_N(C)} \left(\sum_{j=1}^{N} \sigma_j^p(C) \right)^{1/p}.$$

In fact, the above estimates mean the boundedness of the spectral condition number $\nu^{(p)}[(q_0 I - \widetilde{L})^{-1}]$ for $p > 3/2$ by the same constant for any choice of triangulation.

Although the material presented in this chapter is not immediately connected with linear algebra, we find it necessary to show, demonstrating the simplest examples, how matrices satifying the above conditions appear in applications.

CHAPTER 13

Application of the Theory of Functions of Complex Variables

13.1. The Cartan inequality for polynomials

Zeros of a polynomial. The formulation of the Cartan theorem. Inductive construction of coverings. The proof of the Cartan inequality.

Every polynomial $P(z) = (z-z_1)(z-z_2)\dots(z-z_n)$ vanishes at a finite number of points $z_1, z_2, \dots, z_n$, called *zeros* of the polynomial. At the remaining points z of the complex plane, the module $|P(z)|$ is strictly positive ($|P(z)| > 0$ for $z \neq z_j$, $j = 1, 2, \dots, n$). If we cover the zeros by neighborhoods, then outside these neighborhoods $|P(z)| \geq m > 0$, where m depends on the size of the neighborhoods. For example, if $P(z) = (z - z_0)^n$, i.e., if the zeros z_j of the polynomial are located at one point $z_j = z_0$, $j = 1, 2, \dots, n$, then $|P(z)| = |(z - z_0)^n| \geq \rho^n$ outside the circle of radius ρ and center at z_0, i.e., for z such that $|z - z_0| \geq \rho$. It turns out that for any polynomial $P_n(z) = (z - z_1)(z - z_2)\dots(z - z_n)$ and a number $\rho > 0$ there exists a neighborhood of the zeros consisting of disks such that the sum of the radii of disks is equal to ρ. Moreover, outside this neighborhood we can estimate $|P_n(z)|$ from below by a constant depending only on ρ and n. We give an accurate formulation of this assertion.

THEOREM 13.1.1 (Cartan). *For any points $z_1, z_2, \dots, z_n$ in the complex plane and a number $\rho > 0$ there are p, $1 \leq p \leq n$, closed disks of radius ρ_i such that the sum of their radii is equal to ρ ($\rho_1 + \rho_2 + \dots + \rho_p = \rho$, $\rho_i > 0$) and at any point z outside these disks the following inequality holds:*

$$|(z - z_1)(z - z_2)\dots(z - z_n)| > \left(\frac{\rho}{2e}\right)^n.$$

The proof of this theorem is elementary but cumbersome, and requires intensive attention from the reader. This proof can be omitted for the first reading.

PROOF OF THEOREM 13.1.1. Consider the polyhedron that is the convex hull of the points $z_1, z_2, \dots, z_n$ and choose some vertex $z = z_j$. It is obvious that the vertex coincides with one of the points $z_1, z_2, \dots, z_n$ and we can construct a circle that contains the vertex $z = z_j$, but does not contain any point z_m different from z_j (in the geometrical sense). If the multiplicity of z_j is k (i.e., among $z_1, z_2, \dots, z_n$ there are exactly k terms equal to z_j), then we take the circle of radius $\rho k/(2n)$ that contains z_j.

By the above construction, we can conclude that in the complex plane there exists at least one closed disk of radius $l\rho/(2n)$ containing l, $n \geq l \geq 1$, points z_j. Certainly, the number of such disks may be larger than 1.

Denote by l_1 the maximal number of points z_j contained in such a disk and by K_1 one of such disks of radius $r_1 = l_1\rho/(2n)$. Let us show that no disk of radius $r > r_1 = l_1\rho/(2n)$ contains more than $2nr/\rho$ points z_j. Indeed, let the disk $|z-a| \leq r$ contain exactly L points z_j, $r > l_1\rho/(2n)$, $L > l_1$. Three cases are possible.

Case 1. $r > \rho L/(2n) > \rho l_1/(2n)$. Then L is at most $2nr/\rho$.

Case 2. $r = \rho L/(2n) > \rho l_1/(2n)$. Then the disk $|z-a| \leq r$ contains exactly L points z_j and has radius $\rho L/(2n)$; moreover, $L > l_1$, which contradicts the definition of l_1. Case 2 is impossible.

Case 3. $\rho L/(2n) > r > \rho l_1/(2n)$. The disk $|z-a| \leq r'$ of radius $r' = \rho L/(2n) > r > \rho l_1/(2n)$ contains $L' \geq L$ points z_j. For $L' = L$ we are in the situation of Case 2 which is impossible. If $2nr'/\rho = L < L'$, then the disk of radius $r'' = \rho L'/(2n)$ contains $L'' \geq L' > L > l_1$ points z_j. The equality $L'' = L'$ again leads to Case 2 which is impossible. Hence we must assume that $L'' > L' > L > l_1$. Consider the disk $|z-a| \leq \rho L''/(2n)$ and, arguing as above, establish that this disk contains L''' points z_j; moreover, $L''' > L''$. Similarly, the disk $|z-a| \geq \rho L'''/(2n)$ contains $L^{(4)} > L'''$ points z_j and so on:

$$\cdots > L^{(q)} > L^{(q-1)} > \cdots > L^{(4)} > L''' > L'' > L' > L > l_1.$$

The chain of admissible values $L, L', L'', \dots,$ cannot be infinite since we have only n points z_j. The contradiction shows that Case 3 is impossible.

Thus, only Case 1 can occur. Hence the number of points z_j lying in a disk of radius $r > r_1 = \rho l_1/(2n)$ is at most $2nr/\rho$.

Take a closed disk K_1 of radius $r_1 = \rho l_1/(2n)$ that contains exactly l_1 points z_j. Points z_j lying in K_1 are called *points of rank* 1. Eliminate all points of rank 1 from the sequence $\{z_j\}$. Then it contains $n - l_1$ points.

For $n - l_1 > 0$ there exists a closed disk of radius $\rho l/(2n)$, where l is the number of points z_j that lie in the considered disk, but outside K_1. Denote by K_2 such a disk of maximal radius. It contains l_2 points z_j lying outside K_1; moreover, the radius r_2 of K_2 is equal to $\rho l_2/(2n)$.

We show that $l_2 \leq l_1$. For $l_2 > l_1$ we have $r_2 > r_1$ since $r_2 = \rho l_2/(2n)$ and $r_1 = \rho l_1/(2n)$. Furthermore, the number of points z_j lying in the disk K_2 of radius r_2 is less than $2nr_2/\rho = l_2$. The number of points lying in K_2 but not in K_1 is even less. Hence the equality $2nr_2/\rho = l_2$ fails. Thus,

$$l_2 = \frac{2nr_2}{\rho} \leq \frac{2nr_1}{\rho} = l_1.$$

The inequality $l_2 \leq l_1$ is proved.

Every disk $|z-b| \leq r$ of radius r greater than r_2 ($r > r_2 = \rho l_2/(2n)$) contains less than $2nr/\rho$ points z_j lying outside K_1 (hence less than $2nr/\rho$ points lying outside both K_1 and K_2). The proof of this assertion repeats literally the above proof that a disk of radius r_1 contains less than $2nr/\rho$ points z_j ($1 \leq j \leq k$). We recommend that the reader check this assertion.

Points z_j lying in K_2 but not in K_1 are called *points of rank* 2. The number of points of rank 2 is equal to l_2, $l_2 \leq l_1$. Eliminating points of ranks 1 and 2 from the sequence $\{z_j\}$, we obtain a sequence that contains $n - l_2 - l_1$ points.

If $l_1 + l_2 < n$, then there exist closed disks of radii $l\rho/(2n)$, where l is the number of the points z_j lying in such a disk but outside both K_1 and K_2. Denote by K_3 such a disk of the maximal radius $l_3\rho/(2n)$. It contains exactly l_3 points z_j.

Points z_j lying in K_3 and outside both K_1 and K_2 are called *points of rank* 3. We recommend that the reader verify the inequality $l_3 \le l_2$ and show that any disk $|z-c| \le r$ of radius $r > r_3$ contains less than $2nr/\rho$ points z_j lying outside K_1 and K_2. Eliminate all points of K_3 from the sequence $\{z_j\}$. We see that the number of points z_j in the disk $|z-c| \le r$ is less than $2nr/\rho$.

Continuing as above, we construct disks K_m, $m = 1, 2, \dots, p$, such that K_m, $m \ge 2$, contains the maximal number l_m of points from the sequence $z_1, z_2, \dots, z_n$ that do not belong to $K_1, K_2, \dots, K_{m-1}$. The radius r_m of the disk K_m is equal to $r_m = \rho l_m/(2n)$. The number p of such disks is determined from the fact that no point z_j lies outside $K_1, K_2, \dots, K_p$, i.e., p can be found from the equality $l_1 + l_2 + \cdots + l_p = n$.

Analyzing the above procedure we can conclude that $l_m \le l_{m-1}$ for all $m = 2, 3, \dots, p$. For $m = 2$ the detailed proof of this assertion was given above. We recommend the reader to verify similar arguments for $m \ge 3$ and to establish that the disk $|z-g| \le r$ of radius $r > r_m = \rho l_m/(2n)$ contains at most $2nr/\rho$ points z_j lying outside the disks $K_1, K_2, \dots, K_{m-1}$. In particular, for $k > l_m$, $r = \rho k/(2n)$, the last assertion means that any closed disk of radius r contains at most $k-1$ points from the sequence $\{z_j\}$ that are not covered by $K_1, K_2, \dots, K_{m-1}$.

Points z_j lying in K_m but not in $K_1, K_2, \dots, K_{m-1}$ are called *points of rank* m.

Consider a disk $\widehat{K}_m$, $1 \le m \le p$, that is concentric with the disk K_m and has two times greater radius: $\widehat{r}_m = 2r_m = \rho l_m/n$. The sum of the radii of the disks $\widehat{K}_m$, $1 \le m \le p$, is equal to ρ:

$$\widehat{r}_1 + \widehat{r}_2 + \cdots + \widehat{r}_p = \frac{\rho}{n}(l_1 + l_2 + \cdots + l_p) = \frac{\rho}{n} \cdot n = \rho.$$

The interiors of the disks $\widehat{K}_1, \widehat{K}_2, \dots, \widehat{K}_p$ are taken for neighborhoods of the zeros z_j of the polynomial $P(z)$. Outside the neighborhoods we have $|P(z)| > [\rho/(2l)]^n$.

Assume z^* does not belong to any of the disks $\widehat{K}_1, \widehat{K}_2, \dots, \widehat{K}_p$. If z_j lies in some K_m, then $|z^* - z_j| > \rho l_m/(2n)$ by the construction of $\widehat{K}_m$. We construct the disk $\mathcal{K}^s$ such that $|z - z^*| \le \rho s(2n)$, where $z \in \mathcal{K}^s$ and s is a natural number. A point z_j belongs to K_m only if $s > l_m$ or, which is equivalent, $l_m \le s-1$.

Let $l_{M-1} \ge s$ and $s - 1 \ge l_M \ge l_{M+1} \ge \cdots \ge l_p$. As was noted, any disk of radius r greater than $r_M = \rho l_M/(2n)$ cannot contain $2nr/\rho$ points z_j lying outside the disks $K_1, K_2, \dots, K_{M-1}$. In particular, any disk of radius $\rho s/(2n)$ greater than $\rho(s-1)/(2n) \ge \rho l_M/(2n)$ cannot cover more than $s-1$ such points. The disk $\mathcal{K}^s$ has radius $\rho s/(2n)$. Moreover, it does not contain points lying in the disks $K_1, K_2, \dots, K_{M-1}$ since $l_1 \ge l_2 \ge \cdots \ge l_{M-1} \ge s$. Therefore, the number of points from the sequence $z_1, z_2, \dots, z_n$ lying in $\mathcal{K}^s$ ($|z - z^*| \le \rho s/(2n)$) does not exceed $s-1$. We enumerate z_j so that $|z_1 - z^*| \le |z_2 - z^*| \le \cdots \le |z_n - z^*|$. With respect to this numbering we have $|z_j - z^*| > \rho s/(2n)$ for $j \ge s$ (in particular, $|z_s - z^*| > \rho s/(2n)$). Hence we obtain the relation

$$|z_1 - z^*|\,|z_2 - z^*| \dots |z_n - z^*| > \left(\frac{\rho}{2n}\right)^n \cdot 1 \cdot 2 \cdot \dots \cdot n = \frac{n!}{n^n}\left(\frac{\rho}{2}\right)^n,$$

with arbitrary ordering of z_j.

We recall that for z^* we can take any point z outside the disks $\widehat{K}_1, \widehat{K}_2, \dots, \widehat{K}_p$ and the sum of the radii of these disks is ρ. Therefore, if z lies outside a neighborhood of $z_1, z_2, \dots, z_n$, which is formed by the disks $\widehat{K}_i$, then

$$|(z - z_1)(z - z_2) \dots (z - z_n)| \geq \frac{n!}{n^n}\left(\frac{\rho}{2}\right)^n. \tag{13.1.1}$$

Using the inequality $n! > (n/e)^n$, which follows from the Stirling formula $n! = \sqrt{2\pi n}(n/e)^n e^{\theta/(12n)}$, $0 < \theta < 1$, and slightly simplifying (13.1.1), we find that

$$\left|\prod_{j=1}^{n}(z - z_j)\right| > \left(\frac{\rho}{2e}\right)^n \tag{13.1.2}$$

for all points z lying outside a finite number of closed disks such that the sum of the radii of these disks is equal to ρ. The inequality (13.1.2) is the statement of Theorem 13.1.1. □

13.2. The Carathéodory inequality

Harmonic functions. The Poisson formula. The mean value theorem. The logarithm of an analytic function. The Carathéodory inequality.

Let a function $F(x, y)$ be continuous and twice continuously differentiable in some domain. It is said to be *harmonic* if

$$\frac{\partial^2 F}{\partial x^2} + \frac{\partial^2 F}{\partial y^2} = 0$$

everywhere in this domain. For example, the real part $F(x, y)$ and the imaginary part $G(x, y)$ of an analytic function $u(z) = u(x, y) = F + iG$ are harmonic functions. If $F(x, y)$ is harmonic in the disk $x^2 + y^2 < R^2$ and is continuous up to the boundary $x^2 + y^2 = R^2$, then it can be restored at all interior points of the disk from the known boundary values $F(R\cos\theta, R\sin\theta)$ by the Poisson formula $(0 \leq \rho \leq R)$

$$F(\rho\cos\omega, \rho\sin\omega) = -\frac{1}{2\pi}\int_0^{2\pi} F(R\cos\theta, R\sin\theta)\,d\theta$$

$$+ 2\mathrm{Re}\,\frac{1}{2\pi}\int_0^{2\pi} F(R\cos\theta, R\sin\theta)\frac{Re^{i\theta}\,d\theta}{Re^{i\theta} - \rho e^{i\omega}}.$$

The proof of this formula can be found in any textbook on the equations of mathematical physics (see, for example, [**14**]) or on the theory of functions of complex variables (see, for example, [**29**]).

In particular, the Poisson formula implies the equality

$$F(0, 0) = \frac{1}{2\pi}\int_0^{2\pi} F(R\cos\,\theta, R\sin\,\theta)d\theta, \tag{13.2.1}$$

known as the *mean value theorem*. Using (13.2.1), we rewrite the Poisson formula

$$F(\rho\cos\,\omega, \rho\sin\,\omega) + F(0, 0) = 2\mathrm{Re}\frac{1}{2\pi}\int_0^{2\pi} F(\rho\cos\,\theta, \rho\sin\,\theta)\frac{Re^{i\theta}}{Re^{i\theta} - \rho e^{i\omega}}d\theta.$$

Using the elementary relation $|Re^{i\theta} - \rho e^{i\omega}| \geq R - \rho$, we obtain the inequality

$$|F(\rho\cos\omega, \rho\sin\omega) + F(0,0)| \leq \frac{2R}{R-\rho}\max_{\theta}|F(\rho\cos\theta, \rho\sin\theta)|.$$

Let $f(z)$ be an analytic function of complex variable $z = x + iy$. Assume that it is continuous and does not vanish in the disk $|z| \leq R$. The logarithm $\log f(z) = F(x,y) + iG(x,y)$ is also an analytic function in the disk $|z| \leq R$, and the real part $F(x,y) = \log|f(z)|$ is a harmonic function. Consequently, for $|z| < R$ we have

$$\big|\log|f(z)| - \log|f(0)|\big| \leq \frac{2R}{R-|z|}\log M(R),$$

where

$$M(R) = \max_{|z|=R}|f(z)| = \max_{|z|\leq R}|f(z)|.$$

The estimate can be slightly simplified if $f(0) = 1$ ($\log|f(0)| = 0$).

If $f(z)$ is holomorphic and does not vanish for $|z| \leq R$ and $f(0) = 1$, then for $|z| < R$ the following inequality, called the *Carathéodory inequality*, holds:

$$\log|f(z)| \geq -\frac{2R}{R-|z|}\log M(R).$$

13.3. The Jensen inequality

Zeros of a function analytic in a disk. The number of zeros of a function that is analytic in the disk $|z| \leq R$ and does not vanish at the center of the disk. The Jensen theorem. The Jensen inequality.

The goal of this section is to obtain the inequality that connects the number of zeros of a function $f(z)$ analytic in the disk $|z| \leq r$ with the maximal value of the module $|f(z)|$ in the disk of radius $R > r$ ($M(R) = \max_{|z|\leq R}|f(z)| > r$). We begin with the proof of the formula

$$\begin{aligned} I_0(r) &= \frac{1}{2\pi}\int_0^{2\pi} \log|z - z_0|\Big|_{z=re^{i\phi}} d\phi \\ &= \begin{cases} \log|z_0| = \log|r_0| & |z| = r \leq |z_0| = r_0, \\ \log|z| = \log & |z| = r \geq |z_0| = r_0. \end{cases} \end{aligned} \tag{13.3.1}$$

In the domain $|z| < r_0 = |z_0|$, where $z_0 = r_0 e^{i\omega_0}$, the function $z^{-1}\log(z - z_0)$ is analytic everywhere except for the point $z = 0$. At this point, it has the first order pole and is expanded into the Laurent series

$$\begin{aligned} \frac{\log(z - r_0 e^{i\omega_0})}{z} &= \frac{\log[-r_0 e^{i\omega_0}] + \log[1 - \frac{z}{r_0}e^{-i\omega_0}]}{z} \\ &= \frac{\log|r_0| + i[\omega_0 + (2k+1)\pi]}{z} - \frac{e^{-i\omega_0}}{r_0} + \frac{e^{-2i\omega_0}}{2r_0^2}z + \cdots . \end{aligned}$$

The value k depends on the choice of the branch of logarithm. Hence for $r < r_0$ the contour integral

$$J(r) = \oint_{|z|=r<r_0} \frac{\log(z - r_0 e^{i\omega})}{z} dz$$

satisfies the condition

$$J(r) = 2\pi\{i\log|r_0| - [\omega_0 + (2k+1)\pi]\}.$$

Since

$$\log(z - r_0 e^{i\omega}) = \log|z - re^{i\omega_0}| + i\arg(z - r_0 e^{i\omega_0}), \quad 0 \le r < r_0,$$

we obtain

$$I_0(r) = \frac{1}{2\pi}\int_0^{2\pi} \log|re^{i\omega} - r_0 e^{i\omega_0}|d\omega = \frac{1}{2\pi i}\int_0^{2\pi} \frac{\log|re^{i\omega} - r_0 e^{i\omega_0}|}{re^{i\omega}} d(re^{i\omega})$$

$$= \frac{1}{2\pi}\operatorname{Im} J(r) = \log|r_0|.$$

In fact, the integral $I_0(r)$ uniformly converges for $0 \le r \le r_0$. Therefore, $I_0(r)$ is continuous and, consequently, is constant for $0 \le r \le r_0$, i.e.,

$$I_0(r) = \frac{1}{2\pi}\int_0^{2\pi} \log|re^{i\omega} - r_0 e^{i\omega_0}|d\omega = \log|r_0|, \quad 0 \le r \le r_0.$$

PROBLEM 13.3.1. Formulate and prove a proposition about the uniform convergence of the integral. Prove a more general proposition about the uniform convergence of the integral on any segment $0 \le r \le R_0$ ($R_0 > 0$), where the parameter r varies.

We compute $I_0(r)$ for $r > r_0$. We note that

$$|re^{i\omega} - r_0 e^{i\omega_0}| = |r - r_0 e^{-i(\omega-\omega_0)}| = |r - r_0 e^{i(\omega-\omega_0)}| = \left|r - re^{i\omega}\frac{r_0 e^{-i\omega_0}}{r}\right|.$$

Hence

$$I_0(r) = \frac{1}{2\pi}\int_0^{2\pi} \log|re^{i\omega} - r_0 e^{i\omega_0}|d\omega = \frac{1}{2\pi}\int_0^{2\pi} \operatorname{Re}\log\left[r - re^{i\omega}\frac{r_0 e^{-i\omega_0}}{r}\right]d\omega$$

$$= \frac{1}{2\pi}\oint_{|z|=r} \operatorname{Re}\log\left[r - z\frac{r_0 e^{-i\omega_0}}{r}\right]\left(\frac{1}{i}\frac{dz}{z}\right) = \operatorname{Re}\frac{1}{2\pi i}\oint_{|z|=r} \log\left[r - z\frac{r_0 e^{-i\omega_0}}{r}\right]\frac{dz}{z}.$$

In the last integral for $r > r_0$ the function has a single singularity inside the integration contour, namely, the pole $z = 0$ with residue $\log r$. Hence $I_0(r) = \log r$ for $r > r_0$. Thus, we have shown that $I_0(r)$ is a continuous function of r and can be computed according to formula (13.3.1).

Let $P(z) = (z - z_1)(z - z_2)\dots(z - z_n)$ be a polynomial in z and let $z_j \ne 0$, $j = 1, 2, \dots, n$. Since

$$\log|P(z)| = \sum_{j=1}^{n} \log|z - z_j|,$$

it is not hard to obtain the representation

$$I_P(r) = \sum_{|z_j|>r} \log|z_j| + \nu(r)\log r$$

of the integral

$$I_P(r) = \frac{1}{2\pi}\int_0^{2\pi} \log|P(re^{i\omega})|d\omega = \sum_{j=1}^{n} \frac{1}{2\pi}\int_0^{2\pi} \log|re^{i\omega} - z_j|\,d\omega,$$

where $\nu(r)$ is the number of roots z_j of the polynomial $P(z)$ such that the module $|z_j| \leq r$. Therefore,

$$I_P(r) - I_P(0) = \sum_{|z_j|\leq r} (\log r - \log|z_j|) = \sum_{|z_j|\leq r} \int_{z_j}^{r} \frac{dt}{t} = \int_0^r \frac{\nu(t)}{t}dt.$$

If a function $f_0(z)$ is analytic in the disk $|z| \leq R$ and has no zeros there, then the function $\log f_0(z)$ is also analytic in this disk and the real part $\operatorname{Re}\log f_0(z) = \log|f_0(z)|$ is a harmonic function there. By the mean value theorem, for $0 \leq r \leq R$ the quantity

$$I_{f_0}(r) = \frac{1}{2\pi}\int_0^{2\pi} \log|f_0(re^{i\omega})|d\omega = \log|f_0(0)|$$

is independent of r, i.e., $I_{f_0}(r) - I_{f_0}(0) = 0$.

If a function $f(z)$ is analytic in the disk $|z| \leq R$, differs from zero at $z = 0$ ($f(0) \neq 0$), and vanishes only at points $z_1, z_2, \ldots, z_n$, then it can be represented as the product $f(z) = f_0(z) \cdot P(z)$, where the first multiplier is an analytic function and has no zeros in this disk. It is obvious that

$$I_f(r) = \frac{1}{2\pi}\int_0^{2\pi} \log|f(re^{i\omega})|d\omega = \frac{1}{2\pi}\int_0^{2\pi} \log|f_0(re^{i\omega})|d\omega + \frac{1}{2\pi}\int_0^{2\pi} \log|P(re^{i\omega})|d\omega,$$

$$I_f(r) - I_f(0) = \int_0^r \frac{\nu(t)}{t}dt,$$

where $\nu(t)$ is the number of roots of $f(z)$ in the disk $|z| \leq t$.

We note that $I_f(0) = 0$ for $f(0) = 1$ ($\log|f(0)| = 0$). Thus, we have proved the following theorem.

THEOREM 13.3.1 (Jensen). *If a function $f(z)$ is analytic for $|z| \leq R$ and $f(0) = 1$, then for $0 \leq r \leq R$ the number $\nu(r)$ of zeros of the function $f(z)$ for $|z| \leq r$ satisfies the relation*

$$\int_0^r \frac{\nu(t)}{t}dt = \frac{1}{2\pi}\int_0^{2\pi} \log|f(re^{i\omega})|d\omega.$$

As a simple consequence of this theorem, we deduce an inequality that will play an important role later in the volume. Suppose that a function $f(z)$ is analytic for $|z| \leq eR$. Let

$$M(eR) = \max_{|z|\leq eR}|f(z)|, \quad f(0) = 1.$$

In addition,

$$\log M(eR) \geq \frac{1}{2\pi}\int_0^{2\pi} \log|f(eRe^{i\omega})|d\omega = \int_0^{eR} \frac{\nu(t)}{t}dt \geq \int_R^{eR} \frac{\nu(t)}{t}dt \geq \int_R^{eR} \frac{\nu(R)}{t}dt = \nu(R).$$

Thus, we have established the following Jensen inequality. For a function $f(z)$ holomorphic for $|z| \leq eR$ and satisfying the conditions

$$f(0) = 1, \quad |f(eRe^{i\omega})| \leq M(eR)$$

the number of zeros $\nu(R)$ in the disk $|z| \leq R$ is estimated as follows:

$$\nu(R) \leq \log M(eR).$$

13.4. Estimates from below for analytic functions

An estimate from below for the logarithm of the module of a function analytic in a disk. The Blaschke products. An estimate for the module of a function analytic on a circle.

In this section, we prove an important inequality for analytic functions. This inequality will be used in §13.5 to justify the criterion for existence of circles containing no eigenvalues of the operator under consideration. To estimate the norm of the resolvent on these circles from above we need some theorems formulated and proved below.

THEOREM 13.4.1. *Let a function $f(z)$ be holomorphic in the disk $|z| \leq 2eR$ $(R > 0)$, let $f(0) = 1$, and let η be an arbitrary positive number such that $\eta < 3e/2$. Then for $|z| \leq R$, except disks with the common sum of radii less than $4\eta R$, the following inequality holds:*

$$\log|f(z)| > -\left[\log\frac{3e^5}{2\eta}\right]\log M(2eR) \qquad (M(2eR) = \max_{|z|\leq 2eR}|f(z)|).$$

PROOF. [1] Assume that for $|z| \leq 2R$ the only zeroes of $f(z)$ are $z_1, z_2, \dots, z_n$. Consider the ratio

$$\frac{2R(z-z_j)}{(2R)^2 - \overline{z}_j z}, \quad j = 1, 2, \dots, n.$$

If $|z_j| = 2R$, i.e., $z_j = 2Re^{i\phi}$, then for $z \neq z_j$ this ratio is a constant and its module is equal to 1. Indeed,

$$\frac{2R(z-z_j)}{(2R)^2 - \overline{z}_j z} = \frac{2R(z-2Re^{i\phi})}{(2R)^2 - 2Re^{-i\phi}z} = -e^{i\phi}\frac{z-2Re^{i\phi}}{-2Re^{i\phi}+z} = -e^{i\phi}.$$

If $|z_j| < 2R$, then $|\overline{z}_j||z| < (2R)^2$ for $|z| \leq 2R$. Therefore, the denominator does not vanish. In this case, on the circle $z = 2Re^{i\omega}$ we have

$$\left|\frac{2R(z-z_j)}{2R^2 - \overline{z}_j z}\right| = \left|\frac{2R(2Re^{i\omega}-z_j)}{2R^2 - \overline{z}_j 2Re^{i\omega}}\right| = |-e^{i\omega}|\left|\frac{2Re^{i\omega}-z_j}{2Re^{-i\omega}-\overline{z}_j}\right| = 1.$$

[1] The proof of Theorem 13.4.1 follows [**24**].

By the maximum principle for analytic functions, for $|z| \leq 2Re^{i\phi}$ we have

$$\left|\frac{2R(z-z_j)}{(2R)^2 - \overline{z}_j z}\right| \leq 1. \tag{13.4.1}$$

We note that points $z = z_j$ such that $|z_j| = 2R$ are removable singularities. Therefore, we can apply the maximum principle.

Thus, the estimate (13.4.1) holds for all $j = 1, 2, \ldots, n$. Therefore,

$$\left|\prod_{j=1}^{n} \frac{2R(z-z_j)}{(2R)^2 - \overline{z}_j z}\right| \leq 1, \quad |z| \leq 2R.$$

The product under the module sign

$$\pi(z) = \prod_{j=1}^{n} \frac{2R(z-z_j)}{(2R)^2 - \overline{z}_j z}$$

is called the *Blaschke product.* It is obvious that

$$\pi(0) = \prod_{j=1}^{n}\left(-\frac{z_j}{2R}\right), \quad |\pi(0)| \leq 1, \quad |\pi(2Re^{i\omega})| = 1.$$

The function $\phi(z) = \pi(0)\dfrac{f(z)}{\pi(z)}$ has no zeros for $|z| \leq 2R$, $\phi(0) = 1$. On the circle $|z| = 2R$ we have the estimate

$$\begin{aligned}|\phi(z)| = |\phi(2Re^{i\omega})| &= |\pi(0)|\frac{|f(2Re^{i\omega})|}{|\pi(2Re^{i\omega})|} \leq |f(2Re^{i\omega})| \\ &\leq \max_{|z|\leq 2eR} |f(z)| = M(2eR).\end{aligned}$$

By the maximum principle, the same estimate $|\phi(z)| \leq M(2eR)$ holds for $|z| \leq 2R$. By the Carathéodory inequality (for $|z| < 2R$), we have

$$\log|\phi(z)| \geq -\frac{4R}{2R - |z|}\log M(2eR).$$

Consequently,

$$\log|f(z)| \geq -\frac{4R}{2R-|z|}\log M(2eR) + \log\left|\frac{\pi(z)}{\pi(0)}\right|. \tag{13.4.2}$$

Note that for $|z| \leq R$, $|z_j| \leq 2R$ we have

$$\begin{aligned}\left|\frac{\pi(z)}{\pi(0)}\right| \geq |\pi(z)| &= \left|\frac{(2R)^n \prod_{j=1}^{n}(z-z_j)}{\prod_{j=1}^{n}[(2R)^2 - \overline{z}_o z_i]}\right| \geq \frac{(2R)^n\left|\prod_{j=1}^{n}(z-z_j)\right|}{\prod_{j=1}^{n}[(2R)^2 + |z_j|\,|z|]} \\ &\geq \left(\frac{2R}{6R^2}\right)^n \prod_{j=1}^{n}|z-z_j| = \frac{\prod_{j=1}^{n}|z-z_j|}{(3R)^n}.\end{aligned}$$

In addition, by the Cartan theorem, for points z_j there is a neighborhood formed by disks such that the sum of their radii less than $4\eta R$ and outside the neighborhood we have

$$\prod_{j=1}^{n} |z - z_j| > \left(\frac{4\eta R}{2e}\right)^n = \left(\frac{2\eta R}{e}\right)^n.$$

Therefore, for $|z| \leq R$ outside this neighborhood we have

$$\left|\frac{\pi(z)}{\pi(0)}\right| \geq \frac{(2\eta R/e)^n}{(3R)^n} = \left(\frac{2\eta}{3e}\right)^n.$$

By the Jensen inequality, the number n of zeros z_j of $f(z)$ lying in the disk $|z| \leq 2R$ satisfies the estimate $n \leq \log M(2eR)$. Consequently, if $\eta \leq 3e/2$, i.e., $\log 3e/(2\eta) \geq 0$, then

$$\log\left|\frac{\pi(z)}{\pi(0)}\right| \geq -\left[\log\frac{3e}{2\eta}\right]\log M(2eR). \tag{13.4.3}$$

From (13.4.2) and (13.4.3) we obtain the inequalities

$$\begin{aligned}\log|f(z)| &\geq -\frac{4R}{2R-|z|}\log M(2eR) + \log\left|\frac{\pi(z)}{\pi(0)}\right| \\ &\geq -2\log M(2eR) - \left[\log\frac{3e}{2\eta}\right]\log M(2eR) = -\left[\log\frac{3e^5}{2\eta}\right]\log M(2eR)\end{aligned}$$

for $|z| \leq R$. □

Let a function $f(z)$ be holomorphic for all z and let $f(0) = 1$, $|f(z)| < e^{|z|/R_0}$, $0 < \delta < 1$. Setting $\eta = (1-\delta)/16 < 1 < (3e)/2$ and using the relation

$$\max_{|z|\leq 2eR}|f(z)| = M(2eR) < e^{2eR/R_0}$$

and Theorem 13.4.1, we see that in the domain $|z| < R$ it is possible to find a neighborhood of zeros of $f(z)$ that is formed by disks with the sum of radii less than $4\eta R$ and the sum of diameters less $8\eta R = (1-\delta)R/2$ so that outside this neighborhood for $R > R_0/(2e)$ we have

$$\log|f(z)| > -\left[\log\frac{3e^5}{2\eta}\right]\log M(2eR) = -\log\frac{24e^5}{1-\delta}\log M(2eR) > -\log\frac{24e^5}{1-\delta}\frac{2eR}{R_0},$$

$$|f(z)| > \exp(-\frac{R}{R_0}2e\log\frac{24e^5}{1-\delta}).$$

It is obvious that the annulus $R > |z| > \delta R$ contains a circle $z = re^{i\phi}$, $R > r > \delta R$, that has the empty intersection with the disks in which we cannot use the last estimate for $|f(z)|$ (the sum of diameters of these disks is twice less than the width $R - \delta R$ of the annulus). On this circle we have the inequality

$$|f(re^{i\phi})| > \exp\left(-\frac{R}{R_0}2e\log\frac{24e^5}{1-\delta}\right).$$

Thus, we complete the proof of the theorem which was the reason why we turn to the theory of functions of complex variables in §§13.1–13.4.

THEOREM 13.4.2. *If a function $f(z)$ is holomorphic for all z, $f(0) = 1$, and $|f(z)| < e^{|z|/R_0}$ in any annulus $R > |z| > \delta R$ $(0 < \delta < 1, 2eR > R_0)$, then on some circle $z = re^{i\phi}$, $R > r > \delta R$, we have*

$$|f(z)| > \exp(-\frac{R}{R_0}2e\log\frac{24e^5}{1-\delta}).$$

By the Jensen theorem, for $f(z)$ satisfying the assumptions of Theorem 13.4.2 the number of zeros (with multiplicities) in the disk $|z| \leq R$ does not exceed $\log M(2eR)$, i.e., $n(R) \leq 2eR/R_0$.

13.5. Criterion for stratification of spectrum

Estimate for the matrix exponential of a sectorial operator. Relation between the spectral theory of operators and theory of entire analytic functions. The existence of a circle that realizes a dichotomy of the spectrum of an operator.

Consider an operator A given by an $N \times N$-matrix. The main question considered in this section is interesting in the case of very large N. Namely, we analyse whether it is possible to describe the behavior of e^{tA} as $t \to \infty$ or of A^n as $n \to \infty$ by using the exponential or power function of a matrix of significantly smaller size. We explain the matter of the question by a problem about an estimate for the exponential of a sectorial operator A (see §11.5). We recall that in §11.5 we considered an operator A such that all possible values of the quotient $\dfrac{(Au,u)}{(u,u)} = \lambda \equiv \alpha + i\beta$, where $u \neq 0$, satisfy the inequality $|\beta| \leq (q_0 - \alpha)\tan\theta$, $0 < \theta < \pi/2, \alpha \leq q_0$, i.e., we considered an operator whose Hausdorff set lies inside the acute sector bounded by the rays $\alpha = \mathrm{Re}\lambda \leq q_0$, $|\beta| = |\mathrm{Im}\,\lambda| = (q_0 - \mathrm{Re}\lambda)\tan\theta$. Outside the extended open sector $\alpha = \mathrm{Re}\lambda < q$, $|\beta| < (q - \mathrm{Re}\lambda)\tan\theta$ with the same angle for $q > q_0$, the resolvent of the operator A satisfies the estimate

$$\|(\lambda I - A)^{-1}\| \leq \frac{1}{(q - q_0)}\sin\theta.$$

Furthermore, for $t \geq 0$ the matrix exponential of A satisfies the estimate

$$\|e^{tA}\| \leq e^{q_0 t}.$$

As was shown in §11.5, if there exists a circle $|\lambda - q| = r$ centered at the vertex of the extended sector such that $\|(\lambda I - A)^{-1}\| \leq m/r$ on this circle and the arc located inside the sector $|\beta| \leq (q - \mathrm{Re}\lambda)\tan\theta$ $(\mathrm{Re}\lambda < q)$ lies in the left half-plane, then the asymptotic behavior of e^{tA} as $t \to \infty$ is the same as that of e^{tP_rA}, where P_r is the projection

$$P_r = \frac{1}{2\pi i} \oint_{|\lambda - q| = r} (\lambda I - A)^{-1} d\lambda$$

onto the invariant subspace of A corresponding to the eigenvalues of A in the disk $|\lambda - q| < r$. If such a circle exists, we can analyze the stability problems by studying the matrix exponential of the operators P_rA. As a rule, the rank of P_rA is essentially less than N.

The boundedness of the resolvent of an operator A on a circle is equivalent to the assertion that the circle realizes the dichotomy of the spectrum of the operator A. We give sufficient conditions for the existence of circles on which the resolvent is bounded. We require that the operator A satisfies certain conditions formulated using information about singular values.

Recall that the spectral condition number was defined (see §7.8) as follows:

$$\nu^{(k)}(C) = \left(\sum_{j=1}^{N}\left(\frac{\sigma_j(C)}{\sigma_N(C)}\right)^k\right)^{1/k}.$$

We assume that

$$\nu^{(k)}(C^{-1}) = \left(\sum_{j=1}^{N}\left(\frac{\sigma_j(C^{-1})}{\sigma_N(C^{-1})}\right)^k\right)^{1/k} = \sigma_1(C)\left(\sum_{j=1}^{N}\sigma_j^{-k}(C)\right)^{1/k}.$$

We note that for $k > 0$, $a_j = \sigma_j(C^{-1})/\sigma_N(C^{-1}) \leq 1$, $\sum\limits_{j=1}^{N} a_j^k \geq 1$ we have

$$\begin{aligned}\frac{d}{dk}\left(\sum_{j=1}^{N} a_j^k\right)^{1/k} &= \left(\sum_{j=1}^{N} a_j^k\right)^{1/k}\left[\frac{\sum\limits_{j=1}^{N} a_j^k \log a_j}{k\sum\limits_{j=1}^{N} a_j^k} - \frac{\log\sum\limits_{j=1}^{N} a_j^k}{k^2}\right] \\ &= \frac{1}{k^2}\left(\sum_{j=1}^{N} a_j^k\right)^{1/k}\left[\frac{\sum\limits_{j=1}^{N} a_j^k \log a_j^k}{\sum\limits_{j=1}^{N} a_j^k} - \log\sum_{j=1}^{N} a_j^k\right] \\ &= k^{-2}\left(\sum_{j=1}^{N} a_j^k\right)^{1/k-1}\sum_{j=1}^{N} a_j^k \log\frac{a_j^k}{\sum\limits_{j=1}^{N} a_i^k} \leq 0.\end{aligned}$$

Therefore, for $k \geq k_0$ we have

$$\left(\sum_{j=1}^{N} a_j^k\right)^{1/k} \leq \left(\sum_{j=1}^{N} a_j^{k_0}\right)^{1/k}.$$

Consequently, $\nu^{(k)}(C^{-1}) \leq \nu^{(k_0)}(C^{-1})$. We see that if an estimate from above for $\nu_{k_0}(C^{-1})$ holds for some $k_0 > 0$, it also holds for the integer k such that $k_0 + 1 > k \geq k_0$. This remark shows that it suffices to prove these estimates for the resolvent of the operator C that can be obtained from information about the spectral condition number $\nu^{(k)}(C^{-1})$, where k is an integer.

The proof of the estimate presented here is an adaptation to the finite-dimensional case of certain arguments by Lidskii [**25**]. In [**25**] the criterion for the completeness of bases for root invariant subspaces of some operators was established in the case of an infinite-dimensional Hilbert space (the case of noninteger k was also considered in [**25**]).

The results of Lidskii extends important ideas of Keldysh about the relation between the spectral theory of operators and the theory of entire analytic functions (we refer the reader to the talk of Keldysh and Lidskii [**21**] at the All-Union Mathematical Congress in 1961).

We apply the estimate for the resolvent obtained in §8.6 to the matrix C^{-k}, $k = 1, 2, 3, \dots$:

$$\|(I - \xi C^{-k})^{-1}\| \leq \frac{\prod_{j=2}^{N} [1 + |\xi| \sigma_j(C^{-k})]}{\left| \prod_{j=1}^{N} [1 - \xi \lambda_j(C^{-k})] \right|} \leq \frac{e^{|\xi| \sum_{j=1}^{N} \sigma_j(C^{-k})}}{\left| \prod_{j=1}^{N} [1 - \xi \lambda_j(C^{-k})] \right|}.$$

Since (see §5.6)

$$\sum_{j=1}^{N} \sigma_j(C^{-k}) \leq \sum_{j=1}^{N} \sigma_j^k(C^{-1}) = \sum_{j=1}^{N} \frac{1}{\sigma_j^k(C)},$$

we find

$$\|(I - \xi C^{-k})^{-1}\| \leq \frac{e^{|\xi| \sum_{j=1}^{N} \sigma_j^{-k}(C)}}{|f(\xi)|},$$

where the analytic function (polynomial)

$$f(\xi) = \prod_{j=1}^{N} [1 - \xi \lambda_j(C^{-k})]$$

is such that $f(0) = 1$,

$$|f(\xi)| \leq \left| \prod_{j=1}^{N} (1 + |\xi| \, |\lambda_j(C^{-k})|) \right| \leq e^{|\xi| \sum_{j=1}^{N} |\lambda_j(C^{-k})|}.$$

Using the corollary of the Weyl inequality (see §6.3)

$$\sum_{j=1}^{N} |\lambda_j(C^{-k})| \leq \sum_{j=1}^{N} \sigma_j(C^{-k})$$

and the inequality from §5.6, we slightly simplify the estimate for $|f(\xi)|$:

$$|f(\xi)| \leq \exp\left(|\xi| \sum_{j=1}^{N} \sigma_j^{-k}(C) \right).$$

The analytic function (polynomial) $f(\xi)$ satisfies the assumptions of Theorem 13.4.2 for

$$R_0 = \frac{1}{\sum_{j=1}^{N} \sigma_j^{-k}(C)}.$$

Therefore, any annulus $R > |\xi| > \delta R$ ($0 < \delta < 1$ and $2eR > R_0$) contains a circle $\xi = re^{i\phi}$ ($R > r > \delta R$) on which $f(\xi) \neq 0$ and

$$|f(\xi)| > \exp\left(-\frac{R}{R_0} 2e \log \frac{24e^5}{1 - \delta} \right).$$

It is obvious that for $|\xi| = r$ we have

$$\begin{aligned}\|(I - \xi C^{-k})^{-1}\| &< \exp\left(\frac{R}{R_0} 2e\log \frac{24e^5}{1-\delta}\right) \exp\left(R \sum_{j=1}^{N} \sigma_j^{-k}(C)\right) \\ &= \exp\left(\left(1 + 2e\log \frac{24e^5}{1-\delta}\right) \frac{R}{\sigma_1^k(C)} \left(\sigma_1^k(C) \sum_{j=1}^{N} \sigma_j^{-k}(C)\right)\right) \\ &= \exp\left(\left(1 + 2e\log \frac{24e^5}{1-\delta}\right) [\nu^{(k)}(C^{-1})]^k \frac{R}{\sigma_1^k(C)}\right).\end{aligned}$$

Changing the notation

$$\xi = re^{i\phi} \Rightarrow \lambda^k = r^k e^{ik\phi}, \quad R \Rightarrow R^k,$$
$$\delta \Rightarrow \delta^k, \quad R_0 \Rightarrow (R_0/2e)^k, \quad (R_0/2e) \Rightarrow R_0^k,$$

we conclude that any annulus $R > |\lambda| > \delta R$, $0 < \delta < 1$,

$$R > R_0 = \left[2e \sum_{j=1}^{N} \sigma_j^{-k}(C)\right]^{-1/k} = \sigma_1(C)[2e\nu^{(k)}(C^{-1})]^{-1/k}$$

contains a circle $|\lambda| = r$ on which

$$\|(I - \lambda^k C^{-k})^{-1}\| < \exp\left(\left(1 + 2e\log \frac{24e^5}{1-\delta^k}\right)\left(\frac{\nu^{(k)}(C^{-1})R}{\sigma_1(C)}\right)^k\right).$$

Since

$$\begin{aligned}|\lambda|\,\|(\lambda I - C)^{-1}\| &= |\lambda|\,\|C^{-1}(\lambda C^{-1} - I)^{-1}\| \leq [|\lambda|/\sigma_1(C)]\,\|(I - \lambda C^{-1})^{-1}\| \\ &= [|\lambda||\sigma_1(C)|]\,\|(I - \lambda^k C^{-k})^{-1} \sum_{j=0}^{k-1} \lambda^j C^{-j}\| \\ &\leq k[|\lambda/\sigma_1(C)|]^k \|(I - \lambda^k C^{-k})^{-1}\|,\end{aligned}$$

the norm of the resolvent $(\lambda I - C)^{-1}$ on this circle satisfies the estimate

$$|\lambda|\,\|(\lambda I - C)^{-1}\| < k[R/\sigma_1(C)]^k e^{a(k,\delta)[\nu^{(k)}(C^{-1})R/\sigma_1(C)]^k},$$

where

$$a(k, \delta) = 1 + 2e\log \frac{24e^5}{1-\delta^k}.$$

The number $n(R, C)$ of eigenvalues of the operator C, i.e., the number of poles of the resolvent $(\lambda I - C)^{-1}$ in the disk $|\lambda| \leq R$, is k times less than the number of roots of the determinant

$$\det [1 - \lambda^k C^{-k}] = \prod_{j=1}^{N} [1 - \lambda^k \lambda_j(C^{-k})] = f(\lambda^k),$$

in this disk. Equivalently, $n(R, C)$ coincides with the number of zeros of $f(\xi)$ in the disk $|\xi| \leq R^k$. Since for $|\xi| < eR^k$ the module $|f(\xi)|$ does not exceed $e^{e[\nu^{(k)}(C^{-1})R/\sigma_1(C)]^k}$, by the Jensen theorem we have

$$n(R, C) < e[\nu^{(k)}(C^{-1})R/\sigma_1(C)]^k.$$

We complete the proof of the fact that for the operator C with not too large $\nu^{(k)}(C^{-1})$ every annulus $\delta R < |\lambda| < R$ $(0 < \delta < 1,\ R > R_0)$ contains a circle $|\lambda| = r$ that realizes the dichotomy of the spectrum of C: $R_0/\sigma_1(C) = [2e\nu^{(k)}(C^{-1})]^{-1/k}$.

To apply the above results to the sectorial operator A (see the beginning of this section and §11.5), we must put $C = A - qI$.

13.6. Dependence of the dichotomy criterion on the radius

A criterion $\kappa(r) = \|H(r)\|$ for the circle dichotomy. Inequalities for derivatives of a quadratic form whose coefficients are computed from analytic functions. An inequality for $\|H(r)\|$.

In this section, we consider one more application of the theory of functions of complex variable. In Chapter 9 we introduced the integral criteria for the dichotomy quality for the matrix spectrum. In §10.6 we described one-dimensional spectral portraits of matrices, i.e., graphs of the dependence of the dichotomy quality on the location of the line or on the radius of the circle that divide the spectrum. The dichotomy quality for the spectrum by the circle $|\lambda| = r$ is estimated in terms of $\kappa(r) = \|H(r)\|$, where

$$H(r) = \frac{1}{2\pi}\int_0^{2\pi} (re^{-i\phi}I - A^*)^{-1}(re^{i\phi}I - A)^{-1}\,d\phi.$$

In this section, we restrict ourselves to the study of the circle dichotomy which is simpler from the computational point of view. If $r \to |\lambda_j|$, where λ_j is some eigenvalue, then $\kappa(r) \to \infty$. We will show that $\log\kappa(r)$ is a convex function of $\log r$ between two neighboring "infinities."

Let $B = B(\zeta)$ be a nonsingular matrix and let its elements $b_{ij}(\zeta)$ be regular analytic functions of $\zeta = \xi + i\eta$ in some domain. Consider a constant Hermitian positive definite matrix $D = D^* > 0$ and a nonzero vector g. We show that the function

$$\mathcal{E}(\xi,\eta) = (B^*DBg, g) = (DBg, Bg)$$

is positive in the domain $\mathcal{E}(\xi,\eta) > 0$. Moreover,

$$\mathcal{E}_{\xi\xi} + \mathcal{E}_{\eta\eta} \geq 0. \tag{13.6.1}$$

The assertion about positiveness is trivial. Let us prove the inequality (13.6.1). Let the Hermitian matrix D have the form

$$D = U^*\begin{bmatrix} d_1 & & 0 \\ & \ddots & \\ 0 & & d_n \end{bmatrix} U,$$

where U is a unitary matrix. The diagonal elements d_i are positive ($d_i > 0$) since the matrix D is positive definite. Therefore,

$$(DBg, Bg) = \left(\begin{bmatrix} d_1 & & 0 \\ & \ddots & \\ 0 & & d_n \end{bmatrix} Cg, Cg\right),$$

where the matrix $C = UB$ is nonsingular and the elements $c_{ij}(\zeta)$ are regular analytic functions of ζ. Thus,

$$(DBg, Bg) = \sum_{i=1}^{N} d_i \left| \sum_{j=1}^{N} c_{ij} g_j \right|^2.$$

It suffices to show that the quantities

$$|f_i(\zeta)|^2 = \left| \sum_{j=1}^{N} c_{ij}(\zeta) g_j \right|^2$$

satisfy the inequalities

$$\frac{\partial^2 |f_i(\zeta)|^2}{\partial \xi^2} + \frac{\partial^2 |f_i(\zeta)|^2}{\partial \eta^2} \geq 0.$$

If $f(\xi, \eta) = u(\xi, \eta) + iv(\xi, \eta)$ is a regular analytic function of $\zeta = \xi + i\eta$, then $u(\xi, \eta)$ and $v(\xi, \eta)$ are infinitely differentiable functions in the domain where f is regular and satisfy the Cauchy–Riemann equations $u_\xi - v_\eta = 0$ and $u_\eta + v_\xi = 0$. The direct computation shows that

$$|f(\xi, \eta)|^2 = |f^2(\xi, \eta)| = u^2(\xi, \eta) + v^2(\xi, \eta),$$

$$\frac{\partial^2 |f|^2}{\partial \xi^2} + \frac{\partial^2 |f|^2}{\partial \eta^2} = 2u(u_{\xi\xi} + u_{\eta\eta}) + 2v(v_{\xi\xi} + v_{\eta\eta}) + 2(u_\xi^2 + u_\eta^2 + v_\xi^2 + v_\eta^2),$$

$$u_{\xi\xi} + u_{\eta\eta} = (v_\eta)_\xi + (-v_\xi)_\eta = 0,$$

$$v_{\xi\xi} + v_{\eta\eta} = (-u_\eta)_\xi + (u_\xi)_\eta = 0,$$

i.e.,

$$\frac{\partial^2 |f|^2}{\partial \xi^2} + \frac{\partial^2 |f|^2}{\partial \eta^2} = 2(u_\xi^2 + u_\eta^2 + v_\xi^2 + v_\eta^2) = 4(u_\xi^2 + u_\eta^2) \geq 0.$$

The inequality $\mathcal{E}_{\xi\xi} + \mathcal{E}_{\eta\eta} \geq 0$ becomes obvious.

We establish a similar inequality

$$\frac{\partial^2 \log \mathcal{E}}{\partial \xi^2} + \frac{\partial^2 \log \mathcal{E}}{\partial \eta^2} = \frac{1}{\mathcal{E}^2} \left[\mathcal{E} \left(\frac{\partial^2 \mathcal{E}}{\partial \xi^2} + \frac{\partial^2 \mathcal{E}}{\partial \eta^2} \right) - \left(\frac{\partial \mathcal{E}}{\partial \xi} \right)^2 - \left(\frac{\partial \mathcal{E}}{\partial \eta} \right)^2 \right] \geq 0. \tag{13.6.2}$$

Since $\mathcal{E}$ is strictly positive, it suffices to show that

$$\mathcal{E} \left(\frac{\partial^2 \mathcal{E}}{\partial \xi^2} + \frac{\partial^2 \mathcal{E}}{\partial \eta^2} \right) - \left(\frac{\partial \mathcal{E}}{\partial \xi} \right)^2 - \left(\frac{\partial \mathcal{E}}{\partial \eta} \right)^2 \geq 0. \tag{13.6.3}$$

For $f(\zeta) = u(\xi, \eta) + iv(\xi, \eta)$ we have

$$\left(\frac{\partial^2}{\partial \xi^2} + \frac{\partial^2}{\partial \eta^2} \right) |f|^2 = 4(u_\xi^2 + u_\eta^2), \quad |f|^2 = u^2 + v^2,$$

$$\frac{\partial}{\partial \xi} |f|^2 = 2(uu_\xi + vv_\xi) = 2(uu_\xi - vu_\eta),$$

$$\frac{\partial}{\partial \eta} |f|^2 = 2(uu_\eta + vv_\eta) = 2(uu_\eta + vu_\xi),$$

$$\left(\frac{\partial |f|^2}{\partial \xi} \right)^2 + \left(\frac{\partial |f|^2}{\partial \eta} \right)^2 = 4[(uu_\xi - vu_\eta)^2 + (uu_\eta + vu_\xi)^2] = 4(u^2 + v^2)(u_\xi^2 + u_\eta^2).$$

For $\mathcal{E}^{(j)}(\xi,\eta) = |f^{(j)}(\xi,\eta)|^2$ we have

$$\left((\mathcal{E}_\xi^{(j)})^2 + (\mathcal{E}_\eta^{(j)})^2\right)^{1/2} = \sqrt{\mathcal{E}^{(j)}}\sqrt{\mathcal{E}_{\xi\xi}^{(j)} + \mathcal{E}_{\eta\eta}^{(j)}}. \tag{13.6.4}$$

In view of the equalities

$$\mathcal{E} = \sum_{j=1}^{N} |f_j(\zeta)|^2 = \sum_{j=1}^{N} \mathcal{E}^{(j)}(\xi,\eta),$$

the triangle inequality, the Cauchy–Schwarz–Bunyakovskii inequality

$$\sqrt{\mathcal{E}_\xi^2 + \mathcal{E}_\eta^2} \le \sum_{j=1}^{N} \sqrt{(\mathcal{E}_\xi^{(j)})^2 + (\mathcal{E}_\eta^{(j)})^2},$$

$$\sum_{j=1}^{N} \sqrt{\mathcal{E}^{(j)}}\sqrt{\mathcal{E}_{\xi\xi}^{(j)} + \mathcal{E}_{\eta\eta}^{(j)}} \le \sqrt{\mathcal{E}(\mathcal{E}_{\xi\xi} + \mathcal{E}_{\eta\eta})},$$

and the equality (13.6.4), we obtain

$$\sqrt{\mathcal{E}_\xi^2 + \mathcal{E}_\eta^2} \le \sqrt{\mathcal{E}(\mathcal{E}_{\xi\xi} + \mathcal{E}_{\eta\eta})},$$

i.e., (13.6.2) and (13.6.3).

Let $B(\zeta) = A - e^\zeta I$, $D = I$, i.e.,

$$\mathcal{E}(\xi,\eta) = ((A^* - e^{\overline{\zeta}} I)^{-1}(A - e^\zeta I)^{-1} g, g).$$

Assume that the matrix $A - e^\zeta I$ is nonsingular in the strip $\log R_1 \le \operatorname{Re}\zeta \le \log R_2$. In other words, we assume that the annulus $R_1 < |\lambda| < R_2$ contains no eigenvalues $\lambda_j(A)$ of A. For $\log R_1 < \xi < \log R_2$ we consider a strictly positive function

$$\widehat{\mathcal{E}}(\xi,\eta) = \frac{1}{2\pi}\int_0^{2\pi} ((A^* - e^{\xi - i(\phi+\eta)} I)^{-1}(A - e^{\xi + i(\phi+\eta)})^{-1} g, g)\, d\phi$$

and note that $\widehat{\mathcal{E}}(\xi,\eta)$ is strictly positive and is independent of η because the integrand is a periodic function of ϕ, i.e., $\widehat{\mathcal{E}}(\xi,\eta) = \widehat{\mathcal{E}}(\xi)$. Moreover,

$$\frac{d^2 \log \widehat{\mathcal{E}}(\xi)}{d\xi^2} = \frac{1}{\widehat{\mathcal{E}}^2}[\widehat{\mathcal{E}} \cdot \widehat{\mathcal{E}}'' - (\widehat{\mathcal{E}}')^2] \ge 0, \quad |\widehat{\mathcal{E}}_\xi(\xi)| \le \sqrt{\widehat{\mathcal{E}}(\xi) \cdot \widehat{\mathcal{E}}_{\xi\xi}(\xi)}.$$

Indeed,

$$|\widehat{\mathcal{E}}_\xi(\xi)| = \left|\frac{1}{2\pi}\int_0^{2\pi} \mathcal{E}_\xi(\xi,\eta)\,d\eta\right| \le \left[\left|\frac{1}{2\pi}\int_0^{2\pi} \mathcal{E}_\xi(\xi,\eta)\,d\eta\right|^2 + \left|\frac{1}{2\pi}\int_0^{2\pi} \mathcal{E}_\eta(\xi,\eta)\,d\eta\right|^2\right]^{1/2}$$

$$\le \frac{1}{2\pi}\int_0^{2\pi} \sqrt{\mathcal{E}_\xi^2(\xi,\eta)+\mathcal{E}_\eta^2(\xi,\eta)}\,d\eta \le \frac{1}{2\pi}\int_0^{2\pi}\sqrt{\mathcal{E}(\xi,\eta)[\mathcal{E}_{\xi\xi}(\xi,\eta)+\mathcal{E}_{\eta\eta}(\xi,\eta)]}\,d\eta$$

$$\le \left[\frac{1}{2\pi}\int_0^{2\pi}\mathcal{E}(\xi,\eta)\,d\eta\cdot\frac{1}{2\pi}\int_0^{2\pi}[\mathcal{E}_{\xi\xi}(\xi,\eta)+\mathcal{E}_{\eta\eta}(\xi,\eta)]\,d\eta\right]^{1/2}$$

$$= \left[\frac{1}{2\pi}\int_0^{2\pi}\mathcal{E}(\xi,\eta)\,d\eta\right]^{1/2}\cdot\left[\frac{1}{2\pi}\int_0^{2\pi}\mathcal{E}_{\xi\xi}(\xi,\eta)\,d\eta\right]^{1/2} = \sqrt{\widehat{\mathcal{E}}(\xi)\cdot\widehat{\mathcal{E}}_{\xi\xi}}.$$

Thus, for

$$(H(r)g,g) = \frac{1}{2\pi}\int_0^{2\pi}((A^* - re^{-i\phi}I)^{-1}(A-re^{i\phi}I)^{-1}g,g)\,d\phi$$

we have the inequality

$$\frac{d^2\log(H(r)g,g)}{(d\log r)^2} \ge 0,$$

i.e., $\log(H(r)g,g)$ is a convex function of $\log r$. For $R_1 \le r \le R_2$ this implies the inequality

$$\log(H(r)g,g) \le \frac{\log(R_2/r)\cdot\log(H(R_1)g,g)+\log(r/R_1)\cdot\log(H(R_2)g,g)}{\log(R_2/R_1)}.$$

Let g be such that $\|g\|=1$, $(H(r)g,g)=\|H(r)\|$. In addition, let

$$(H(R_1)g,g)\le\|H(R_1)\|,\quad (H(R_2)g,g)\le\|H(R_2)\|.$$

We arrive at the inequality

$$\log\|H(r)\| \le \frac{\log(R_2/r)\cdot\log\|H(R_1)\|+\log(r/R_1)\cdot\log\|H(R_2)\|}{\log(R_2/R_1)},$$

which means that $\log\|H(r)\|$ is a convex function of $\log r$ in any annulus $R_1 \le |\lambda| \le R_2$ containing no eigenvalues $\lambda_j(A)$.

Denoting $r = R_1^{1-\alpha}R_2^{\alpha}$ $(0\le\alpha\le1)$, it is convenient to rewrite the last inequality in the following equivalent form:

(13.6.5) $$\|H(R_1^{1-\alpha}R_2^{\alpha})\| \le \|H(R_1)\|^{1-\alpha}\|H(R_2)\|^{\alpha}.$$

We recall our assumption that the annulus $R_1 \le |\lambda| \le R_2$ contains no eigenvalues of the matrix A. This assumption was essentially used in the arguments that lead to the proof of the elegant estimate (13.6.5).

13.7. Logarithmic subharmonicity of the resolvent

A harmonic majorant for the logarithm of the norm of the resolvent. Subharmonic and logarithmically harmonic functions.

As was shown in §13.6, if an $N \times N$-matrix $B(\zeta)$ is nonsingular and bounded in a domain of the complex plane $\zeta = \xi + i\eta$ and the elements of $B(\zeta)$ are analytic functions of ζ, then the expression

$$\mathcal{E}(\xi, \eta) = \log (B(\zeta)g, B(\zeta)g) = 2\log \|B(\zeta)g\|,$$

where g is any nonzero constant vector, has continuous second order derivatives and satisfies the inequality

$$\frac{\partial^2 \mathcal{E}}{\partial \xi^2} + \frac{\partial^2 \mathcal{E}}{\partial \eta^2} \geq 0. \tag{13.7.1}$$

In particular, if $B(\zeta) = (\zeta I - A)^{-1}$, then (13.7.1) holds at all points of the complex plane except for the points $\zeta = \lambda_j(A)$ of the spectrum of A. In this case, omitting the constant coefficient 2, we set

$$\mathcal{E}(\xi, \eta) = \log \|\zeta I - A)^{-1} g\|.$$

The change of the coefficients does not affect the validity of (13.7.1).

Let G be a bounded domain in the complex plane that contains no eigenvalues $\lambda_j(A)$ of the matrix A and the boundary of G is sufficiently regular. For example, we can assume that the boundary ∂G is a nonselfintersecting closed curve composed of a finite number of smooth arcs with common endpoints. Nonsimply connected domains are also admissible provided that the boundary is composed by several closed curves without common points.

Let a function $F(\xi, \eta)$ be harmonic in G and continuous up to the boundary ∂G. Assume that it coincides with $\mathcal{E}(\xi, \eta)$ on ∂G, i.e.,

$$\frac{\partial^2 F}{\partial \xi^2} + \frac{\partial^2 F}{\partial \eta^2} = 0, \quad \zeta = \xi + i\eta \in G,$$
$$F(\xi, \eta) = \mathcal{E}(\xi, \eta) = \log \|(\zeta I - A)^{-1} g\|, \quad \zeta = \xi + i\eta \in \partial G.$$

It is known that such a function exists and is unique.

It is obvious that the difference $\mathcal{E} - F$ vanishes on ∂G; moreover,

$$\begin{gathered} \frac{\partial^2 (\mathcal{E} - F)}{\partial \xi^2} + \frac{\partial^2 (\mathcal{E} - F)}{\partial \eta^2} \geq 0, \quad \zeta \in G, \\ \mathcal{E} - F = 0, \quad \phi \in \partial G. \end{gathered} \tag{13.7.2}$$

By (13.7.2), everywhere in G we have

$$\mathcal{E}(\xi, \eta) - F(\xi, \eta) \leq 0, \quad \zeta \in G. \tag{13.7.3}$$

This assertion is usually established in the proof of the uniqueness theorem for a solution to the Dirichlet problem for the Laplace equation. We give this simple proof here.

Suppose that (13.7.3) fails, i.e., there exists an interior point $\zeta_0 = \xi_0 + i\eta_0$ of G at which $\mathcal{E}(\xi_0, \eta_0) - F(\xi_0, \eta_0) = M > 0$. Constructing an auxiliary function

$$V(\xi, \eta) = \mathcal{E}(\xi, \eta) - F(\xi, \eta) + \frac{M}{2d^2}[(\xi - \xi_0)^2 + (\eta - \eta_0)^2], \tag{13.7.4}$$

where d is the diameter of G, we see that

$$V(\xi,\eta) \le M/2 \quad \zeta = \xi + i\eta \in \partial G, \quad V(\xi_0,\eta_0) = M.$$

Therefore, in the closure $\overline{G} = G + \partial G$ of G the function $V(\xi,\eta)$ attains the maximal value which cannot be less than M. Since $V(\xi,\eta) \le M/2$ on the boundary, there exists at least one interior point $\zeta_1 = \xi_1 + i\eta_1$ at which V attains the maximal value.

It is known that at the maximum point we have $V_\xi = 0$, $V_\eta = 0$, $V_{\xi\xi} \le 0$, $V_{\eta\eta} \le 0$. Therefore,

$$\left[\frac{\partial^2 V}{\partial \xi^2} + \frac{\partial^2 V}{\partial \eta^2}\right]_{\substack{\xi=\xi_1\\ \eta=\eta_1}} \le 0. \tag{13.7.5}$$

But the inequality (13.7.5) cannot be valid since (13.7.2) and (13.7.4) imply that

$$\frac{\partial^2 V}{\partial \xi^2} + \frac{\partial^2 V}{\partial \eta^2} = \frac{\partial^2(\mathcal{E} - F)}{\partial \xi^2} + \frac{\partial^2(\mathcal{E} - F)}{\partial \eta^2} + 2\frac{M}{d^2} \ge \frac{M}{d^2} > 0$$

everywhere in G. Thereby, the assumption that at some interior point the equality $\mathcal{E}(\xi_0,\eta_0) - F(\xi_0,\eta_0) = M > 0$ holds is wrong, i.e., the inequality (13.7.3) is proved. Thus, if we have constructed a function $F_g(\xi,\eta)$ that is harmonic in G, continuous up to the boundary of G, and satisfies the condition

$$F_g(\xi,\eta) = \log \|(\zeta I - A)^{-1} g\|, \quad \zeta \in \partial G,$$
$$\frac{\partial^2 F_g(\xi,\eta)}{\partial \xi^2} + \frac{\partial^2 F_g(\xi,\eta)}{\partial \eta^2} = 0, \quad \zeta \in G,$$

we, in fact, have proved that $\log \|\zeta I - A)^{-1} g\| \le F_g(\xi,\eta)$ everywhere in G. Let $\zeta^* = \xi^* + i\eta^*$ be a point of the domain G ($\zeta^* \in G$) and let g^* be a vector of unit norm ($\|g^*\| = 1$) such that $\|(\zeta^* I - A)^{-1} g^*\| = \|(\zeta^* I - A)^{-1}\|$. Then $\log \|(\zeta^* I - A)^{-1}\| \le F_{g^*}(\xi,\eta)$.

Let $F_0(\xi,\eta)$ be harmonic in the domain G and continuous up to the boundary of G. Assume that it takes the values $F_0(\xi,\eta)\big|_{\partial G} = \log \|(\zeta I - A)^{-1}\|$ on ∂G. Since

$$\begin{aligned} F_0(\xi,\eta)\big|_{\partial G} &= \log \|(\zeta I - A)^{-1}\| = \sup_{g \ne 0} \log \|(\zeta I - A)^{-1} g\| / \|g\| \\ &\ge \log \|(\zeta I - A)^{-1} g^*\| = F_{g^*}(\xi,\eta)\big|_{\partial G}, \end{aligned}$$

on the boundary ∂G we have

$$F_0(\xi,\eta) - F_{g^*}(\xi,\eta) \ge 0, \quad \zeta = \xi + i\eta \in \partial G.$$

The difference $F_0 - F_{g^*}$ is a harmonic function in G, i.e.,

$$\frac{\partial^2(F_0 - F_{g^*})}{\partial \xi^2} + \frac{\partial^2(F_0 - F_{g^*})}{\partial \eta^2} = 0.$$

By the maximum principle for harmonic functions, $F_{g^*}(\xi,\eta) \le F_0(\xi,\eta)$ everywhere in G. In particular, $\log \|(\zeta^* I - A)^{-1}\| \le F_0(\xi^*,\eta^*)$. Since a point $\zeta^* \in G$ was taken arbitrarily, we arrive at the following important assertion.

THEOREM 13.7.1 (on a harmonic majorant). *Let G be a bounded domain in the complex plane with sufficiently smooth boundary. If G contains no eigenvalues of a matrix A, then the logarithm of the norm of the resolvent* $\log \|(\zeta I - A)^{-1}\|$ *admits* *a harmonic majorant* $F_0(\xi, \eta)$ *which is continuous up to the boundary and satisfies the conditions*

$$\frac{\partial^2 F_0}{\partial \xi^2} + \frac{\partial^2 F_0}{\partial \eta^2} = 0, \quad F_0(\xi, \eta) \geq \log \|(\zeta I - A)^{-1}\|, \quad \zeta = \xi + i\eta \in G,$$

in the domain G*; moreover,* $F_0(\xi, \eta) = \log \|(\zeta I - A)^{-1}\|$ *on the boundary* ∂G.

REMARK 13.7.1. Unlike the case of $\log \|(\zeta I - A)^{-1} g\|$, for a fixed vector g the logarithm of the norm of the resolvent $\log \|(\zeta I - A)^{-1}\|$ does not necessarily have continuous second order derivatives. The reader can verify this fact by checking the relation

$$\log \|(\zeta I - A)^{-1}\| = -\frac{1}{2} \log [(1 - |\xi|)^2 + \eta^2]$$

for the second order matrix

$$A = \begin{bmatrix} 1 & 0 \\ 0 & -1 \end{bmatrix}.$$

A function $W(\xi, \eta)$ in a domain is called *subharmonic* if in every bounded subdomain with sufficiently smooth boundary the function admits a majorant $u(\xi, \eta)$ which is harmonic in the subdomain, is continuous up to the boundary of the subdomain, and satisfies the conditions $u(\xi, \eta) = W(\xi, \eta)$ on the boundary and $u(\xi, \eta) \geq W(\xi, \eta)$ in the subdomain.

If $\log W(\xi, \eta)$ is a subharmonic function, then $W(\xi, \eta)$ is said to be *logarithmically subharmonic.*

REMARK 13.7.2. One can show that a logarithmically subharmonic function $W(\xi, \eta)$ is subharmonic. We do not prove this fact here.

Using the definition of a logarithmically subharmonic function, we can reformulate Theorem 13.7.1 as follows.

THEOREM 13.7.2. *The norm of the resolvent* $\|(\zeta I - A)^{-1}\|$ *is a logarithmically subharmonic function in any domain in the complex plane that contains no eigenvalues of the matrix* A.

CHAPTER 14

Computational Algorithms of Spectral Analysis

14.1. The computation of solutions to matrix Lyapunov equations

The simplest case of the "discrete equation." A series represention of a solution. Recurrence formulas for even sums. An algorithm. An error estimate. Application to the classical Lyapunov equation.

We begin with the discrete Lyapunov equation (see §4.4)

$$X - A^*XA = C. \tag{14.1.1}$$

Suppose that (14.1.1) is solvable for any right-hand side C. More precisely, we suppose that the eigenvalues of the matrix A lie in the unit disk. We restrict ourselves (although this is not essential) to the case of a positive definite Hermitian matrix C ($C = C^* > 0$) on the right-hand side of (14.1.1).

As was shown in §9.3, a solution X to equation (14.1.1) can be represented as the following infinite matrix series:

$$X = C + \sum_{j=1}^{\infty} (A^*)^j C A^j. \tag{14.1.2}$$

Denote by S_k the partial sum

$$S_k = \sum_{j=0}^{2^k} (A^*)^j C A^j. \tag{14.1.3}$$

We note that

$$\begin{aligned} S_k &= \sum_{j=0}^{2^{k-1}} (A^*)^j C A^j + (A^*)^{2^{k-1}} \Bigg[\sum_{j=0}^{2^{k-1}} (A^*)^j C A^j \Bigg] A^{2^{k-1}} \\ &= S_{k-1} + (A^*)^{2^{k-1}} S_{k-1} A^{2^{k-1}}, \\ A^{2^k} &= (A^{2^{k-1}})^2. \end{aligned}$$

We obtain the following computational scheme:

$$\begin{gathered} S_0 = C, \quad A_0 = A, \\ S_k = S_{k-1} + A_{k-1}^* S_{k-1} A_{k-1}, \quad A_k = A_{k-1}^2, \quad k = 1, 2, \ldots. \end{gathered} \tag{14.1.4}$$

Comparing (14.1.2) with (14.1.3) and taking into account the relations

$$X - S_k = \sum_{j=2^k+1}^{\infty} (A^*)^j C A^j, \quad \|X - S_k\| \le \|C\| \sum_{j=2^k+1}^{\infty} \|A^j\|^2,$$

we see that to estimate the convergence rate of the above computational process we must estimate the rate of the decrease of $\|A^j\|$ as j increases. Such an estimate was obtained in §9.3 for the case where $C = C^* > 0$:

$$\|A^j\| \leq \sqrt{\|X\|\,\|X^{-1}\|}\left[1 - \frac{1}{\|C^{-1}\|\,\|X\|}\right]^{j/2}. \tag{14.1.5}$$

Using this estimate, we find that

$$\begin{aligned}\|X - S_k\| &\leq \|C\|\,\|X\|\,\|X^{-1}\| \sum_{j=2^{k+1}}^{\infty} \left[1 - \frac{1}{\|C^{-1}\|\,\|X\|}\right]^{j} \\ &\leq \|C\|\,\|X\|\,\|X^{-1}\| \sum_{j=2^{k+1}}^{\infty} \left[1 + \frac{1}{\|C^{-1}\|\,\|X\|}\right]^{-j} \\ &= \|C\|\,\|X\|\,\|X^{-1}\| \left[\frac{\|C^{-1}\|\,\|X\|}{1 + \|C^{-1}\|\,\|X\|}\right]^{2^{k+1}-1},\end{aligned}$$

i.e., the norm of the error in the approximate solution S_k satisfies the estimate

$$\frac{\|X - S_k\|}{\|X\|} \leq \|C\|\,\|X^{-1}\| \left[\frac{\|C^{-1}\|\,\|X\|}{1 + \|C^{-1}\|\,\|X\|}\right]^{2^{k+1}-1}. \tag{14.1.6}$$

At the first glance, such an estimate is useless since the error is expressed in terms of an unknown solution. However, a solution X to the Lyapunov equation is not suitable for applications if the norm $\|X\|$ is too large. Indeed, the large norm $\|X\|$ means that either the matrix A has eigenvalues $\lambda_p(A)$ whose module $|\lambda_p(A)|$ is close to 1 (consequently, the decrease of $\|A^j\|$ is very slow as j increases), or eigenvalues are multiple, or angles between eigenvectors are small, which again leads to the slow decrease of $\|A^j\|$. Therefore, in the case $\|X\| > M$ for sufficiently large M, it is natural to regard $X - A^*XA = C$ as a practically unsolvable equation because we cannot obtain an estimate for the decrease of the powers A^j.

If we are interested in only those solutions X that $\|X\| \leq M$, then the estimate (14.1.6) allows us to find the number of steps which is necessary to obtain solutions with guaranteed accuracy. After k steps we can verify if the result S_k satisfies the condition

$$\|S_k\| \leq M + \text{an admissible error.}$$

We can also verify whether or not the Lyapunov equation (14.1.1) holds with the required accuracy. If at least one such verification shows a negative result, then we conclude that a solution X such that $\|X\| \leq M$ does not exist.

It is also necessary to take into account the influence of unavoidable errors appearing as the result of arithmetic operations performed by a computer. These arguments can serve as a basis of a computational process which either leads to the conclusion that the Lyapunov equations have no solutions X such that $\|X\| \leq M$ (for example, $\|X\| \leq 10^{10}$) or computes the solutions X with any guaranteed accuracy. Such processes were developed and justified in [**4, 5**], where the author mainly studied the Lyapunov equations $XA + A^*X + D = 0$ applied in the theory of stability of continuous processes.

Let a matrix $S = S^*$ be an approximate solution to the equation $X - A^*XA = C$, i.e.,

$$S - A^*SA = C + \Delta, \quad \|C^{-1}\|\,\|\Delta\| < 1. \tag{14.1.7}$$

In this case,

$$S \geq C + \Delta \geq (\|C^{-1}\|^{-1} - \|\Delta\|)I = \frac{1 - \|C^{-1}\|\,\|\Delta\|}{\|C^{-1}\|}I.$$

In other words, the matrix S is positive definite and satisfies the inequality

$$\|S\|\,\|S^{-1}\| \leq \frac{\|S\|\,\|C^{-1}\|}{1 - \|C^{-1}\|\,\|\Delta\|}.$$

Hence we can apply the estimate (14.1.5) with the corresponding replacement to the matrix A from equation (14.1.7):

$$\begin{aligned}
\|A^j\| &\leq \sqrt{\|S\|\,\|S^{-1}\|}\left[1 - \frac{1}{\|(C+\Delta)^{-1}\|\,\|S\|}\right]^{j/2} \\
&\leq \sqrt{\frac{\|S\|\,\|C^{-1}\|}{1 - \|C^{-1}\|\,\|\Delta\|}}\left[1 - \frac{1 - \|C^{-1}\|\,\|\Delta\|}{\|C^{-1}\|\,\|S\|}\right]^{j/2},
\end{aligned}$$

$$\begin{aligned}
\sum_{j=0}^{\infty} \|(A^*)^j\|\,\|A^j\| = \sum_{j=0}^{\infty} \|A^j\|^2 &\leq \frac{\|S\|\,\|C^{-1}\|}{1 - \|C^{-1}\|\,\|\Delta\|} \sum_{j=0}^{\infty}\left[1 - \frac{1 - \|C^{-1}\|\,\|\Delta\|}{\|C^{-1}\|\,\|S\|}\right]^j \\
&= \frac{\|S\|^2\,\|C^{-1}\|^2}{(1 - \|C^{-1}\|\,\|\Delta\|)^2}.
\end{aligned}$$

The difference $X - S$ satisfies the equation $(X - S) + A^*(X - S)A = -\Delta$ and, consequently, is represented as the series

$$X - S = -\sum_{j=0}^{\infty} (A^*)^j \Delta A^j,$$

which implies the estimate

$$\|X - S\| \leq \|\Delta\| \sum_{j=0}^{\infty} \|A^j\|^2 \leq \frac{\|S\|^2\,\|C^{-1}\|^2}{(1 - \|C^{-1}\|\,\|\Delta\|)^2}\|\Delta\|. \tag{14.1.8}$$

Analyzing the proof of (14.1.8), we can verify that from the existence of an approximate solution S such that the residual Δ in (14.1.7) satisfies the condition $\|\Delta\|\,\|C^{-1}\| < 1$, it follows that there exists an exact solution X to the discrete Lyapunov equation $X - A^*XA = C$ $(C = C^* > 0)$ and the estimate (14.1.8) for the error $X - S$ holds.

The classical Lyapunov equation

$$XB + B^*X + D = 0 \tag{14.1.9}$$

is used for the analysis of the stability of solutions to the system of equations $\dot{x} = Bx$. It can be reduced to the discrete Lyapunov equation. Hence we can apply the above iteration procedure.

We begin by computing the auxiliary matrices

$$\begin{aligned} A &= e^{\tau B} = I + \tau B + \frac{\tau^2}{2!} B^2 + \cdots, \\ C &= \int_0^\tau e^{tB^*} D e^{tB}\, dt = \tau D_0 + \frac{\tau^2}{2!} D_1 + \frac{\tau^3}{3!} D_2 + \cdots, \end{aligned} \tag{14.1.10}$$

where $D_0 = D$ and $D_j = D_{j-1} B + B^* D_{j-1}$. The parameter τ is chosen to be not too large so that the Taylor series of A and C will converge quickly. For example, we can take $\tau = 1/(2\|B\|)$.

PROBLEM 14.1.1. Verify (14.1.10).

It turns out that the solution

$$X = \int_0^\infty e^{tB^*} D e^{tB} dt \tag{14.1.11}$$

to the Lyapunov equation (14.1.9) is simultaneously a solution to the discrete Lyapunov equation $X - A^*XA = C$ (see §9.3). Indeed, $\lambda_j(A) = e^{\tau\lambda_j(B)}$ are eigenvalues of $A = e^{\tau B}$ (see Problem 8.5.3). Therefore, if all $\lambda_j(B)$'s lie in the left half-plane (this is the necessary condition for the existence of a positive definite solution X to equation (14.1.9) in the case $D = D^* > 0$), then all $\lambda_j(A)$'s lie in the unit disk. Therefore, the equation $X - A^*XA = C$ is uniquely solvable and the solution X is represented as the series

(14.1.12)

$$\begin{aligned} X &= C + \sum_{j=1}^\infty (A^*)^j C A^j = \int_0^\tau e^{tB^*} D e^{tB} dt + \sum_{j=1}^\infty e^{j\tau B^*} \left[\int_0^\tau e^{tB^*} D e^{tB} dt \right] e^{j\tau B} \\ &= \sum_{j=0}^\infty \int_{j\tau}^{(j+1)\tau} e^{tB^*} D e^{tB} dt = \int_0^\infty e^{tB^*} D e^{tB} dt. \end{aligned}$$

Comparing (14.1.12) with (14.1.11), we see that the solution X to the equation $X - A^*XA = C$ coincides with a solution to equation (14.1.9) if A and C are connected with B and D by formulas (14.1.10). The algorithm presented is simple; moreover, it converges quickly. The algorithm was suggested by Davison and Man [7] in 1968.

14.2. Computation of the spectrum dichotomy of a regular pencil

The statement of the problem (reminder). The canonical form of the condition of the problem. Auxiliary ribbon systems of equations and elimination of unknowns. Representation of solutions. The convergence of the iteration process.

We recall (see §10.3) that a matrix pencil $\lambda A + B$ regular on the unit circle consists of square $(N \times N)$-matrices A and B which are simultaneously reduced to the canonical forms

$$A = T \begin{bmatrix} A_1 & 0 \\ 0 & I_{N_2} \end{bmatrix} Q, \quad B = T \begin{bmatrix} I_{N_1} & 0 \\ 0 & B_2 \end{bmatrix} Q \tag{14.2.1}$$

for some T and Q ($\det T \neq 0$, $\det Q \neq 0$). Furthermore, the eigenvalues of the $N_1 \times N_1$-block A_1 and the $N_2 \times N_2$-block B_2, where $N_1 + N_2 = N$, lie in the unit disk.

The study of the regularity of a matrix pencil $\lambda A + B$ on the unit circle $|\lambda| = 1$ is reduced to the construction of the projections P and $I - P$ onto the invariant subspaces corresponding to the eigenvalues inside and outside the unit disk, respectively.

In the adopted notation, the integral representation (10.3.5) of the projection P is written as follows:

$$P = Q^{-1} \begin{bmatrix} 0 & 0 \\ 0 & I \end{bmatrix} Q = \frac{1}{2\pi} \int_0^{2\pi} (A + e^{-i\varphi} B)^{-1} A \, d\varphi. \tag{14.2.2}$$

If the projection P is somehow found, one of the possible representations (14.2.1) can be constructed by using forumlas in §10.3. Recall that Q^{-*} means $(Q^*)^{-1}$ (see the end of §10.1). We recommend the that reader read once again the end of §10.3 and review the construction of the matrices Q and T suggested there. In order to obtain orthonormal bases for the reducing subspaces, we can apply the singular value decomposition of P and $I - P$. The singular value decomposition, as well as the analysis of influences of computational errors, can be found in [**15**].

Suppose that the projection P is found. Computing the matrices T and Q and choosing a positive definite Hermitian matrix $C = C^* > 0$, we pass to the solution of the equations

$$\begin{aligned} AHA^* - BHB^* &= B[PCP^* - (I-P)C(I-P^*)]B^* \\ &\quad + A[PCP^* - (I-P)C(I-P^*)]A^*, \\ PH &= HP^*, \end{aligned} \tag{14.2.3}$$

which, together with the equalities

$$P^2 = P, \quad P(A-B)^{-1}A = (A-B)^{-1}AP \tag{14.2.4}$$

form the system of *matrix equations of the problem about the circle dichotomy.* These equations were studied in §10.4 (note that other notation was used there). As was shown, if a solution (P, H) to the system (14.2.3), (14.2.4) exists and the matrix H is positive definite, then the solution is unique. The existence of a solution guarantees the regularity of the pencil. Conversely, if the pencil is regular, then there exists a solution (P, H) such that $H = H^* > 0$.

It turns out that for known P, T, and Q the computation of H is reduced to the computation of solutions to the discrete Lyapunov equations. This means that we can apply the Davison–Man algorithm (see §14.1).

In view of the uniqueness of a solution to the system (14.2.3) and (14.2.4), it is not hard to obtain the representation

$$H = PHP^* + (I-P)H(I-P^*),$$

where the terms on the right-hand side satisfy the equations

$$\begin{aligned} &A(PHP^*)A^* - B(PHP^*)B^* = A(PCP^*)A^* + B(PCP^*)B^*, R \\ &B((I-P)H(I-P^*))B^* - A((I-P)H(I-P^*))A^* \\ &\qquad = A((I-P)C(I-P^*))A^* + B((I-P)C(I-P^*))B^*. \end{aligned} \tag{14.2.5}$$

From (14.2.2) it follows that the matrices PHP^*, $(I-P)H(I-P^*)$, PCP^*, and $(I-P)C(I-P^*)$ have the following form:

$$PHP^* = Q^{-1}\begin{bmatrix}0 & 0\\ 0 & H_{22}\end{bmatrix}Q^{-*},$$
$$PCP^* = Q^{-1}\begin{bmatrix}0 & 0\\ 0 & C_{22}\end{bmatrix}Q^{-*},$$
$$(I-P)H(I-P^*) = Q^{-1}\begin{bmatrix}H_{11} & 0\\ 0 & 0\end{bmatrix}Q^{-*},$$
$$(I-P)C(I-P^*) = Q^{-1}\begin{bmatrix}C_{11} & 0\\ 0 & 0\end{bmatrix}Q^{-*}.$$

Using (14.2.1), we can reduce equations (14.2.5) for A and B to the discrete Lyapunov equations

$$\begin{aligned}H_{11} - A_1H_{11}A_1^* &= C_{11} + A_1C_{11}A_1^*,\\ H_{22} - B_2H_{22}A_2^* &= C_{22} + A_2C_{22}A_2^*.\end{aligned} \tag{14.2.6}$$

For given P, T, and Q it is easy to find A_1, B_1, C_{11}, and C_{22} from the representations

$$\begin{gathered}\begin{bmatrix}A_1 & 0\\ 0 & I_{N_2}\end{bmatrix} = T^{-1}AQ^{-1}, \quad \begin{bmatrix}I_{N_1} & 0\\ 0 & B_2\end{bmatrix} = T^{-1}BQ^{-1},\\ QCQ^* = \begin{bmatrix}C_{11} & C_{12}\\ C_{21} & C_{22}\end{bmatrix}.\end{gathered} \tag{14.2.7}$$

After that we can use the algorithm from §14.1 to find H_{11} and H_{22} as solutions to the system (14.2.6). Now, we can compute the matrix H by the formula

$$H = Q\begin{bmatrix}H_{11} & 0\\ 0 & H_{22}\end{bmatrix}Q^*.$$

We recall (see §10.4) that H is represented as the matrix integral

$$\begin{aligned}H &= \frac{1}{2\pi}\int_0^{2\pi}(e^{i\varphi}A+B)^{-1}(ACA^*+BCB^*)(e^{i\varphi}A+B)^{-*}\,d\varphi\\ &\equiv \frac{1}{2\pi}\int_0^{2\pi}(A-e^{i\varphi}B)^{-1}(ACA^*+BCB^*)(A-e^{i\varphi}B)^{-*}d\varphi.\end{aligned}$$

Thus, if the projection P is found, then the problem (14.2.3), (14.2.4) about the circle dichotomy is reduced to the known procedures.

We pass to an algorithm for computing the projection P. First, we note that if in (14.2.7) the matrices A_1, B_2, T, Q are replaced with the matrices

$$\widetilde{A}_1 = (\sqrt{H_{11}})^{-1}A_1\sqrt{H_{11}}, \quad \widetilde{B}_2 = (\sqrt{H_{22}})^{-1}B_2\sqrt{H_{22}},$$
$$\widetilde{T} = T\begin{bmatrix}\sqrt{H_{11}} & 0\\ 0 & \sqrt{H_{22}}\end{bmatrix}, \quad \widetilde{Q} = \begin{bmatrix}(\sqrt{H_{11}})^{-1} & 0\\ 0 & (\sqrt{H_{22}})^{-1}\end{bmatrix}Q,$$

leaving A and B the same, then we obtain the representations

$$A = \widetilde{T}\begin{bmatrix}\widetilde{A}_1 & 0\\ 0 & I_{N_2}\end{bmatrix}\widetilde{Q}, \quad B = \widetilde{T}\begin{bmatrix}I_{N_1} & 0\\ 0 & \widetilde{B}_2\end{bmatrix}\widetilde{Q},$$

where the eigenvalues of $\widetilde{A}_1$ and $\widetilde{B}_2$ are of module less than 1 (they coincide with the eigenvalues of the initial blocks A_1 and B_2); moreover, $\|\widetilde{A}_1\| < 1$ and $\|\widetilde{B}_2\| < 1$. Let a positive definite Hermitian matrix H_{ii} be reduced to the diagonal form

$$H_{ii} = U^* \begin{bmatrix} h_1 & & & 0 \\ & h_2 & & \\ & & \ddots & \\ 0 & & & h_{N_i} \end{bmatrix} U, \quad h_j > 0, \tag{14.2.8}$$

with the help of a unitary matrix

$$\sqrt{H_{ii}} = U^* \begin{bmatrix} \sqrt{h_1} & & & 0 \\ & \sqrt{h_2} & & \\ & & \ddots & \\ 0 & & & \sqrt{h_{N_i}} \end{bmatrix} U, \quad \sqrt{h_i} > 0. \tag{14.2.9}$$

Then

$$(\sqrt{H_{ii}})^{-1} H_{ii} (\sqrt{H_{ii}})^{-1} = I_{N_i}, \quad \sqrt{H_{ii}}\,\sqrt{H_{ii}} = H_{ii}.$$

In view of the last equalities, the Lyapunov equation (14.2.6) transforms to the equations

$$\begin{aligned} I_{N_1} - \widetilde{A}_1 \widetilde{A}_1^* &= (\sqrt{H_{11}})^{-1} [C_{11} + A_1 C_{11} A_1^*] (\sqrt{H_{11}})^{-1} = F_{11}, \\ I_{N_2} - \widetilde{B}_2 \widetilde{B}_2^* &= (\sqrt{H_{22}})^{-1} [C_{22} + B_2 C_{22} B_2^*] (\sqrt{H_{22}})^{-1} = F_{22}, \end{aligned} \tag{14.2.10}$$

where the matrices $F_{11} = F_{11}^* > 0$ and $F_{22} = F_{22}^* > 0$ are positive definite since the matrices C_{11} and C_{22} are positive definite because of the positive definiteness of the matrix C and the representations (14.2.7). The equalities (14.2.10) lead to the estimates

$$\begin{aligned} \|\widetilde{A}_1\|^2 &\leq 1 - \lambda_{\min}(F_{11}) = 1 - \|F_{11}^{-1}\|^{-1} < 1, \\ \|\widetilde{A}_2\|^2 &\leq 1 - \lambda_{\min}(F_{22}) = 1 - \|F_{22}^{-1}\|^{-1} < 1, \end{aligned}$$

which shows that $\|\widetilde{A}_1\| < 1$ and $\|\widetilde{A}_2\| < 1$. Later when describing the computation of the projections P and $I - P$ for matrices A and B forming a pencil regular on the unit circle we use the canonical decompositions

$$A = T \begin{bmatrix} \Lambda & 0 \\ 0 & I_{N_2} \end{bmatrix} Q, \quad B = T \begin{bmatrix} I_{N_1} & 0 \\ 0 & M \end{bmatrix} Q,$$

where $\|\Lambda\| \leq \alpha < 1$, $\|M\| \leq \alpha < 1$. The above arguments mean that our assumption does not restrict generality.

We introduce a square $N \times N$-matrix X_1 as a solution to the matrix equation

$$A(I + X_1) + B(I - X_1) = 0.$$

Under the above assumptions, this equation is uniquely solvable:

$$\begin{aligned} X_1 &= (B - A)^{-1}(A + B) \\ &= Q^{-1} \begin{bmatrix} (I_{N_1} - \Lambda)^{-1}(I_{N_1} + \Lambda) & 0 \\ 0 & -(I_{N_2} - M)^{-1}(I_{N_2} + M) \end{bmatrix} Q. \end{aligned} \tag{14.2.11}$$

Further construction is based on the study of the matrix equations

$$\begin{aligned}
AZ_0 + BZ_1 &= 0, \\
AZ_1 + BZ_2 &= 0, \\
&\cdots\cdots \\
AZ_{2^k-1} + BZ_{2^k} &= 0, \\
Z_0 - Z_{2^k} &= I_N,
\end{aligned} \tag{14.2.12}$$

which, in view of the equalities

$$A = -\frac{1}{2}(B - A)(I - X_1), \quad B = \frac{1}{2}(B - A)(I + X_1)$$

can be rewritten in the following convenient form:

$$\begin{aligned}
(I - X_1)Z_0 - (I + X_1)Z_1 &= 0, \\
(I - X_1)Z_1 - (I + X_1)Z_2 &= 0, \\
&\cdots\cdots \\
(I - X_1)Z_{2^k-1} - (I + X_1)Z_{2^k} &= 0, \\
Z_0 - Z_{2^k} &= I.
\end{aligned} \tag{14.2.13}$$

It is not hard to eliminate the matrices Z_{2j+1} with odd indices from each pair of equalities

$$\begin{aligned}
(I - X_1)Z_{2j} - (I + X_1)Z_{2j+1} &= 0, \\
(I - X_1)Z_{2j+1} - (I + X_1)Z_{2j+2} &= 0.
\end{aligned}$$

We obtain the equation

$$(I - X_1)^2 Z_{2j} - (I + X_1)^2 Z_{2j+2} = 0,$$

which can be rewritten in the following equivalent form:

$$[(I + X_1^2) - 2X_1]Z_{2j} - [(I + X_1^2) + 2X_1]Z_{2j+2} = 0$$

or, setting $X_2 = 2(I + X_1^2)^{-1}X_1$, in the form

$$(I - X_2)Z_{2j} - (I + X_2)Z_{2j+2} = 0.$$

We note that the matrix $I + X_1^2$ is nonsingular and, consequently invertible. The nonsingularity follows from the representation

$$I + X_1^2 = Q^{-1}\begin{bmatrix} 2(I_{N_1} - \Lambda)^{-2}(I_{N_1} + \Lambda^2) & 0 \\ 0 & 2(I_{N_2} - M)^{-2}(I_{N_2} + M^2) \end{bmatrix} Q, \tag{14.2.14}$$

which is valid in view of (14.2.11) and the inequalities $\|\Lambda\| \leq \alpha < 1$, $\|M\| \leq \alpha < 1$. From (14.2.11) and (14.2.14) we find the canonical decomposition

$$X_2 = Q^{-1}\begin{bmatrix} (I_{N_1} + \Lambda^2)^{-1}(I_{N_1} - \Lambda^2) & 0 \\ 0 & -(I_{N_2} + M^2)^{-1}(I_{N_2} - M^2) \end{bmatrix} Q, \tag{14.2.15}$$

which differs from (14.2.11) for X_1 only in that Λ and M are replaced with $-\Lambda^2$ and $-M^2$. In addition, $\| - \Lambda^2\| \leq \|\Lambda\|^2 \leq \alpha^2 < 1$ and $\| - M^2\| \leq \alpha^2 < 1$. Eliminating all unknown Z_{2j+1}'s with odd indices from (14.2.13), we obtain the system

$$
\begin{gathered}
(I - X_2)Z_0 - (I + X_2)Z_2 = 0, \\
(I - X_2)Z_2 - (I + X_2)Z_4 = 0, \\
\dots\dots\dots\dots\dots\dots\dots\dots \\
(I - X_2)Z_{2^k-2} - (I + X_2)Z_{2^k} = 0, \\
Z_0 - Z_{2^k} = I
\end{gathered}
$$

with fewer equations. Similarly, eliminating all unknown Z_{2j}'s with indices that are not divisible by 4 we arrive at the following shortened system:

$$
\begin{gathered}
(I - X_3)Z_{4j} - (I + X_3)Z_{4(j+1)} = 0, \\
X_3 = 2(I + X_2^2)^{-1}X_2.
\end{gathered}
$$

We can continue the elimination procedure, setting

$$X_{r+1} = 2(I + X_r^2)^{-1}X_r, \quad r = 1, 2, \dots, k. \tag{14.2.16}$$

and obtaining

$$(I - X_{r+1})Z_{2^r j} - (I + X_{r+1})Z_{2^r(j+1)} = 0,$$

until we arrive at the system consisting of two matrix equations

$$
\begin{gathered}
(I - X_{k+1})Z_0 - (I + X_{k+1})Z_{2^k} = 0, \\
Z_0 - Z_{2^k} = I.
\end{gathered} \tag{14.2.17}
$$

It is obvious that the matrices

$$Z_0 = \frac{1}{2}(I + X_{k+1}^{-1}), \quad Z_{2^k} = -\frac{1}{2}(I - X_{k+1}^{-1})$$

form a solution to the system (14.2.17). Acting as in the proof of (14.2.15) for X_2 and using the recurrence formulas (14.2.16), by induction we establish the equality

$$X_k = Q^{-1} \begin{bmatrix} \Psi_{11} & 0 \\ 0 & \Psi_{22} \end{bmatrix}, \tag{14.2.18}$$

where

$$
\begin{aligned}
\Psi_{11} &= (I + (-1)^k \Lambda^{2^k})^{-1}(I - (-1)^k \Lambda^{2^k}), \\
\Psi_{22} &= -(I + (-1)^k M^{2^k})^{-1}(I - (-1)^k M^{2^k}).
\end{aligned}
$$

In view of the conditions $\|\Lambda\| \le \alpha < 1$ and $\|M\| \le \alpha < 1$, we obtain

$$\lim_{k\to\infty} X_k = \lim_{k\to\infty} X_k^{-1} = Q^{-1} \begin{bmatrix} I_{N_1} & 0 \\ 0 & -I_{N_2} \end{bmatrix} Q = X.$$

The limit X of the sequence $\{X_k\}$ satisfies the equalities $X^2 = I$ and $X^{-1} = X$. The first and the last matrices of the sequence $Z_0, Z_1, \dots, Z_{2^k}$, i.e., the matrices Z_0 and Z_{2^k}, turn out to be close to the projection $I - P = \frac{1}{2}(I + X)$ and the matrix $-P = -\dfrac{1}{2}(I - X)$ for large k. In other words, $X = I_N - 2P$.

Thus, we have shown that the iteration process

$$
\begin{aligned}
X_1 &= (B - A)^{-1}(A + B), \\
X_{r+1} &= 2(I_N + X_r^2)^{-1}X_r
\end{aligned} \tag{14.2.19}
$$

converges as $r \to \infty$ if the pencil $\lambda A + B$ is regular on the unit circle $|\lambda| = 1$. Furthermore, it converges to the matrix $X = I_N - 2P$, i.e., $\lim_{r\to\infty} X_r = I_N - 2P$.

Consequently, we can apply this process to the computation of the projection P. Moreover, for large k the coefficient matrices $(I - X_k)$ and $(I + X_k)$ in (14.2.17) almost coincide with the corresponding projections multiplied by 2:

$$\lim_{k\to\infty}(I + X_k) = 2(I - P), \quad \lim_{k\to\infty}(I - X_k) = 2P. \tag{14.2.20}$$

At each step of this process we find the inverse matrices to $I_N + X_r^2$. It turns out that it is not necessary to deal with the inverse matrices if we eliminate the unknowns from the initial system (14.2.12) in the same order as above, but using orthogonal transformations instead of the inverse matrices. In §14.3 we will describe an algorithm based on this idea. Applying it, we can find a solution to one of the possible systems of matrix equations which appear in the criterion of the dichotomy of a matrix pencil; moreover, this can be done without additional computations.

To facilitate the understanding of the scheme of the orthogonal elimination algorithm, we connected the iteration process (14.2.19) with the solution of the system (14.2.12). However, to justify (14.2.19) it suffices to give iteration formulas and establish the convergence by proving the representation of X_k in the canonical form (14.2.18). This fact was proved above with the help of only formulas (14.2.19).

We consider the problem about linear dichotomy of the spectrum by the imaginary axis. It is reduced to the construction of the projections P and $I - P$ onto the subspaces that reduce from the right the pencil $\lambda A + B$ and correspond to the eigenvalues in the right and left half-plane, respectively.

If matrices A and B form a pencil regular on the imaginary axis, then they admit canonical representations

$$A = T\begin{bmatrix}\widetilde{\Lambda} & 0\\ 0 & I_{N_2}\end{bmatrix}Q, \quad B = T\begin{bmatrix}I_{N_1} & 0\\ 0 & \widetilde{M}\end{bmatrix}Q, \qquad \det T \neq 0, \ \det Q \neq 0,$$

where the eigenvalues of $\widetilde{\Lambda}$ lie in the left half-plane $\operatorname{Re}\lambda < 0$; whereas the eigenvalues of $\widetilde{M}$ are located in the right half-plane $\operatorname{Re}\lambda > 0$. The projection P has the form

$$P = Q^{-1}\begin{bmatrix}0 & 0\\ 0 & I_{N_2}\end{bmatrix}Q. \tag{14.2.21}$$

The pencil $\lambda(A+B) + (A-B)$ is formed by the matrices $A+B$ and $A-B$, which admit the canonical decomposition

$$A + B = \widehat{T}\begin{bmatrix}(I_{N_1} - \widetilde{\Lambda})^{-1}(I_{N_1} + \widetilde{\Lambda}) & 0\\ 0 & I_{N_2}\end{bmatrix}Q,$$

$$A - B = \widehat{T}\begin{bmatrix}I_{N_1} & 0\\ 0 & (I_{N_2} + \widetilde{M})^{-1}(I_{N_2} - \widetilde{M})\end{bmatrix}Q,$$

where

$$\widehat{T} = T\begin{bmatrix}I_{N_1} - \widetilde{\Lambda} & 0\\ 0 & I_{N_2} + \widetilde{M}\end{bmatrix}.$$

We used the fact that the matrices $I_{N_1} - \widetilde{\Lambda}$ and $I_{N_2} + \widetilde{M}$ are nonsingular, which follows from the assumptions on the eigenvalues of $\widetilde{\Lambda}$ and $\widetilde{M}$. In view of the same assumptions, the eigenvalues of the blocks $(I_{N_1} - \widetilde{\Lambda})^{-1}(I_{N_1} + \widetilde{\Lambda})$ and $(I_{N_2} + \widetilde{M})^{-1}(I_{N_2} - \widetilde{M})$ lie in the unit disk, i.e., the pencil $\lambda(A+B) + (A-B)$

is regular on the unit circle. The projection P onto the right reducing subspace corresponding to the eigenvalues in the unit disk can be represented in the form

$$P = Q^{-1} \begin{bmatrix} 0 & 0 \\ 0 & I_{N_2} \end{bmatrix} Q.$$

This coincides with the representation (14.2.21).

Thus, we can compute P applying the above iteration procedure to the pencil $\lambda(A+B)+(A-B)$.

14.3. The orthogonal elimination algorithm

The solution of the ribbon system by the orthogonal elimination method. The preliminary normalization of a matrix pencil. An announcement about the invariance of the integral dichotomy criterion. Canonical representations. Identities connected canonical decompositions.

We describe a procedure based on the orthogonal transformations or, more exactly, on the QR decomposition (see §3.1). The procedure eliminates the unknowns from the chain of matrix equations used for obtaining the projections P and $I-P$ onto the subspaces reducing from the right a matrix pencil $\lambda A+B$ and corresponding to the eigenvalues inside and outside the unit disk $|\lambda| \leq 1$. In §14.2, we already considered this chain, which consists of equations

$$AZ_{j-1} + BZ_j = 0.$$

To eliminate Z_j from two successive equations

$$\begin{aligned} AZ_{j-1} + BZ_j &= 0, \\ AZ_j + BZ_{j+1} &= 0, \end{aligned}$$

we rewrite them as the matrix equality

$$\begin{bmatrix} B & A & 0 \\ A & 0 & B \end{bmatrix} \begin{pmatrix} Z_j \\ Z_{j-1} \\ Z_{j+1} \end{pmatrix} = \begin{pmatrix} 0 \\ 0 \end{pmatrix} \tag{14.3.1}$$

and apply the QR decomposition

$$\begin{bmatrix} B & A \\ A & 0 \end{bmatrix} = Q_1 \begin{bmatrix} C_1 & D_1 \\ 0 & A_1 \end{bmatrix},$$

i.e., we represent the matrix on the left-hand side as the product of a unitary $(2N \times 2N)$-matrix Q_1 and an upper triangular matrix consisting of the zero block and the blocks D_1, C_1, and A_1. We note that C_1 and A_1 are triangular matrices. However, this is not essential later. The QR decomposition is a well developed and widely used computational procedure. We mentioned it in §3.4 and refer the reader to [**15**] for detailed information.

Usually, we assume that the pencil consists of real matrices. Sometimes we will emphasize necessary modifications in the complex case. For definiteness, we will speak about orthogonal but not unitary eliminations. Acting from the left on both sides of (14.3.1) by the matrix Q_1^* ($Q_1^*Q_1 = I_{2N}$), we find

$$Q_1^* \begin{bmatrix} B & A & 0 \\ A & 0 & B \end{bmatrix} \begin{pmatrix} Z_j \\ Z_{j-1} \\ Z_{j+1} \end{pmatrix} = \begin{bmatrix} C_1 & D_1 & E_1 \\ 0 & A_1 & B_1 \end{bmatrix} \begin{pmatrix} Z_j \\ Z_{j-1} \\ Z_{j+1} \end{pmatrix} = \begin{pmatrix} 0 \\ 0 \end{pmatrix}. \tag{14.3.2}$$

We obtain the equality $A_1Z_{j-1} + B_1Z_{j+1} = 0$ which does not contain Z_j. We emphasize that if the QR decomposition is realized by a series of orthogonal (unitary) reflections, then the matrix Q_1 itself is not computed in this process. However this fact is not essential for the further consideration.

Acting as before, we eliminate Z_j from the equations

$$A_1Z_{j-2} + B_1Z_j = 0,$$
$$A_1Z_j + B_1Z_{j+2} = 0.$$

We obtain the equalities

$$A_2Z_{j-2} + B_2Z_{j+2} = 0,$$

where the difference between the indices of unknown matrices is divisible by 4. Similarly, continuing the process of successive eliminations, we obtain

$$A_rZ_j + B_rZ_{j+2^r} = 0.$$

To compute A_r and B_r for given matrices A_{r-1} and B_{r-1}, i.e., to eliminate Z_j from the equations

$$A_{r-1}Z_{j-2^{r-1}} + B_{r-1}Z_j = 0,$$
$$A_{r-1}Z_j + B_{r-1}Z_{j+2^{r-1}} = 0,$$

it suffices to choose a unitary $2N \times 2N$-matrix Q_r ($Q_r^*Q_r = I_{2N}$) such that the matrix

$$Q_r^* \begin{bmatrix} B_{r-1} & A_{r-1} & 0 \\ A_{r-1} & 0 & B_{r-1} \end{bmatrix} = \begin{bmatrix} C_r & D_r & E_r \\ 0 & A_r & B_r \end{bmatrix} \tag{14.3.3}$$

contains the zero $N \times N$-block in the left lower corner, i.e., it is necessary that Q_r appears in the QR decomposition:

$$\begin{bmatrix} B_{r-1} & A_{r-1} \\ A_{r-1} & 0 \end{bmatrix} = Q_r \begin{bmatrix} C_r & D_r \\ 0 & A_r \end{bmatrix}.$$

We will show later that for the initial matrices A and B forming the pencil $\lambda A + B$ regular on the unit circle, the pencils $\lambda A_r + B_r$, $r = 1, 2, \dots$, are regular on the same circle, which happens only if the rows of the matrix $[A_r \vdots B_r]$ are linearly independent. Recall that in a similar situation (see §14.2) we arrived at the equations $(I - X_r)Z_j - (I + X_r)Z_{j+2^{r-1}} = 0$. Hence there exists a nonsingular matrix K_r ($\det K_r \neq 0$) such that

$$A_{r-1} = K_r(I - X_r), \quad B_{r-1} = -K_r(I + X_r).$$

To prove this assertion, we use the linear independence of the rows of the matrix

$$\frac{1}{2}[I + X_r \vdots I - X_r] = [I \vdots X_r] \begin{bmatrix} \frac{1}{2}I & -\frac{1}{2}I \\ \frac{1}{2}I & \frac{1}{2}I \end{bmatrix}.$$

Thus,

$$(A_r - B_{r-1})^{-1}A_{r-1} = \frac{1}{2}(I - X_r).$$

As was shown in §14.2 (see (14.2.20)),

$$\lim_{r\to\infty} \frac{1}{2}(I - X_r) = P = \lim_{r\to\infty} (A_r - B_r)^{-1}A_r. \tag{14.3.4}$$

Another proof of the second equality in (14.3.4) will be given below.

We see that in the elimination process based on the QR decomposition, the matrix pairs $A_1, B_1; A_2, B_2; \ldots$ are successively computed without computing the inverse matrices. Only after a sufficiently large number of eliminations, obtaining A_r and B_r for a sufficiently large r, we may try to compute the projection P by taking the inverse matrix to the difference $A_r - B_r$ and multiplying the result $(A_r - B_r)^{-1}$ by A_r from the right.

Establishing the regularity of the pencils $\lambda A_r + B_r$, we thereby complete the justification of the simplified algorithm for computing the projection P. In addition, one more fine property of the algorithm becomes clear in the proof. Without much additional computation, we can find the positive definite matrix H in the integral form

$$\begin{aligned} H &= \frac{1}{2\pi}\int_0^{2\pi} (e^{i\varphi}A + B)^{-1}(ACA^* + BCB^*)(e^{i\varphi}A + B)^{-*}d\varphi \\ &\equiv \frac{1}{2\pi}\int_0^{2\pi} (A - e^{i\varphi}B)^{-1}(ACA^* + BCB^*)(A - e^{i\varphi}B)^{-*}d\varphi, \end{aligned} \tag{14.3.5}$$

where $C = C^*$ is a given positive definite matrix.

We emphasize that (14.3.5) holds only for real matrices A and B. In order that it remains valid in the complex case, it is necessary to replace A and B with $\overline{A}$ and $\overline{B}$ on the right-hand side of this equality.

As was shown in §10.4, H and P satisfy the system of equations

$$\begin{aligned} BHB^* - AHA^* = A[PCP^* &- (I-P)C(I-P^*)]A^* \\ &+ B[PCP^* - (I-P)C(I-P^*)]B^*, \end{aligned}$$

$$PH = HP^*, \quad P^2 = P, \quad P(A+B)^{-1}B = (A+B)^{-1}BP. \tag{14.3.6}$$

Thus, indicating the method for computing approximations of the matrices H and P, we simultaneously obtain the method for solving equations (14.3.6).

To obtain a matrix H satisfying (14.3.6), we need to modify the first step of the computation of the matrix pairs $A_1, B_1; A_2, B_2; \ldots$. First we find a matrix L such that

$$LL^* = ACA^* + BCB^* = G.$$

To this end, computing $G = ACA^* + BCB^*$, we set, for example, $L^* = L = \sqrt{G}$. The square root of a positive definite matrix can be found after the reduction of the matrix to the diagonal form by a similar unitary transformation (see §14.2, formulas (14.2.8) and (14.2.9)). It is simpler to represent G as the product of two conjugate triangular matrices L^* and L according to the well-known Holetskii algorithm (see [**33**]). Having computed L, we replace the initial matrices A and B with the matrices $A_0 = L^{-1}A$ and $B_0 = L^{-1}B$. Hence

$$A_0 C A_0^* + B_0 C B_0^* = I.$$

The integral representation of H becomes as follows:

$$H = \frac{1}{2\pi}\int_0^{2\pi} (A_0 - e^{i\varphi}B_0)^{-1}(A_0 - e^{i\varphi}B_0)^{-*}\,d\varphi.$$

Taking A_0 and B_0 in (14.3.2) instead of A and B, we compute A_1 and B_1. The further computation of the matrix pairs $A_2, B_2; A_3, B_3; \ldots; A_r, B_r$ is performed as above. In §14.4, we will show that the integral

$$H_r = \frac{1}{2\pi}\int_0^{2\pi} (A_r - e^{i\varphi}B_r)^{-1}(A_r - e^{i\varphi}B_r)^{-*}\,d\varphi$$

is independent of r ($H_r = H$) and prove the equality

$$H = \lim_{r\to\infty} \widetilde{A}_r\widetilde{A}_r^* + \widetilde{B}_r\widetilde{B}_r^*,$$

where

$$\widetilde{A}_r = (A_r - B_r)^{-1}A_r(A_r - B_r)^{-1}, \quad \widetilde{B}_r = (A_r - B_r)^{-1}B_r(A_r - B_r)^{-1}.$$

Passing to a preliminary part of the proof, we denote by $\Omega_{ij}^{(r)}$ the blocks of the matrix

$$Q_r^* = \begin{bmatrix} \Omega_{11}^{(r)} & \Omega_{12}^{(r)} \\ \Omega_{21}^{(r)} & \Omega_{22}^{(r)} \end{bmatrix} \equiv \begin{bmatrix} \Omega_{11}^{(r)} & \Omega_{12}^{(r)} \\ \Omega_{1}^{(r)} & \Omega_{2}^{(r)} \end{bmatrix}.$$

In further considerations, only the blocks $\Omega_{2j}^{(r)}$ are mainly used; therefore, we omit the subscript 2 in the notation. Since $Q_r^*Q_r = I_{2N}$ is unitary, the matrix equality (14.3.3) written in detail as

$$\begin{bmatrix} \Omega_{11}^{(r)} & \Omega_{12}^{(r)} \\ \Omega_{1}^{(r)} & \Omega_{2}^{(r)} \end{bmatrix}\begin{bmatrix} B_{r-1} & A_{r-1} & 0 \\ A_{r-1} & 0 & B_{r-1} \end{bmatrix} = \begin{bmatrix} C_r & D_r & E_r \\ 0 & A_r & B_r \end{bmatrix}$$

implies the relations

$$\begin{gathered} \Omega_1^{(r)}[\Omega_1^{(r)}]^* + \Omega_2^{(r)}[\Omega_2^{(r)}]^* = I_N, \\ \Omega_1^{(r)}B_{r-1} + \Omega_2^{(r)}A_{r-1} = 0, \\ \Omega_1^{(r)}A_{r-1} = A_r, \\ \Omega_2^{(r)}B_{r-1} = B_r. \end{gathered} \tag{14.3.7}$$

As was shown in §14.2, the matrices A_0 and B_0 forming a regular pencil admit the canonical decompositions

$$\begin{gathered} A_0 = T_0\begin{bmatrix} \Lambda & 0 \\ 0 & I_{N_2} \end{bmatrix}Q, \quad B_0 = T_0\begin{bmatrix} I_{N_1} & 0 \\ 0 & M \end{bmatrix}Q, \\ \det T_0 \neq 0, \quad \det Q \neq 0, N_1 + N_2 = N, \\ \|\Lambda\| \le \alpha, \quad \|M\| \le \alpha, \quad 0 \le \alpha < 1. \end{gathered} \tag{14.3.8}$$

Hence any solution of the difference equations

$$A_0 Z_{j-1} + B_0 Z_j = 0, \quad j = 1, 2, \ldots, 2^k,$$

can be written in the form

$$Z_j = Q^{-1} \begin{bmatrix} (-\Lambda)^j & 0 \\ 0 & (-M)^{2^k - j} \end{bmatrix} \Phi_k. \tag{14.3.9}$$

The matrix Φ_k is arbitrary: for any choice of Φ_k in (14.3.9) the equations hold. The relation $A_k Z_0 + B_k Z_{2^k} = 0$ holds for any solution if

$$\begin{aligned} -A_k &= T_k \begin{bmatrix} (-\Lambda)^{2^k} & 0 \\ 0 & -I_{N_2} \end{bmatrix} Q, \\ -B_k &= T_k \begin{bmatrix} -I_{N_1} & 0 \\ 0 & (-M)^{2^k} \end{bmatrix} Q. \end{aligned} \tag{14.3.10}$$

Let us prove by induction that $\det T_k \neq 0$ for the matrices A_k and B_k obtained in the elimination process. Assume that this is true for some k (for $k = 0$ this assertion is contained in (14.3.8)). We find some $\Omega_1^{(k+1)}$ and $\Omega_2^{(k+1)}$ from the first two relations in (14.3.7). It is not necessary that they are uniquely determined. Denote by $\Phi_{ij}^{(k+1)}$ and $\Psi_{ij}^{(k+1)}$ the blocks of the matrices $\Omega_1^{(k+1)} T_k$ and $\Omega_2^{(k+1)} T_k$:

$$\begin{bmatrix} \Phi_{11}^{(k+1)} & \Phi_{12}^{(k+1)} \\ \Phi_{21}^{(k+1)} & \Phi_{22}^{(k+1)} \end{bmatrix} = \Omega_1^{(k+1)} T_k, \quad \begin{bmatrix} \Psi_{11}^{(k+1)} & \Psi_{12}^{(k+1)} \\ \Psi_{21}^{(k+1)} & \Psi_{22}^{(k+1)} \end{bmatrix} = \Omega_2^{(k+1)} T_k.$$

Since the matrix Q is nonsingular, from the second equality in (14.3.7) we find that

$$\begin{bmatrix} \Phi_{11}^{(k+1)} & \Phi_{12}^{(k+1)} \\ \Phi_{21}^{(k+1)} & \Phi_{22}^{(k+1)} \end{bmatrix} \begin{bmatrix} -I_{N_1} & 0 \\ 0 & (-M)^{2^k} \end{bmatrix} + \begin{bmatrix} \Psi_{11}^{(k+1)} & \Psi_{12}^{(k+1)} \\ \Psi_{21}^{(k+1)} & \Psi_{22}^{(k+1)} \end{bmatrix} \begin{bmatrix} (-\Lambda)^{2^k} & 0 \\ 0 & -I_{N_2} \end{bmatrix} = \begin{bmatrix} 0 & 0 \\ 0 & 0 \end{bmatrix},$$

which is equivalent to the relations

$$\begin{aligned} \Phi_{11}^{(k+1)} &= \Psi_{11}^{(k+1)} (-\Lambda)^{2^k}, \quad \Psi_{12}^{(k+1)} = \Phi_{12}^{(k+1)} (-M)^{2^k}, \\ \Phi_{21}^{(k+1)} &= \Psi_{21}^{(k+1)} (-\Lambda)^{2^k}, \quad \Psi_{22}^{(k+1)} = \Phi_{22}^{(k+1)} (-M)^{2^k}, \end{aligned}$$

$$\begin{bmatrix} \Phi_{11}^{(k+1)} & \Phi_{12}^{(k+1)} \\ \Phi_{21}^{(k+1)} & \Phi_{22}^{(k+1)} \end{bmatrix} = \begin{bmatrix} \Psi_{11}^{(k+1)} & \Phi_{12}^{(k+1)} \\ \Psi_{21}^{(k+1)} & \Phi_{22}^{(k+1)} \end{bmatrix} \begin{bmatrix} (-\Lambda)^{2^k} & 0 \\ 0 & I_{N_2} \end{bmatrix} \equiv F_{k+1} \begin{bmatrix} (-\Lambda)^{2^k} & 0 \\ 0 & I_{N_2} \end{bmatrix},$$

$$\begin{bmatrix} \Psi_{11}^{(k+1)} & \Psi_{12}^{(k+1)} \\ \Psi_{21}^{(k+1)} & \Psi_{22}^{(k+1)} \end{bmatrix} = \begin{bmatrix} \Psi_{11}^{(k+1)} & \Phi_{12}^{(k+1)} \\ \Psi_{22}^{(k+1)} & \Phi_{22}^{(k+1)} \end{bmatrix} \begin{bmatrix} I_{N_1} & 0 \\ 0 & (-M^2)^k \end{bmatrix} \equiv F_{k+1} \begin{bmatrix} I_{N_1} & 0 \\ 0 & (-M)^{2^k} \end{bmatrix}.$$

Thus, we have proved the existence of a matrix F_{k+1} such that

$$\begin{aligned} \Omega_1^{(k+1)} &= F_{k+1} \begin{bmatrix} (-\Lambda)^{2^k} & 0 \\ 0 & I_{N_2} \end{bmatrix} T_k^{-1}, \\ \Omega_2^{(k+1)} &= F_{k+1} \begin{bmatrix} I_{N_1} & 0 \\ 0 & (-M)^{2^k} \end{bmatrix} T_k^{-1}. \end{aligned} \tag{14.3.11}$$

The matrix F_{k+1} is nonsingular because the equalities

$$I_N = \Omega_1^{(k+1)}(\Omega_1^{(k+1)})^* + \Omega_2^{(k+1)}(\Omega_2^{(k+1)})^*$$
$$= F_{k+1}\left\{ \begin{bmatrix} (-\Lambda)^{2^k} & 0 \\ 0 & I_{N_2} \end{bmatrix} T_k^{-1}(\Omega_1^{(k+1)})^* + \begin{bmatrix} I_{N_1} & 0 \\ 0 & (-M)^{2^k} \end{bmatrix} T_k^{-1}(\Omega_2^{(k+1)})^* \right\}$$

are impossible if F_{k+1} is singular. From (14.3.10), (14.3.11), and (14.3.7) we obtain the following representations of A_{k+1} and B_{k+1}:

(14.3.12)
$$-A_{k+1} = F_{k+1} \begin{bmatrix} (-\Lambda)^{2^k} & 0 \\ 0 & -I_{N_2} \end{bmatrix} Q,$$
$$-B_{k+1} = F_{k+1} \begin{bmatrix} -I_{N_1} & 0 \\ 0 & (-M)^{2^{k+1}} \end{bmatrix} Q.$$

To complete induction, it remains to replace F_{k+1} with T_{k+1} and verify that (14.3.12) differs from (14.3.10) only by k replaced with $k+1$. We note that (14.3.11) can be rewritten in the form

$$\Omega_1^{(r)} = T_r \begin{bmatrix} (-\Lambda)^{2^{r-1}} & 0 \\ 0 & I_{N_2} \end{bmatrix} T_{r-1}^{-1},$$
$$\Omega_2^{(r)} = T_r \begin{bmatrix} I_{N_1} & 0 \\ 0 & (-M)^{2^{r-1}} \end{bmatrix} T_{r-1}^{-1}.$$

Using the equality $\Omega_1^{(r)}[\Omega_1^{(r)}]^* + \Omega_2^{(r)}[\Omega_2]^* = I_N$ from (14.3.7), we obtain the recurrence relation

(14.3.13)
$$T_r^{-1}T_r^{-*} = \begin{bmatrix} (-\Lambda)^{2^{r-1}} & 0 \\ 0 & I_{N_2} \end{bmatrix} T_{r-1}^{-1}T_{r-1}^{-*} \begin{bmatrix} (-\Lambda)^{2^{r-1}} & 0 \\ 0 & I_{N_2} \end{bmatrix}^*$$
$$+ \begin{bmatrix} I_{N_1} & 0 \\ 0 & (-M)^{2^{r-1}} \end{bmatrix} T_{r-1}^{-1}T_{r-1}^{-*} \begin{bmatrix} I_{N_1} & 0 \\ 0 & (-M)^{2^{r-1}} \end{bmatrix}^*.$$

To avoid cumbersome computations, it is convenient to introduce the notation

$$G_{-j} = Q^{-1} \begin{bmatrix} -\Lambda^j & 0 \\ 0 & 0 \end{bmatrix} Q, \quad G_{+j} = Q^{-1} \begin{bmatrix} 0 & 0 \\ 0 & -M^j \end{bmatrix} Q, \quad j = 0, 1, \dots .$$

We emphasize that

$$G_{-0} = Q^{-1} \begin{bmatrix} I_{N_1} & 0 \\ 0 & 0 \end{bmatrix} Q = (I-P), \quad G_{+0} = Q^{-1} \begin{bmatrix} 0 & 0 \\ 0 & I_{N_0} \end{bmatrix} Q = P.$$

Denote $S_r = (T_rQ)^{-1}$ ($\det S_r \neq 0$) and rewrite the recurrence relations (14.3.13) as follows:

$$S_rS_r^* = (P + G_{-2^{r-1}})S_{r-1}S_{r-1}^*(P + G_{-2^{r-1}})^*$$
$$+ [(I-P) + G_{+2^{r-1}}]S_{r-1}S_{r-1}^*[(I-P) + G_{+2^{r-1}}]^*.$$

The representations (14.3.10) take the form

(14.3.14)
$$A_r = S_r^{-1}[P - G_{-2^r}],$$
$$B_r = S_r^{-1}[(I-P) - G_{+2^r}].$$

Note that

(14.3.15)
$$G_{+j}G_{+k} = G_{+(j+k)}, \quad G_{-j}G_{-k} = G_{-(j+k)},$$
$$G_{+j}G_{-k} = 0, \quad PG_{+j} = G_{+j}, \quad (I-P)G_{-j} = G_{-j}.$$

In view of the assumption that $\|\Lambda\| \leq \alpha < 1$ and $\|M\| \leq \alpha < 1$, we have

$$\|G_{-j}\| \leq \|Q\| \, \|Q^{-1}\|\alpha^j, \quad \|G_j\| \leq \|Q\| \, \|Q^{-1}\|\alpha^j.$$

It is easy to see that (14.3.14) and (14.3.15) imply the equality

$$(A_r - B_r)^{-1}A_r = P + \sum_{k=1}^{\infty} G_{+2^k r} + \sum_{k=1}^{\infty} G_{-2^k r}, \tag{14.3.16}$$

and the estimates (14.3.8) lead to the inequality

$$\|(A_r - B_r)^{-1}A_r - P\| \leq 2\|Q\| \, \|Q^{-1}\| \frac{\alpha^{2^r}}{1-\alpha^{2^r}}, \tag{14.3.17}$$

which completes the justification of the orthogonal elimination algorithm.

Simultaneously, we prepare a number of important relations which will be used in §14.4, where we will study the matrix integral H (see (14.3.5)) and obtain estimates for the convergence rate in terms of $\|H\|$. Namely, we will deduce the estimate

$$\|G_{\pm k}\| \leq \sqrt{\|C^{-1}\| \, \|H\|} \, e^{-k/(1+\|C^{-1}\| \, \|H\|)}.$$

Using this estimate in (14.3.16), instead of (14.3.17) we obtain

$$\|(A_r - B_r)^{-1}A_r - P\| \leq 2\sqrt{\|C^{-1}\| \, \|H\|} e^{-\frac{2^r}{1+\|C^{-1}\| \, \|H\|}} \left(1 - e^{-\frac{2^r}{1+\|C^{-1}\| \, \|H\|}}\right)^{-1}.$$

In §14.4, we will indicate a rule for computing the matrix integral H on the basis of simple formulas for A_r and B_r without integration if r is sufficiently large.

14.4. Properties of the orthogonal elimination algorithm

An infinite sum representing the integral in the dichotomy criterion. An estimate for the summands in terms of the norm of the integral. Applications of this estimate and of identities from §14.3 for the proof of the convergence of the orthogonal elimination method and the computation of the dichotomy parameter. Application to the linear dichotomy problem.

In this section, we complete the analysis of the orthogonal elimination algorithm. Using this algorithm, we can verify the regularity of a matrix pencil $\lambda A + B$ on the unit circle. For a regular pencil the algorithm yields the projections P and $I - P$ onto subspaces reducing the pencil from the right and corresponding to the eigenvalues inside and outside the unit disk. Hereinafter we assume that the pencil is regular.

We recall that the original matrices A_0 and B_0 differ from the matrices A and B forming the pencil only by the left nonsingular factor L^{-1}, i.e.,

$$A_0 = L^{-1}A, \quad B_0 = L^{-1}B,$$

which yields the equality

$$A_0 C A_0^* + B_0 C B_0^* = I_N, \tag{14.4.1}$$

where C is a given positive definite Hermitian matrix. The matrices A_0 and B_0 admit the canonical decompositions

$$A_0 = T_0 \begin{bmatrix} \Lambda & 0 \\ 0 & I_{N_2} \end{bmatrix} Q, \quad B_0 = T_0 \begin{bmatrix} I_{N_1} & 0 \\ 0 & M \end{bmatrix} Q,$$
$$\det T_0 \neq 0, \quad \|\Lambda\| \leq \alpha < 1, \quad \|M\| \leq \alpha < 1.$$

The matrices A_r and B_r computed in the realization of the algorithm are also represented in the similar form ($r \geq 1$)

$$A_r = T_r \begin{bmatrix} -\Lambda^{2^r} & 0 \\ 0 & I_{N_2} \end{bmatrix} Q, \quad B_r = T_r \begin{bmatrix} I_{N_1} & 0 \\ 0 & -M^{2^r} \end{bmatrix} Q. \tag{14.4.2}$$

We introduce the auxiliary matrices

$$\begin{gathered} G_{-0} \equiv I - P = Q^{-1} \begin{bmatrix} I_{N_1} & 0 \\ 0 & 0 \end{bmatrix} Q, \\ G_{+0} \equiv P = Q^{-1} \begin{bmatrix} 0 & 0 \\ 0 & I_{N_2} \end{bmatrix} Q, \\ G_{-j} = Q^{-1} \begin{bmatrix} \Lambda^j & 0 \\ 0 & 0 \end{bmatrix} Q, \quad G_{+j} = Q^{-1} \begin{bmatrix} 0 & 0 \\ 0 & M^j \end{bmatrix} Q. \end{gathered} \tag{14.4.3}$$

Setting $S_r = (T_r Q)^{-1}$, we find

$$\begin{aligned} A_r &= S_r^{-1}[P - G_{-2^r}], \\ B_r &= S_r^{-1}[(I - P) - G_{+2^r}], \end{aligned} \tag{14.4.4}$$

where $S_r = (T_r Q)^{-1}$. In view of the inequality $\|G_{\pm j}\| \leq \|Q^{-1}\| \, \|Q\| \, \alpha^j$, we can represent the positive definite Hermitian matrix $H = H^* > 0$ as an infinite series:

$$H = G_{+0} C G_{+0}^* + G_{-0} C G_{-0}^* + 2 \sum_{j=1}^{\infty} (G_{+j} C G_{+j}^* + G_{-j} C G_{-j}^*). \tag{14.4.5}$$

It is convenient to represent H as the sum of two positive semidefinite matrices H_+ and H_-:

$$\begin{gathered} H_+ = G_{+0} C G_{+0}^* + 2 \sum_{j=1}^{\infty} G_{+j} C G_{+j}^* = P H P^*, \\ H_- = G_{-0} C G_{-0}^* + 2 \sum_{j=1}^{\infty} G_{-j} C G_{-j}^* = (I - P) H (I - P^*), \\ H = H_+ + H_-, \quad \|H_+\| \leq \|H\|, \quad \|H_-\| \leq \|H\|, \\ H = P H P^* + (I - P) H (I - P^*). \end{gathered} \tag{14.4.6}$$

The following theorem establishes a very important estimate.

THEOREM 14.4.1. *We have*

$$\|G_{\pm j}\| \leq \sqrt{\|C^{-1}\| \, \|H\|} \left[1 - \frac{2}{1 + \|C^{-1}\| \, \|H\|} \right]^{j/2}. \tag{14.4.7}$$

PROOF. Starting with an arbitrary vector f, we construct a sequence $f_0, f_1, \dots, f_k, \dots$ by the formulas

$$\begin{gathered} f_0 = G^*_{+0} f, \\ f_1 = G^*_{+1} f = G^*_{+1} f_0, \\ f_2 = G^*_{+2} f = (G^*_{+1})^2 f = (G^*_{+1})^2 f_0, \\ \dots\dots\dots\dots\dots\dots\dots\dots\dots\dots \\ f_k = G^*_{+k} f = (G^*_{+1})^k f = (G^*_{+1})^k f_0, \\ \dots\dots\dots\dots\dots\dots\dots\dots\dots\dots \end{gathered}$$

It is easy to verify that different vectors on the right-hand sides of the above equalities coincide if they define the same f_k. Indeed, we can use the relations

$$\begin{gathered} G_{-j} G_{+k} = G_{+k} G_{-j} = 0, \\ G_{+k} G_{+j} = G_{+(k+j)}, \\ G_{-k} G_{-j} = G_{-(k+j)}, \end{gathered} \tag{14.4.8}$$

which follow from (14.4.3). Formulas (14.4.5), (14.4.6), and (14.4.8) imply that

$$\begin{gathered} (H_+ f_{k+1}, f_{k+1}) = (H_+ f_k, f_k) - (C f_k, f_k) - (C f_{k+1}, f_{k+1}), \\ ([C + H_+] f_{k+1}, f_{k+1}) = ([C + H_+] f_k, f_k) - 2(C f_k, f_k), \\ (H_+ f_{k+1}, f_{k+1}) = (H f_{k+1}, f_{k+1}), \\ (H_+ f_k, f_k) = (H f_k, f_k), \\ ((C + H) f_{k+1}, f_{k+1}) = ((C + H) f_k, f_k) - 2(C f_k, f_k), \\ ((C + H) f_{k+1}, f_{k+1}) = \left[1 - \frac{2(C f_k, f_k)}{((C + H) f_k, f_k)}\right] ((C + H) f_k, f_k). \end{gathered}$$

Note that

$$\begin{aligned} \frac{(C f_k, f_k)}{((C + H) f_k, f_k)} &= \frac{(C^{1/2} f_k, C^{1/2} f_k)}{([I + C^{-1/2} H C^{-1/2}] C^{1/2} f_k, C^{1/2} f_k)} \\ &\geq \frac{1}{(1 + \|C^{-1/2} H C^{-1/2}\|)} \geq (1 + \|C^{-1}\| \, \|H\|)^{-1}. \end{aligned}$$

Therefore,

$$\begin{aligned} ((C + H) f_{k+1}, f_{k+1}) &\leq \left[1 - \frac{2}{1 + \|C^{-1}\| \, \|H\|}\right] ((C + H) f_k, f_k) \\ &\leq \left[1 - \frac{2}{1 + \|C^{-1}\| \, \|H\|}\right]^{k+1} ((C + H) f_0, f_0). \end{aligned}$$

Further,

$$\begin{gathered} (C f_0, f_0) = (G_{+0} C G^*_{+0} f_0, f_0) \leq (H_+ f_0, f_0) = (H_+ f, f) \leq (H f, f), \\ (C f_k, f_k) = (G_{+0} C G^*_{+0} f_k, f_k) \leq (H_+ f_k f_k) \leq (H f_k, f_k), \\ ((C + H) f_0, f_0) \leq 2(H f, f) \leq 2\|H\| \, \|f\|^2 \end{gathered}$$

and

$$\frac{2}{\|C^{-1}\|}\|f_k\|^2 \le 2(Cf_k, f_k) \le ((C+H)f_k, f_k)$$
$$\le \left[1 - \frac{2}{1+\|C^{-1}\|\,\|H\|}\right]^k ((C+H)f_0, f_0) \le 2\|H\|\,\|f\|^2.$$

Thus, we have established the relation

$$\|G_{+k}^* f\| = \|f_k\| \le \sqrt{\|C^{-1}\|\,\|H\|}\left[1 - \frac{2}{1+\|C^{-1}\|\,\|H\|}\right]^{j/2} \|f\|.$$

Thereby, we have proved that

$$\|G_{+k}\| = \|G_{+k}^*\| \le \sqrt{\|C^{-1}\|\,\|H\|}\left[1 - \frac{2}{1+\|C^{-1}\|\,\|H\|}\right]^{k/2}. \tag{14.4.9}$$

The following inequality is proved in a similar way:

$$\|G_{-k}\| \le \sqrt{\|C^{-1}\|\,\|H\|}\left[1 - \frac{2}{1+\|C^{-1}\|\,\|H\|}\right]^{k/2}. \tag{14.4.10}$$

Combining (14.4.9) and (14.4.10), we obtain (14.4.7). □

To clarify the meaning of Theorem 14.4.1, we explain why the Hermitian matrix H attracts our attention. It turns out that H can be defined by not only the exotic formula (14.4.5), but by an elegant integral representation.

Using (14.4.2) or (14.4.4), it is easy to prove that

$$\begin{aligned}(A_0 - e^{i\varphi}B_0)^{-1} &= Q^{-1}\begin{bmatrix}[\Lambda - e^{i\varphi}I_{N_1}]^{-1} & 0\\ 0 & [I_{N_2} - e^{i\varphi}M]^{-1}\end{bmatrix} QS_0,\\ (A_0 - e^{i\varphi}B_0)^{-1}A_0 &= Q^{-1}\begin{bmatrix}[\Lambda - e^{i\varphi}I_{N_1}]^{-1}\Lambda & 0\\ 0 & [I_{N_2} - e^{i\varphi}M]^{-1}\end{bmatrix} Q.\end{aligned} \tag{14.4.11}$$

We note that the series on the right-hand sides of the equalities

$$\begin{aligned}\Lambda[\Lambda - e^{i\varphi}I_{N_1}]^{-1} &= -\sum_{k=1}^{\infty} e^{-ik\varphi}\Lambda^k,\\ [I_{N_2} - e^{i\varphi}M]^{-1} &= \sum_{l=0}^{\infty} e^{il\varphi}M^l\end{aligned} \tag{14.4.12}$$

are uniformly convergent since $\|\Lambda\| \le \alpha < 1$ and $\|M\| \le \alpha < 1$. Using the matrices $G_{\pm j}$ defined by (14.4.3), from (14.4.11) and (14.4.12) we can obtain the representation

$$(A_0 - e^{ik\varphi}B_0)^{-1}A_0 = -\sum_{k=1}^{\infty} e^{-ik\varphi}G_{-k} + \sum_{l=0}^{\infty} e^{il\varphi}G_{+l}$$

and compute the integral

$$\begin{aligned}&\frac{1}{2\pi}\int_0^{2\pi}(A_0 - e^{i\varphi}B_0)^{-1}A_0CA_0^*(A_0 - e^{i\varphi}B_0)^{-*}\,d\varphi\\ &\qquad = G_{+0}CG_{+0}^* + \sum_{k=1}^{\infty} G_{-k}CG_{-k}^* + \sum_{l=1}^{\infty} G_{+l}CG_{+l}^*.\end{aligned} \tag{14.4.13}$$

The integral

$$\begin{aligned}(14.4.14)\qquad &\int_0^{2\pi} (A_0 - e^{i\varphi}B_0)^{-1}B_0CB_0^*(A_0 - e^{i\varphi}B_0)^{-*}d\varphi \\ &\qquad = G_{-0}CG_{-0}^* + \sum_{k=1}^{\infty} G_{-k}CG_{-k}^* + \sum_{l=1}^{\infty} G_{+l}CG_{+l}^*\end{aligned}$$

is computed in a similar way. Adding (14.4.13) to (14.4.14) and recalling the definition of the Hermitian matrix H (see (14.4.5)), we obtain the representation

$$H = \frac{1}{2\pi}\int_0^{2\pi} (A_0 - e^{i\varphi}B_0)^{-1}(A_0CA_0^* + B_0CB_0^*)(A_0 - e^{i\varphi}B_0)^{-*}d\varphi,$$

where the factor $A_0CA_0^* + B_0CB_0^*$ can be omitted because it is equal to the identity matrix in the chosen normalization (see (14.4.1)), i.e.,

$$H = \frac{1}{2\pi}\int_0^{2\pi} (A_0 - e^{i\varphi}B_0)^{-1}(A_0 - e^{i\varphi}B_0)^{-*}d\varphi.$$

If we know the original pencil $\lambda A + B$ from which the first pencil $\lambda A_0 + B_0$ in the orthogonal elimination algorithm can be obtained by normalization, we can find

$$(14.4.15)\qquad H = \frac{1}{2\pi}\int_0^{2\pi} (A - e^{i\varphi}B)^{-1}(ACA^* + BCB^*)(A - e^{i\varphi}B)^{-*}d\varphi.$$

Our next goal is to prove that the integral

$$H_r = \frac{1}{2\pi}\int_0^{2\pi} (A_r - e^{i\varphi}B_r)^{-1}(A_r - e^{i\varphi}B_r)^{-*}d\varphi$$

is invariant with respect to the iterations. In other words, we prove the equalities

$$H = H_0 = H_1 = H_2 = \cdots = H_{r-1} = H_r = \cdots.$$

Our arguments are based on the representations

$$(14.4.16)\qquad \begin{aligned} A_r &= S_r^{-1}[P - G_{-2^r}] = S_r^{-1}[G_{+0} - G_{-2^r}], \\ B_r &= S_r^{-1}[(I - P) - G_{+2^r}] = S_r^{-1}[G_{-0} - G_{+2^r}] \end{aligned}$$

and the recurrence relations

$$(14.4.17)\qquad \begin{aligned} S_rS_r^* = {}&(G_{+0} + G_{-2^{r-1}})S_{r-1}S_{r-1}^*(G_{+0} + G_{-2^{r-1}})^* \\ &+ (G_{-0} + G_{+2^{r-1}})S_{r-1}S_{r-1}^*(G_{-0} + G_{+2^{r-1}})^* \end{aligned}$$

established in §14.3. From (14.4.16) we find

$$(14.4.18)\qquad (A_r - e^{i\varphi}B_r)^{-1} = [G_{+0} + e^{i\varphi}G_{+2^r} - e^{i\varphi}G_{-0} - G_{-2^r}]^{-1}S_r.$$

Since

$$
\begin{aligned}
&[G_{+0} + e^{i\varphi} G_{+2^r} - e^{i\varphi} G_{-0} - G_{-2^r}]^{-1} \\
&\quad = Q^{-1} \begin{bmatrix} [-e^{i\varphi} I_{N_1} - \Lambda^{2^r}]^{-1} & 0 \\ 0 & [I_{N_2} + e^{i\varphi} M^{2^r}]^{-1} \end{bmatrix} Q \\
&\quad = Q^{-1} \begin{bmatrix} -\sum\limits_{k=0}^{\infty} (-1)^k e^{-i(k+1)\varphi} \Lambda^{k\cdot 2^r} & 0 \\ 0 & \sum\limits_{l=0}^{\infty} (-1)^l e^{il\varphi} M^{l\cdot 2^r} \end{bmatrix} Q \\
&\quad = -\sum_{k=0}^{\infty} (-1)^k e^{-i(k+1)\varphi} G_{-k\cdot 2^r} + \sum_{l=0}^{\infty} (-1)^l e^{il\varphi} G_{-l\cdot 2^r},
\end{aligned}
$$

it is not hard to verify that

$$
\begin{aligned}
H_r &= \frac{1}{2\pi} \int_0^{2\pi} (A_r - e^{i\varphi} B_r)^{-1} (A_r - e^{i\varphi} B_r)^{-*} d\varphi \\
&= \sum_{k=0}^{\infty} G_{-k\cdot 2^r} S_r S_r^* G_{-k\cdot 2^r}^* + \sum_{l=0}^{\infty} G_{+l\cdot 2^r} S_r S_r^* G_{+l\cdot 2^r}^*.
\end{aligned}
\tag{14.4.19}
$$

The right-hand side of (14.4.19) contains two sums. Using the recurrence relation (14.4.17) for $S_r S_r^*$, we transform the first sum as follows:

$$
\begin{aligned}
&\sum_{k=0}^{\infty} G_{-k\cdot 2^r} S_r S_r^* G_{-k\cdot 2^r}^* \\
&\quad = \sum_{k=0}^{\infty} [G_{-k\cdot 2^r} (G_{+0} + G_{-2^{r-1}}) S_{r-1} S_{r-1}^* (G_{+0}^* + G_{-2^{r-1}}^*) G_{-k\cdot 2^r}^* \\
&\qquad\qquad + G_{-k\cdot 2^r} (G_{-0} + G_{+2^{r-1}}) S_{r-1} S_{r-1}^* (G_{-0}^* + G_{+2^{r-1}}^*) G_{-k\cdot 2^r}^*] \\
&\quad = \sum_{k=0}^{\infty} [G_{-(2k+1)\cdot 2^{r-1}} S_{r-1} S_{r-1}^* G_{-(2k+1)\cdot 2^{r-1}}^* + G_{-2k\cdot 2^{r-1}} S_{r-1} S_{r-1}^* G_{-2k\cdot 2^{r-1}}^*] \\
&\quad = \sum_{k=0}^{\infty} G_{-k\cdot 2^{r-1}} S_{r-1} S_{r-1}^* G_{-k\cdot 2^{r-1}}^*.
\end{aligned}
\tag{14.4.20}
$$

We used the rule (14.4.8), in view of which the terms $G_{-k\cdot 2^r}$, G_{+0}, $G_{2^{r-1}}^*$, and $G_{-k\cdot 2^r}^*$ as well as the similar terms vanish.

The second sum is transformed in a similar way:

$$
\sum_{l=0}^{\infty} G_{+l\cdot 2^r} S_r S_r^* G_{+l\cdot 2^r}^* = \sum_{l=0}^{\infty} G_{+l\cdot 2^{r-1}} S_{r-1} S_{r-1}^* G_{+l\cdot 2^{r-1}}^*.
\tag{14.4.21}
$$

To complete the proof of the following theorem, it remains to compare (14.4.19) with (14.4.20) and (14.4.21).

THEOREM 14.4.2. *The integrals*

$$H_r = \frac{1}{2\pi}\int_0^{2\pi}(A_r - e^{i\varphi}B_r)^{-1}(A_r - e^{i\varphi}B_r)^{-*}d\varphi$$

are independent of the number r *if the pencil* $\lambda A_0 + B_0$ *is regular on the unit circle and the pencils* $\lambda A_{j-1} + B_{j-1}$ *and* $\lambda A_j + B_j$ *are connected by the orthogonal elimination transformation from* §14.3.

Since the integral H_r is invariant and for large r in the representations (14.4.19), (14.4.16), and (14.4.18) the main contribution is given by the terms that contains only $G_{\pm 0}$ and $G^*_{\pm 0}$, we suggest the following approximate formula for computing H:

(14.4.22)

$$\begin{aligned} H = H_r \approx{}& (A_r - B_r)^{-1}A_r(A_r - B_r)^{-1}[(A_r - B_r)^{-1}A_r(A_r - B_r)^{-1}]^* \\ &+ (A_r - B_r)^{-1}B_r(A_r - B_r)^{-1}[(A_r - B_r)^{-1}B_r(A_r - B_r)^{-1}]^*. \end{aligned}$$

We sketch the proof of (14.4.22). We will explain only principal steps and omit some intermediate computations. First, we estimate the norms of the matrices $G_{-0}S_r$ and $G_{+0}S_r$ for large r. To this end, we rewrite (14.4.19) in the form

$$\begin{aligned} H = G_{-0}S_rS_r^*G_{-0} + G_{+0}S_rS_r^*G_{+0} &+ \sum_{k=1}^{\infty} G_{-k\cdot 2^r}G_{-0}S_rS_r^*G^*_{-0}G^*_{-k\cdot 2^r} \\ &+ \sum_{l=1}^{\infty} G_{+l\cdot 2^r}G_{+0}S_rS_r^*G^*_{+0}G^*_{+l\cdot 2^r} \end{aligned}$$

and, using (14.4.7), deduce the relation

$$\begin{aligned} &\|H - G_{-0}S_rS_r^*G^*_{-0} - G_{+0}S_rS_r^*G_{+0}\| \\ &\le \frac{\sqrt{\|C^{-1}\|\,\|H\|}\left[1 - \frac{2}{1+\|C^{-1}\|\,\|H\|}\right]^{2^r}}{1 - \left[1 - \frac{2}{1+\|C^{-1}\|\,\|H\|}\right]^{2^r}}(\|G_{-0}S_rS_r^*G^*_{-0}\| + \|G_{+0}S_rS_r^*G^*_{+0}\|) \\ &\le \frac{2\sqrt{\|C^{-1}\|\,\|H\|}\left[1 - \frac{2}{1+\|C^{-1}\|\,\|H\|}\right]^{2^r}}{1 - \left[1 - \frac{2}{1+\|C^{-1}\|\,\|H\|}\right]^{2^r}}\|G_{-0}S_rS_r^*G^*_{-0} + G_{+0}S_rS_r^*G^*_{+0}\| \\ &= \varepsilon_r\|G_{-0}S_rS_r^*G_{-0} + G_{+0}S_rS_r^*G_{+0}\|, \end{aligned}$$

$$\varepsilon_r = \frac{2\sqrt{\|C^{-1}\|\,\|H\|}\left[1 - \frac{2}{1+\|C^{-1}\|\,\|H\|}\right]^{2^r}}{1 - \left[1 - \frac{2}{1+\|C^{-1}\|\,\|H\|}\right]^{2^r}},$$

which, in turn, yields the estimates

$$\begin{aligned} \|G_{-0}S_rS_r^*G_{-0}^* + G_{+0}S_rS_r^*G_{+0}^* - H\| &\le \frac{\varepsilon_r}{1-\varepsilon_r}\|H\|, \\ \|G_{\pm 0}S_rS_r^*G_{\pm 0}\| &\le \frac{1}{1-\varepsilon_r}\|H\|, \\ \|G_{\pm 0}S_r\| &\le \frac{1}{\sqrt{1-\varepsilon_r}}\sqrt{\|H\|}, \\ \|S_r\| = \|G_{+0}S_r + G_{-0}S_r\| &\le \frac{2}{\sqrt{1-\varepsilon_r}}\sqrt{\|H\|}. \end{aligned} \tag{14.4.23}$$

Using (14.4.2) and (14.4.4), we can show that

$$\begin{aligned} (A_r - B_r)^{-1}A_r(A_r - B_r)^{-1} &= Q^{-1}\begin{bmatrix} \Phi & 0 \\ 0 & (I + M^{2^r})^{-2} \end{bmatrix} QS_r \\ &= G_{+0}S_r + Q^{-1}\begin{bmatrix} \Phi & 0 \\ 0 & -(2M^{2^r} + M^{2^{r+1}})(I + M^{2^r})^{-2} \end{bmatrix} QS_r, \end{aligned}$$

$$(A_r - B_r)^{-1}B_r(A_r - B_r)^{-1} = G_{-0}S_r + Q^{-1}\begin{bmatrix} \Psi & 0 \\ 0 & -M^{2^r}(I + M^{2^r})^{-2} \end{bmatrix} QS_r,$$

where

$$\begin{aligned} \Phi &= -\Lambda^{2^r}(I + \Lambda^{2^r})^{-2}, \\ \Psi &= -(2\Lambda^{2^r} + \Lambda^{2^{r+1}})(I + \Lambda^{2^r})^{-2}, \end{aligned}$$

i.e., we justify the approximate equalities

$$\begin{aligned} (A_r - B_r)^{-1}A_r(A_r - B_r)^{-1} &\approx G_{+0}S_r, \\ (A_r - B_r)^{-1}B_r(A_r - B_r)^{-1} &\approx G_{-0}S_r. \end{aligned}$$

It is not hard to estimate the remainders. We only give their orders:

$$\begin{aligned} &\frac{\|(A_r - B_r)^{-1}A_r(A_r - B_r)^{-1} - G_{+0}S_r\|}{\sqrt{\|H\|}} \\ &\qquad = O\Big[\sqrt{\|C^{-1}\|\,\|H\|}\Big(1 - \frac{2}{1 + \|C^{-1}\|\,\|H\|}\Big)^{2^{r-1}}\Big], \\ &\frac{\|(A_r - B_r)^{-1}B_r(A_r - B_r)^{-1} - G_{-0}S_r\|}{\sqrt{\|H\|}} \\ &\qquad = O\left[\sqrt{\|C^{-1}\|\,\|H\|}\left(1 - \frac{2}{1 + \|C^{-1}\|\,\|H\|}\right)^{2^{r-1}}\right]. \end{aligned} \tag{14.4.24}$$

The comparison of (14.4.23) and (14.4.24) leads to the approximate equality (14.4.22). Moreover, it is possible to obtain an estimate for the remainder, but we do not deduce it here.

The possibility of finding H by analyzing the sequence of the pencils $\lambda A_r + B_r$ is very important if we want to construct an algorithm with guaranteed accuracy. Such algorithms can be applied to the study of the spectrum dichotomy for regular matrix pencils.

We formulate the rule of definition H in the following theorem.

THEOREM 14.4.3. *If the matrices A_0, B_0, A_r, and B_r are constructed from a pencil $\lambda A + B$ regular on the unit circle and a positive definite matrix $C = C^* > 0$ is constructed from $A_0 = L^{-1}A$ and $B_0 = L^{-1}B$ by normalization so that $A_0CA_0^* + B_0CB_0^* I_N$ and the further work of the orthogonal elimination algorithm is possible* (*see* §14.3), *then there exists the limit*

$$H = \lim_{r\to\infty} \{(A_r - B_r)^{-1}A_r(A_r - B_r)^{-1}(A_r - B_r)^{-*}A_r^*(A_r - B_r)^{-*} \\ + (A_r - B_r)^{-1}B_r(A_r - B_r)^{-1}(A_r - B_r)^{-*}B_r^*(A_r - B_r)^{-*}\},$$

which can be expressed in terms of A, B, and C as the matrix integral

$$H = \frac{1}{2\pi}\int_0^{2\pi} (A - e^{i\varphi}B)^{-1}(ACA^* + BCB^*)(A - e^{i\varphi}B)^{-*}d\varphi. \tag{14.4.25}$$

The projections P and $I - P$ can be found by the formulas

$$P = \lim_{r\to\infty}(A_r - B_r)^{-1}A_r, \quad I - P = \lim_{r\to\infty}(A_r - B_r)^{-1}B_r.$$

We write formulas from §10.3 in another notation:

$$\begin{aligned} P &= \frac{1}{2\pi}\int_0^{2\pi} (A + e^{-i\varphi}B)^{-1}A\,d\varphi, \\ I - P &= \frac{1}{2\pi}\int_0^{2\pi} (B + e^{-i\varphi}A)^{-1}B\,d\varphi, \end{aligned} \tag{14.4.26}$$

Recall that in §§10.4 and 14.3 we considered the system of matrix equations for H and P

$$\begin{aligned} AHA^* - BHB^* &= B[PCP^* - (I - P)C(I - P)^*]B^* \\ &\quad + A[PCP^* - (I - P)C(I - P^*)]A^*, \\ PH = HP^*, \quad P^2 &= P, \quad P(B - A)^{-1}A = (B - A)^{-1}BP. \end{aligned} \tag{14.4.27}$$

The system is solvable if the pencil $\lambda A + B$ is regular on the unit circle. In this case, there exists a unique solution such that $H = H^* > 0$ is positive definite. Conversely, the existence of a solution (H, P) such that $H = H^* > 0$ implies the regularity of the pencil and the representation of H, P, and $I - P$ in the integral forms (14.4.25) and (14.4.26).

To verify the accuracy of the result of the algorithm, one can use equations (14.4.27) in the same way as we verified the accuracy of the solution to the discrete Lyapunov equation in §14.1. We do not discuss this question here and refer the reader to works of Malyshev [**26**]–[**28**] who suggested the algorithm discussed here.

We note that the algorithm can be applied to the computation of the matrix integrals

$$\widetilde{H} = \frac{1}{2\pi}\int_0^{+2\pi} (A - re^{i\varphi})^{-1}C(A^* - re^{-i\varphi}I)^{-1}d\varphi.$$

To this end, we consider the pencil $\lambda rI + A$ and, using the orthogonal elimination algorithm, compute the projection

$$P = \frac{1}{2\pi}\int_0^{2\pi}(rI + e^{-i\varphi}A)^{-1}rI\,d\varphi$$

onto the invariant subspace of this pencil that corresponds to the eigenvalues in the disk of radius r. We also consider the integral

$$H = \frac{1}{2\pi}\int_0^{2\pi}(A - re^{i\varphi}I)^{-1}(r^2C + ACA^*)(A^* - re^{-i\varphi})^{-1}d\varphi.$$

After that it is not hard to find $\widetilde{H}$ (see Problem 10.4.3) by the formula

$$\widetilde{H} = \frac{1}{2}(H + C - PC - CP^*).$$

We described in detail and studied the solution of matrix equations that generalize the discrete Lyapunov equation to the case of the circle dichotomy. The detection of the linear dichotomy of the spectrum of D by the imaginary axis is reduced (see §§10.2 and 10.5) to the solution of the system of the matrix equations

$$\begin{gathered} XD + D^*X + P^*FP - (I - P^*)F(I - P) = 0, \\ PX = XP^*, \quad PD = DP, \quad P^2 = P, \quad X = X^*, \end{gathered} \tag{14.4.28}$$

where the "right-hand side" $F = F^* > 0$ is a given positive definite matrix. Here only a solution given by a positive definite Hermitian matrix $X = X^* > 0$ is of interest. Such a solution exists (and is unique) if and only if the matrix D has no purely imaginary eigenvalues.

As was shown in §10.5, the solution of equation (14.4.28) can be reduced to the solution of the above discrete equations. Namely, setting

$$A = e^{\tau D}, \quad C = \int_0^{\tau} e^{sD^*}Fe^{sD}\,ds,$$

where τ is a parameter (for example, $\tau = \dfrac{1}{2\|D\|}$), we first solve the system

$$\begin{gathered} S - A^*SA = P^*CP - (I - P)^*C(I - P) + A^*[P^*CP + (I - P)^*C(I - P)] = 0, \\ P^*S = SP, \quad PA = AP, \quad P^2 = P, \quad S = S^*, \end{gathered}$$

and find a solution such that $S = S^* > 0$. After that we put

$$X = \frac{1}{2}(S + P^*CP - (I - P)^*C(I - P)). \tag{14.4.29}$$

The projections P and $I - P$ do not change. The system (14.4.29) is a special case of the system (14.4.27) and, consequently, can be solved by the orthogonal elimination algorithm.

14.5. Approximations of invariant subspaces

The canonical form of a matrix admitting the circle dichotomy of the spectrum. Indefinite form and a criterion for the closeness between a subspace and an invariant subspace. Transformations of almost invariant subspace in iteration process. The orthogonal-power method.

Consider an $(N \times N)$-matrix A and suppose that no eigenvalue of A lies on the circle $|\lambda| = r$, i.e. the Hermitian matrix

$$H = \frac{1}{2\pi} \int_0^{2\pi} (A^* - re^{i\varphi} I_N)^{-1} (A^* A + r^2 I_N)(A - re^{i\varphi} I_N)^{-1} \, d\varphi \tag{14.5.1}$$

is defined and the norm $\omega(r) = \|H(r)\|$ can be taken for a parameter characterizing the dichotomy quality for the spectrum of the matrix A by the circle $|\lambda| = r$. Moreover, the projections P and $I - P$ onto the invariant subspaces of A corresponding to the eigenvalues $\lambda_j(A)$ outside and inside the disk $|\lambda| < r$ can be computed as the integral

$$I_N - P = \frac{1}{2\pi} \int_0^{2\pi} (I_N - \frac{1}{r} e^{-i\varphi} A)^{-1} d\varphi.$$

By the Schur theorem (see §4.2), there exists a unitary transformation U that reduces the matrix A to the upper triangular form

$$UAU^* = \begin{bmatrix} rB & D \\ 0 & rC \end{bmatrix},$$

where the eigenvalues of the triangular blocks rB and rC are located outside and inside the disk $|\lambda| < r$, respectively. The fact that the blocks B and C are triangular is not essential. Only the location of their eigenvalues in the complex plane is important. Under the above condition, the Sylvester equation

$$rBL - L(rC) = D$$

has a solution, i.e., there exists a rectangular matrix L such that

$$\begin{bmatrix} rB & D \\ 0 & rC \end{bmatrix} = \begin{bmatrix} I_{N_0} & -L \\ 0 & I_{N-N_0} \end{bmatrix} \begin{bmatrix} rB & 0 \\ 0 & rC \end{bmatrix} \begin{bmatrix} I_{N_0} & L \\ 0 & I_{N-N_0} \end{bmatrix}.$$

Thus, the matrix A admits the canonical decomposition

$$A = U^* \begin{bmatrix} I_{N_0} & -L \\ 0 & I_{N-N_0} \end{bmatrix} \begin{bmatrix} rB & 0 \\ 0 & rC \end{bmatrix} \begin{bmatrix} I_{N_0} & L \\ 0 & I_{N-N_0} \end{bmatrix} U = T^{-1} \begin{bmatrix} rB & 0 \\ 0 & rC \end{bmatrix} T,$$

where

$$T = \begin{bmatrix} I_{N_0} & L \\ 0 & I_{N-N_0} \end{bmatrix} U, \quad U^* U = I_N$$

and the spectra of the blocks B and C are located outside and inside the disk $|\lambda| < 1$, respectively. In addition,

$$P = T^{-1} \begin{bmatrix} I_{N_0} & 0 \\ 0 & 0 \end{bmatrix} T, \quad N_0 = \operatorname{tr} P.$$

In §14.4, where we described the orthogonal elimination algorithm for the study of the dichotomy of the spectrum of the pencil $\lambda A + B$ by the circle $|\lambda| = 1$, we used the norm of the integral (see (14.4.15))

$$\widetilde{H} = \frac{1}{2\pi} \int_0^{2\pi} (A - e^{i\varphi} B)^{-1} (ACA^* + BCB^*)(A^* - e^{-i\varphi} B^*)^{-1} d\varphi$$

as the dichotomy criterion and denoted by P the projection on the subspace reducing the pencil $\lambda A + B$ from the right and corresponding to the eigenvalues λ_i in the unit disk. To apply the results from §14.4, we introduce new notation

$$A \to rI_N, \quad B \to A^*, \quad C \to I_N, \quad P^* \to I - P.$$

In this notation, $\widetilde{H}$ coincides with H and the representation (14.4.5) for $\widetilde{H}$, again after the change of notation $G_{\pm j} \to G^*_{\pm j}$, becomes the following equation:

$$\begin{aligned} H &= P^* H P + (I - P^*) H (I - P) \\ &= G^*_{+0} G_{+0} + G^*_{-0} G_{-0} + 2 \sum_{j=1}^{\infty} (G^*_{+j} G_{+j} + G^*_{-j} G_{-j}). \end{aligned}$$

In this notation, (14.4.3) can be written in the form

$$\begin{aligned} G_{+j} &= T^{-1} \begin{bmatrix} 0 & 0 \\ 0 & C^j \end{bmatrix} T = U^* \begin{bmatrix} I_{N_0} & -L \\ 0 & I_{N-N_0} \end{bmatrix} \begin{bmatrix} 0 & 0 \\ 0 & C^j \end{bmatrix} \begin{bmatrix} I_{N_0} & L \\ 0 & I_{N-N_0} \end{bmatrix} U \\ &= U^* \begin{bmatrix} 0 & -LC^j \\ 0 & C^j \end{bmatrix} U, \end{aligned}$$

$$\begin{aligned} G_{-j} &= T^{-1} \begin{bmatrix} B^{-j} & 0 \\ 0 & 0 \end{bmatrix} T = U^* \begin{bmatrix} I_{N_0} & -L \\ 0 & I_{N-N_0} \end{bmatrix} \begin{bmatrix} B^{-j} & 0 \\ 0 & 0 \end{bmatrix} \begin{bmatrix} I_{N_0} & L \\ 0 & I_{N-N_0} \end{bmatrix} U \\ &= U^* \begin{bmatrix} B^{-j} & B^{-j} L \\ 0 & 0 \end{bmatrix} U. \end{aligned}$$

Since the matrix U is unitary, these formulas imply the inequalities

$$(14.5.2) \qquad \begin{aligned} \|C^j\| \le \|G_{+j}\| &= \left\| \begin{pmatrix} -L \\ I_{N-N_0} \end{pmatrix} C^j \right\| \le \sqrt{1 + \|L\|^2} \|C^j\|, \\ \|B^{-j}\| \le \|G_{-j}\| &= \|(B^{-j} \vdots B^{-j} L)\| \le \sqrt{1 + \|L\|^2} \|B^{-j}\| \end{aligned}$$

and the equalities

$$(14.5.3) \qquad \begin{aligned} \|P\| &= \|G_{-0}\| = \sqrt{1 + \|L\|^2}, \\ \|I - P\| &= \|G_{+0}\| = \sqrt{1 + \|L\|^2}. \end{aligned}$$

Now we can use Theorem 14.4.1 which asserts that

(14.5.4)

$$\|G_{\pm j}\| \le \sqrt{\|H\|} \left(1 - \frac{2}{1 + \|H\|}\right)^{j/2} \le \sqrt{\|H\|} e^{-j/(1+\|H\|)} = \sqrt{\omega(r)} e^{-j/[1+\omega(r)]}.$$

(Since $\|G_{\pm j}\| = \|G^*_{\pm j}\|$, it is not necessary to use the new notation $G^*_{\pm j}$ instead of $G_{\pm j}$ in the formulation of Theorem 14.4.4.) Comparing (14.5.4) with (14.5.2) and

(14.5.3), we conclude that

$$\begin{aligned} \|L\| \le \|P\| = \|I - P\| = \sqrt{1 + \|L\|^2} \le \sqrt{\omega(r)}, \\ \|C^j\| \le \sqrt{\omega(r)} e^{-j/[1+\omega(r)]}, \\ \|B^{-j}\| \le \sqrt{\omega(r)} e^{-j/[1+\omega(r)]}. \end{aligned} \tag{14.5.5}$$

Thus, we arrive at the following conclusion.

THEOREM 14.5.1. *If for some $r > 0$ the parameter $\omega(r) = \|H\|$, where*

$$H = \frac{1}{2\pi} \int_0^{2\pi} (A^* - re^{i\varphi} I_N)^{-1} (A^*A + r^2 I_N)(A - re^{i\varphi} I_N)^{-1} \, d\varphi,$$

is defined and is finite, then the matrix A admits the canonical decomposition

$$\frac{1}{r} A = U^* \begin{bmatrix} I_{N_0} & -L \\ 0 & I_{N-N_0} \end{bmatrix} \begin{bmatrix} B & 0 \\ 0 & C \end{bmatrix} \begin{bmatrix} I_{N_0} & L \\ 0 & I_{N-N_0} \end{bmatrix} U,$$

where the matrix U is unitary ($U^*U = I_N$) and the blocks L, B, and C satisfy the estimates (14.5.5).

In the spectral analysis, Theorem 14.5.1 plays an important role in the description and study of the orthogonal-power method and its modifications.

Consider an N_0-dimensional subspace such that the coordinates of the basis vectors form the columns (linearly independent) of the rectangular $N \times N_0$-matrix W_0. This subspace coincides with the invariant subspace of A corresponding to the eigenvalues $\lambda_j(A)$ such that $|\lambda_j| > r$ if $PW_0 = W_0$ and $(I - P)W_0 = 0$. The set of vectors such that for some $\gamma > 0$ we have $\|(I - P)x\| \le \gamma \|Px\|$ or, equivalently,

$$([P^*P - \frac{1}{\gamma^2}(I - P^*)(I - P)]x, x) \ge 0,$$

form a cone that contains the invariant subspace. It is clear that any N_0-dimensional subspace located in this cone can be regarded as an approximation of the invariant subspace and the smaller γ is, the more exact is this approximation.

The linear subspace of the vectors $x = W_0 y$, i.e., the subspace spanned by the basis vectors whose coordinates form the columns of W_0, approximates the invariant subspace if

$$([P^*P - (1/\gamma^2)(I - P^*)(I - P)]W_0 y, W_0 y) \ge 0, \tag{14.5.6}$$

i.e., if the matrix

$$W_0^*[P^*P - (1/\gamma^2)(I - P^*)(I - P)]W_0$$

is positive semidefinite.

We give another interpretation of this criterion. For vectors $x = W_0 y$ we use the following representations and inequalities:

$$Px = U^* \begin{bmatrix} I_{N_0} & L \\ 0 & 0 \end{bmatrix} U W_0 y,$$

$$\|Px\| = \|[I_{N_0} \vdots L] U W_0 y\|,$$

$$(I - P)x = U^* \begin{bmatrix} 0 & -L \\ 0 & I_{N-N_0} \end{bmatrix} U W_0 y,$$

$$\|(I-P)x\| \le \sqrt{1 + \|L\|^2} \left\| (0 \vdots I_{N-N_0}) U W_0 y \right\|.$$

Introducing the notation $Z^{(0)}$ and $Z^{(1)}$ for the blocks of the matrix

$$\text{(14.5.7)} \qquad \begin{matrix} N_0 \text{ rows} \\ N - N_0 \text{ rows} \end{matrix} \underbrace{\begin{bmatrix} Z^{(0)} \\ Z^{(1)} \end{bmatrix}}_{N_0 \text{ columns}} = U W_0.$$

We obtain the inequalities

$$\text{(14.5.8)} \qquad \begin{aligned} \frac{\|Px\|}{\|(I-P)x\|} &\ge \frac{\|[Z^{(0)} + LZ^{(1)}]y\|}{\sqrt{1 + \|L\|^2} \|Z^{(1)} y\|} \\ &\ge \frac{1}{\sqrt{1 + \|L\|^2} \|Z^{(1)} [Z^{(0)} + LZ^{(1)}]^{-1} \|)}. \end{aligned}$$

In other words, the condition (14.5.6) is satisfied if

$$\text{(14.5.9)} \qquad \gamma \ge \sqrt{1 + \|L\|^2} l_0, \quad l_0 = \|Z^{(1)} (Z^{(0)} + LZ^{(1)})^{-1}\|.$$

The parameter l_0 computed by formula (14.5.9) in terms of $Z^{(0)}$ and $Z^{(1)}$ in (14.5.7) will be used together with the parameter γ to characterize the closeness between the subspace of vectors $x = W_0 y$ and the invariant subspace with the projection P.

We note that l_0 does not change if we multiply W_0 from the right by any nonsingular $N_0 \times N_0$-matrix, i.e., the parameter l_0 is determined by the subspace itself and is independent of the choice of the basis.

Since l_0 is finite, which is equivalent to the fact that $Z^{(0)} + LZ^{(1)}$ is nonsingular, the columns of W_0 are linearly independent.

Together with the subspace spanned by the columns of W_0, we consider the subspace spanned by the columns of the matrix W_k obtained from W_0 by the linear transformation A^k, i.e., $W_k = A^k W_0$. We decompose UW_k into blocks in the same way as UW_0 (see (14.5.7)):

$$U W_k = \underbrace{\begin{bmatrix} Z_k^{(0)} \\ Z_k^{(1)} \end{bmatrix}}_{N_0 \text{ columns}} \begin{matrix} N_0 \text{ rows} \\ N - N_0 \text{ rows.} \end{matrix}$$

Similarly to l_0 in (14.5.9), we introduce and estimate the parameter

$$l_k = \|Z_k^{(0)} (Z_k^{(0)} + LZ_k^{(1)})^{-1}\|.$$

This parameter l_k characterizes the location of the subspace of vectors $y = W_k x$, with respect to the subspace on which P projects. Since

$$\begin{pmatrix} Z_k^{(0)} \\ Z_k^{(1)} \end{pmatrix} = UW_k = UA^kW_0 = UA^kU^*UW_0 = UA^kU^* \begin{pmatrix} Z^{(0)} \\ Z^{(1)} \end{pmatrix}$$

$$= r^k \begin{bmatrix} I_{N_0} & -L \\ 0 & I_{N-N_0} \end{bmatrix} \begin{bmatrix} B^k & 0 \\ 0 & C^k \end{bmatrix} \begin{bmatrix} I_{N_0} & L \\ 0 & I_{N-N_0} \end{bmatrix} \begin{pmatrix} Z^{(0)} \\ Z^{(1)} \end{pmatrix},$$

we have

$$\begin{pmatrix} Z_k^{(0)} + LZ_k^{(1)} \\ Z_k^{(1)} \end{pmatrix} = r^k \begin{bmatrix} B^k & 0 \\ 0 & C^k \end{bmatrix} \begin{pmatrix} Z^{(0)} + LZ^{(1)} \\ Z^{(1)} \end{pmatrix},$$

$$Z_k^{(1)}[Z_k^{(0)} + LZ_k^{(1)}]^{-1} = C^k Z^{(1)}[Z^{(0)} + LZ^{(1)}]^{-1}B^{-k}.$$

Consequently, $l_k \leq \|C^k\| \, \|B^{-k}\| l_0$. In view of the estimates (14.5.5) for $\|C^k\|$ and $\|B^{-k}\|$, we conclude that

$$l_k \leq \omega(r)e^{-2k/[1+\omega(r)]};$$

moreover, any vector $x = W_k y$ satisfies the inequality

$$\|Px\| \geq \frac{1}{\gamma_k}\|(I-P)x\|$$

for $\gamma_k \geq \omega^{3/2}(r)e^{-2k/[1+\omega(r)]} \geq \sqrt{\omega(r)}l_k$.

Thus, if we can choose a subspace of dimension N_0 such that the basis is described by W_0, and l_0, as well as the dichotomy parameter $\omega(r)$, is finite, then for a sufficiently large k the subspace spanned by the columns of $W_k = A^k W_0$, is a good approximation of the invariant subspace of A corresponding to the eigenvalues $\lambda_j(A)$ such that $|\lambda_j| > r$. Since the parameters $l_0, l_1, \ldots, l_k$ are finite, the columns of each of the matrices $W_0, W_1, \ldots, W_k$ are linearly independent.

The above arguments form a basis of a number of algorithms that yield arbitrarily accurate approximations of the invariant subspaces of a matrix A corresponding to some eigenvalues with the maximal module. However, when constructing algorithms one must keep in mind that not every basis is suitable for practical description of subspaces. It is preferable to use orthonormal (or almost orthonormal) bases. A basis composed of vectors whose coordinates form the columns of a matrix W with N_0 columns is orthonormal if $W^*W = I_{N_0}$. As a rule, the bases $W_k = A^k W_0$ are not orthonormal even if the initial basis W_0 is. Using the fact that the multiplication from the right of the basis matrix W_k by any nonsingular $N_0 \times N_0$-matrix yields again a basis matrix for the same subspaces, we can indicate the recurrence procedure for constructing orthonormal matrices Q_k $(Q_k^* Q_k = I_{N_0})$ whose columns form bases for the same subspaces as W_k.

We describe a procedure called the *orthogonal-power method.*

Take some basis matrix W_0 whose columns are not necessarily orthonormal. Consider the QR decomposition, i.e., consider the representation $W_0 = Q_0 R_0$, where the matrix Q_0 has N_0 orthonormal columns and the matrix R_0 is upper triangular. The QR decomposition was briefly considered in §3.1. Here, we only note that since the columns of the matrix W_0 are linearly independent (they form a basis), the triangular matrix R_0 is nonsingular, i.e., the diagonal elements of R_0 differ from zero.

The matrix Q_0 obtained as the result of QR decomposition consists of columns that contain coordinates of some orthonormal basis vectors. The linear span of these basis vectors coincides with the linear span of the vectors whose coordinates form the columns of W_0. It is obvious that the linear span of the columns of $W_1 = AW_0$, the linear span of the columns of AQ_0, and the linear span of the orthonormal columns of Q_1 ($Q_1^* Q_1 = I_{N_0}$) obtained from the QR decomposition coincide. Furthermore,

$$W_1 = AW_0 = AQ_0 R_0 = Q_1 R_1 R_0.$$

Since the columns of W_1 are linearly independent, the matrix $R_1 R_0$ is nonsingular. Hence triangular matrices R_1 and R_0 are also nonsingular.

We continue the recurrence procedure for constructing matrices Q_j composed of N_0 orthonormal columns ($Q_j^* Q_j = I_{N_0}$) by successive QR decompositions $AQ_{j-1} = Q_j R_j$. Arguing as above, we see that

$$W_j = Q_j R_j R_{j-1} \dots R_1 R_0;$$

moreover, the upper triangular matrices $R_0, R_1, \dots, R_j$ are nonsingular, i.e., their diagonal elements do not vanish.

The linear spans of vectors whose coordinates form the columns of Q_j and W_j are the same subspace of dimension N_0.

The parameter l_j becomes smaller as the subspace spanned by the columns of W_j approximates the invariant subspace better. This parameter can be found from any basis, not only from W_j. In particular, for this purpose the basis formed by the columns of Q_j can be used. Indeed, it suffices to represent UQ_j in the block form

$$UQ_j = UW_k = \underbrace{\begin{bmatrix} Q_j^{(0)} \\ Q_j^{(1)} \end{bmatrix}}_{N_0 \text{ columns}} \begin{matrix} N_0 \text{ rows} \\ N - N_0 \text{ rows} \end{matrix}$$

and set $l_j = \|Q_j^{(1)} [Q_j^{(0)} + LQ_j^{(1)}]^{-1}\|$. The qualitative assertion that the triangular matrices R_j are nonsingular can be specified as a quantitative estimate for the condition number. We use the equalities

$$\begin{bmatrix} I_{N_0} & -L \\ 0 & I_{N-N_0} \end{bmatrix}^{-1} = \begin{bmatrix} I_{N_0} & L \\ 0 & I_{N-N_0} \end{bmatrix},$$

$$\left\| \begin{bmatrix} I_{N_0} & -L \\ 0 & I_{N-N_0} \end{bmatrix}^{-1} \right\|^{-1} = \left\| \begin{bmatrix} I_{N_0} & L \\ 0 & I_{N-N_0} \end{bmatrix} \right\|^{-1} = \frac{1}{\sqrt{1 + \|L\|^2}}$$

and the relation ($x = Q_{j-1} y$)

$$\begin{aligned} \|Ax\| = \|UAx\| &\geq \left\| \begin{bmatrix} I_{N_0} & -L \\ 0 & I_{N_0} \end{bmatrix}^{-1} \right\|^{-1} \left\| \begin{bmatrix} I_{N_0} & -L \\ 0 & I_{N-N_0} \end{bmatrix} UAx \right\| \\ &\geq \frac{r}{\sqrt{1 + \|L\|^2}} \left\| \begin{pmatrix} B(Q_j^{(0)} + LQ_{j-1}^{(1)}) y \\ CQ_{j-1}^{(1)} y \end{pmatrix} \right\| \\ &\geq \frac{r \|B^{-1}\|^{-1}}{\sqrt{1 + \|L\|^2}} \|(Q_{j-1}^{(0)} + LQ_{j-1}^{(1)}) y\|. \end{aligned}$$

It is obvious that

$$\begin{aligned}
\|x\| = \|UQ_{j-1}y\| &= \sqrt{\|Q_{j-1}^{(0)}y\|^2 + \|Q_{j-1}^{(1)}y\|^2} \\
&= \sqrt{\|(Q_{j-1}^{(0)} + LQ_{j-1}^{(1)})y - LQ_{j-1}^{(1)}y\|^2 + \|Q_{j-1}^{(1)}y\|^2} \\
&\le \sqrt{(\|(Q_{j-1}^{(0)} + LQ_{j-1}^{(1)})y\| + \|L\|\,\|Q_{j-1}y\|)^2 + \|Q_{j-1}^{(1)}y\|^2} \\
&\le \sqrt{2\|(Q_{j-1}^{(0)} + LQ_{j-1}^{(1)})y\|^2 + (1+2\|L\|^2)\|Q_{j-1}^{(1)}y\|^2}.
\end{aligned}$$

Therefore, for vectors $x = Q_{j-1}y$ we have

$$\begin{aligned}
\frac{\|Ax\|}{\|x\|} &\ge \frac{r\|B^{-1}\|^{-1}}{\sqrt{1+\|L\|^2}} \frac{\|(Q_{j-1}^{(0)} + LQ_{j-1}^{(1)})y\|}{\sqrt{2\|Q_{j-1}^{(0)} + LQ_{j-1}^{(1)})y\|^2 + (1+2\|L\|^2)\|Q_{j-1}^{(1)}y\|^2}} \\
&\ge \frac{r\|B^{-1}\|^{-1}}{\sqrt{(1+\|L\|^2)[2+(1+2\|L\|^2)l_{j-1}^2]}}.
\end{aligned}$$

The equalities

$$AQ_{j-1} = Q_jR_j, \quad Q_j^*Q_j = I_N$$

imply

$$R_jy = Q_j^*Q_jR_jy = Q_j^*AQ_{j-1}y.$$

Therefore,

$$\begin{aligned}
\|(Q_j^*AQ_{j-1})^{-1}\|^{-1} = \|R_j^{-1}\|^{-1} &= \min_{y\neq 0}\frac{\|R_jy\|}{\|y\|} = \min_{y\neq 0}\frac{\|Q_jR_jy\|}{\|Q_{j-1}y\|} \\
&= \min_{y\neq 0}\frac{\|AQ_{j-1}y\|}{\|Q_{j-1}y\|} = \min_{x=Q_{j-1}y,\,y\neq 0}\frac{\|Ax\|}{\|x\|} \\
&\ge \frac{r\|B^{-1}\|^{-1}}{\sqrt{(1+\|L\|^2)[2+(1+2\|L\|^2)l_{j-1}^2]}}.
\end{aligned} \tag{14.5.10}$$

On the other hand,

$$\|R_j\| = \|Q_j^*AQ_{j-1}\| \le \|A\|. \tag{14.5.11}$$

Comparing (14.5.10) and (14.5.11), we estimate the conditionality of the triangular matrix R_j:

$$\|R_j\|\,\|R_j^{-1}\| \le \frac{\|A\|}{r}\|B^{-1}\|\sqrt{(1+\|L\|^2)[2+(1+2\|L\|^2)l_{j-1}^2]}.$$

Note that $\|B^{-1}\|^{-1}$ and $\|L\|$ on the right-hand side can be estimated in terms of $\omega(r)$ using (14.5.5). In other words, the condition number of the triangular matrix R_j is estimated in terms of $\frac{\|A\|}{r}$, $\omega(r)$, and l_{j-1}.

PROBLEM 14.5.1. If the matrix

$$Q_{j-1}^*\left[P^*P - \frac{1}{\gamma^2}(I-P^*)(I-P)\right]Q_{j-1} \ge 0$$

is positive semidefinite, then $l_{j-1} \le \gamma$.

Hint: Analyze and modify the proof of the inequalities (14.5.8) and (14.5.9).

14.6. Stability of the orthogonal-power algorithm

Simulation of errors. Ribbon systems for the bases of successive approximations of invariant subspaces. Estimates for the influence of errors in terms of estimates for solutions to ribbon systems.

Applying the orthogonal-power algorithm (see §14.5)

$$\begin{gathered} AQ_{j-1} = Q_j R_j, \quad Q_j^* Q_j = I_{N_0}, \\ R_j \text{ is an upper triangular matrix,} \end{gathered} \tag{14.6.1}$$

we must keep in mind that the relations (14.6.1) between the matrices obtained in computations are not valid because of unavoidable round-off errors. In particular, the columns of the matrices Q_j are not strictly orthonormal and the left- and right-hand sides of the equalities $AQ_{j-1} = Q_j R_j$ are not really equal. Moreover, the product AQ_j is found with error.

To demonstrate the technique applied for the deduction of estimates, we consider the influence of only this error. The exposition becomes cumbersome if we take into account the influence of other errors. We refer the reader to [**18**] for a general case. This section presents a simplified variant of the article [**18**].

Suppose that

$$\begin{gathered} (A + \Delta_j) Q_{j-1} = Q_j R_j, \quad Q_j^* Q_j = I_{N_0}, \\ R_j \text{ is an upper triangular matrix} \end{gathered} \tag{14.6.2}$$

$$\|\Delta_j\| < \varepsilon \|A\|, \tag{14.6.3}$$

where ε is a sufficiently small positive parameter characterizing the admissible accuracy of computation. Assume that $\varepsilon < 1$.

Let a matrix A be the same as in §14.5. We assume that A admits the canonical decomposition

$$A = U^* \begin{bmatrix} I_{N_0} & -L \\ 0 & I_{N-N_0} \end{bmatrix} \begin{bmatrix} rB & 0 \\ 0 & rC \end{bmatrix} \begin{bmatrix} I_{N_0} & L \\ 0 & I_{N-N_0} \end{bmatrix} U,$$

where

$$\begin{gathered} U^* U = I_N, \\ \sqrt{1 + \|L\|^2} \le \sqrt{\omega(r)}, \quad 1 + \|L\| \le \sqrt{2\omega(r)}, \\ \|C^j\| \le \sqrt{\omega(r)} e^{-j/[1+\omega(r)]}, \quad j \ge 1, \\ \|B^{-j}\| \le \sqrt{\omega(r)} e^{-j/[1+\omega(r)]}, \quad j \ge 1. \end{gathered} \tag{14.6.4}$$

It is convenient to represent the products UQ_j as the block matrices

$$UQ_j = \underbrace{\begin{bmatrix} Q_j^{(0)} \\ Q_j^{(1)} \end{bmatrix}}_{N_0 \text{ columns}} \begin{matrix} N_0 \text{ rows} \\ N - N_0 \text{ rows.} \end{matrix}$$

It is obvious that

$$[Q_j^{(0)}]^* Q_j^{(0)} + [Q_j^{(1)}]^* [Q_j^{(1)}] = I_{N_0}.$$

The relation $(A+\Delta_j)Q_{j-1} = Q_j R_j$ can be rewritten in the form

$$\begin{bmatrix} rB & 0 \\ 0 & rC \end{bmatrix} \begin{pmatrix} Q_{j-1}^{(0)} + LQ_{j-1}^{(1)} \\ Q_{j-1}^{(1)} \end{pmatrix} + r \begin{bmatrix} \Delta_{11}^{(j)} & \Delta_{12}^{(j)} \\ \Delta_{21}^{(j)} & \Delta_{22}^{(j)} \end{bmatrix} \begin{pmatrix} Q_{j-1}^{(0)} + LQ_{j-1}^{(1)} \\ Q_{j-1}^{(1)} \end{pmatrix} = \begin{pmatrix} Q_j^{(0)} + LQ_j^{(1)} \\ Q_j^{(1)} \end{pmatrix} R_j, \tag{14.6.5}$$

where

$$\begin{bmatrix} \Delta_{11}^{(j)} & \Delta_{12}^{(j)} \\ \Delta_{21}^{(j)} & \Delta_{22}^{(j)} \end{bmatrix} = \frac{1}{r} \begin{bmatrix} I_{N_0} & L \\ 0 & I_{N-N_0} \end{bmatrix} U\Delta_j.$$

It is obvious that

$$\|\Delta_{km}^{(j)}\| \le \varepsilon \frac{\|A\|}{r}(1+\|L\|) \le \varepsilon \frac{\sqrt{2\omega(r)}\|A\|}{r}, \quad k,m=1,2. \tag{14.6.6}$$

LEMMA 14.6.1. *Let the matrix $Q_{j-1}^{(0)} + LQ_{j-1}^{(0)}$ be nonsingular and, consequently the parameter*

$$l_{j-1} = \|Q_{j-1}^{(1)}(Q_{j-1}^{(0)} + LQ_{j-1}^{(0)})^{-1}\|$$

be finite, and let

$$\varepsilon < \frac{r}{\sqrt{2}\|A\|\omega(1+l_{j-1})}. \tag{14.6.7}$$

Then the matrices $Q_j^{(0)} + LQ_j^{(1)}$ and R_j are nonsingular.

PROOF. From (14.6.5) it follows that

$$\begin{aligned} &[Q_j^{(0)} + LQ_j^{(1)}]R_j[Q_{j-1}^{(0)} + LQ_{j-1}^{(1)}]^{-1} \\ &\quad = r\{B + \Delta_{11}^{(j)} + \Delta_{12}^{(j)} Q_{j-1}^{(1)}[Q_{j-1}^{(0)} + LQ_{j-1}^{(1)}]^{-1}\} \\ &\quad = rB\{I + B^{-1}\Delta_{11}^{(j)} + B^{-1}\Delta_{12}^{(j)} Q_{j-1}^{(1)}[Q_{j-1}^{(0)} + LQ_{j-1}^{(1)}]^{-1}\}. \end{aligned} \tag{14.6.8}$$

The conditions (14.6.4) and (14.6.6) imply the estimates

$$\|B^{-1}\| \le \sqrt{\omega} e^{-1/(1+\omega)} < \sqrt{\omega},$$

$$\|B^{-1}\Delta_{11}^{(j)} + B^{-1}\Delta_{12}^{(j)} Q_{j-1}^{(1)}[Q_{j-1}^{(0)} + LQ_{j-1}^{(1)}]^{-1}\| < \sqrt{\omega}\varepsilon \frac{\sqrt{2\omega}\|A\|}{r}(1+l_{j-1}) < 1,$$

which show that the right-hand side of (14.6.8) is nonsingular. Hence all the factors on the left-hand side of (14.6.8) are nonsingular. □

Now we have the following question. Can we fix a small ε so that for a given l_0 it is possible to prove that every term of the sequence $l_0, l_1, l_2, \ldots$ is bounded by the same constant? We show that such an ε exists; moreover, if the estimate (14.6.3) holds with this ε, then the sequence $l_0, l_1, l_2, \ldots$ is bounded. In addition, the elements of $l_0, l_1, l_2, \ldots$, starting with some element, decrease until l_j becomes of order $O(\varepsilon)$.

Suppose that we can choose ε so that the matrices $Q_m^{(0)} + LQ_m^{(1)}$, $R_1, R_2, \ldots, R_m$ are nonsingular. Then we can find the matrices Y_m, Z_m; $Y_{m-1}, Z_{m-1}; \ldots; Y_1, Z_1$;

Y_0, Z_0 by the formulas

$$\begin{aligned}
&Y_m = Q_m^{(1)}[Q_m^{(0)} + LQ_m^{(1)}]^{-1},\\
&Z_m = I_{N_0};\\
&Y_{m-1} = Q_{m-1}^{(1)} R_m^{-1}[Q_m^{(0)} + LQ_m^{(1)}]^{-1},\\
&Z_{m-1} = [Q_{m-1}^{(0)} + LQ_{m-1}^{(1)}]R_m^{-1}[Q_m^{(0)} + LQ_m^{(1)}]^{-1},\\
&\cdots\cdots\cdots\cdots\cdots\cdots\cdots\cdots\cdots\cdots\\
&Y_j = Q_j^{(1)} R_{j+1}^{-1} R_{j+2}^{-1} \dots R_m^{-1}[Q_m^{(0)} + LQ_m^{(1)}]^{-1},\\
&Z_j = [Q_j^{(0)} + LQ_j^{(1)}]R_{j+1}^{-1} R_{j+2}^{-1} \dots R_m^{-1}[Q_m^{(0)} + LQ_m^{(1)}]^{-1},\\
&\cdots\cdots\cdots\cdots\cdots\cdots\cdots\cdots\cdots\cdots\\
&Y_0 = Q_0^{(1)} R_1^{-1} R_2^{-1} \dots R_m^{-1}[Q_m^{(0)} + LQ_m^{(1)}]^{-1},\\
&Z_0 = [Q_0^{(0)} + LQ_0^{(1)}]R_1^{-1} R_2^{-1} \dots R_m^{-1}[Q_m^{(0)} + LQ_m^{(1)}]^{-1}.
\end{aligned}$$

It is easy to check that equations (14.6.5) are equivalent to the system

$$\tag{14.6.9}\begin{aligned}
Z_m &= I_{N_0},\\
Z_j &= BZ_{j-1} + \Delta_{11}^{(j)} Z_{j-1} + \Delta_{12}^{(j)} Y_{j-1}, \quad m \ge j \ge 1,\\
Y_j &= CY_{j-1} + \Delta_{21}^{(j)} Z_{j-1} + \Delta_{22}^{(j)} Y_{j-1}, \quad m \ge j \ge 1,\\
Y_0 &= \mathcal{L} Z_0 \quad (\mathcal{L} = Q_0^{(1)}[Q_0^{(0)} + LQ_0^{(1)}]^{-1}, \quad \|\mathcal{L}\| = l_0).
\end{aligned}$$

Studying this system, we can estimate the quantities

$$\mu_m = \max_{0 \le j \le m} \|Y_j\|, \quad \nu_m = \max_{0 \le j \le m} \|Z_j\|,$$

and, consequently, estimate l_m:

$$\|Y_m\| = \|Q_m^{(1)}[Q_m^{(0)} + LQ_m^{(1)}]^{-1}\| = l_m \le \mu_m.$$

From (14.6.9) we obtain

$$\tag{14.6.10}\begin{aligned}
Z_m &= I_{N_0},\\
Z_p &= B^{-(m-p)} - \sum_{i=p}^{m-1} B^{-(m-i)}[\Delta_{11}^{(i+1)} Z_i + \Delta_{12}^{(i+1)} Y_i], \quad 0 \le p \le m-1,\\
Y_q &= C^q \mathcal{L} Z_0 + \sum_{i=0}^{q-1} C^i [\Delta_{21}^{(q-i)} Z_{q-i-1} + \Delta_{22}^{(q-i)} Y_{q-i-1}], \quad 1 \le q \le m,\\
Y_0 &= \mathcal{L} Z_0.
\end{aligned}$$

From these equalities and the estimates (14.6.4) and (14.6.6) we obtain the inequalities

$$\tag{14.6.11}\begin{aligned}
\nu_m &\le \sqrt{\omega}\left[1 + \sum_{k=0}^{\infty} e^{-k/(1+\omega)} \frac{\varepsilon\sqrt{2\omega}\|A\|}{r}(\nu_m + \mu_m)\right],\\
\mu_m &\le \sqrt{\omega}\left[l_0 \nu_m + \sum_{k=0}^{\infty} e^{-k/(1+\omega)} \frac{\varepsilon\sqrt{2\omega}\|A\|}{r}(\nu_m + \mu_m)\right].
\end{aligned}$$

Since

$$\sum_{k=0}^{\infty} e^{-k/(1+\omega)} = \frac{1}{1-e^{-1/(1+\omega)}} \leq \frac{1}{1-e^{-1/2}} = \frac{\sqrt{e}}{\sqrt{e}-1},$$

$$\sqrt{\omega}\sqrt{2}\sum_{k=0}^{\infty} e^{-k/(1+\omega)} = \frac{\sqrt{2\omega}}{1-e^{-1/(1+\omega)}} < \frac{\sqrt{2e\omega}}{\sqrt{e}-1},$$

the inequalities (14.6.11) can be written in the following compact form:

$$\nu_m \leq \sqrt{\omega} + \varepsilon\frac{\|A\|}{r}\frac{\sqrt{2e\omega}}{\sqrt{e}-1}(\nu_m+\mu_m),$$

$$\mu_m \leq \sqrt{\omega}l_0\nu_m + \varepsilon\frac{\|A\|}{r}\frac{\sqrt{2e\omega}}{\sqrt{e}-1}(\nu_m+\mu_m).$$

Under the condition that

$$\varepsilon < \frac{1}{2(3+2\omega\sqrt{\omega}l_0)}\frac{r}{\|A\|}\frac{\sqrt{e}-1}{\sqrt{2e\omega}} < \frac{1}{2}\frac{r}{\|A\|}\frac{\sqrt{e}-1}{\sqrt{2e\omega}} \tag{14.6.12}$$

we have

$$\nu_m \leq \sqrt{\omega} + \frac{1}{2}\nu_m + \varepsilon\frac{\|A\|}{r}\frac{\sqrt{2e\omega}}{\sqrt{e}-1}\mu_m,$$

$$\nu_m \leq 2\sqrt{\omega} + 2\varepsilon\frac{\|A\|}{r}\frac{\sqrt{2e\omega}}{\sqrt{e}-1}\mu_m,$$

$$\begin{aligned}
\mu_m &\leq \sqrt{\omega}l_0\left[2\sqrt{\omega} + 2\varepsilon\frac{\|A\|}{r}\frac{\sqrt{2e\omega}}{\sqrt{e}-1}\mu_m\right] \\
&\quad + \frac{1}{2}\left[2\sqrt{\omega} + 2\varepsilon\frac{\|A\|}{r}\frac{\sqrt{2e\omega}}{\sqrt{e}-1}\mu_m\right] + 2\varepsilon\frac{\|A\|}{r}\frac{\sqrt{2e\omega}}{\sqrt{e}-1} \\
&= \sqrt{\omega}(2\sqrt{\omega}l_0+1) + \varepsilon\frac{\|A\|}{r}\frac{\sqrt{2e\omega}}{\sqrt{e}-1}(2\omega\sqrt{\omega}l_0+3)\mu_m \\
&< \sqrt{\omega}(2\sqrt{\omega}l_0+1) + \frac{1}{2}\mu_m.
\end{aligned}$$

Consequently,

$$\mu_m < 2\sqrt{\omega}(2\sqrt{\omega}l_0+1),$$

$$\nu_m \leq 2\sqrt{\omega} + 2\varepsilon\frac{\|A\|}{r}\frac{\sqrt{2e\omega}}{\sqrt{2e}-1}\mu_m < 2\sqrt{\omega} + 2\frac{1}{2}\mu_m < 4\sqrt{\omega}(\sqrt{\omega}l_0+1).$$

Hence, under the condition (14.6.12) on ε, the following estimates hold:

$$\begin{aligned}
&\nu_m < 4\sqrt{\omega}(\sqrt{\omega}l_0+1), \\
&\mu_m + \nu_m < 2\sqrt{\omega}(4\sqrt{\omega}l_0+3), \\
&l_m \leq \mu_m < 2\sqrt{\omega}(2\sqrt{\omega}l_0+1).
\end{aligned} \tag{14.6.13}$$

It is essential that the right-hand side of these estimates is independent of m. If

$$\varepsilon < \min\left\{\frac{1}{2(3+2\sqrt{\omega}l_0)}\frac{r}{\|A\|}\frac{\sqrt{e}-1}{\sqrt{2e\omega}}, \frac{1}{\sqrt{2}[1+2\sqrt{\omega}(2\sqrt{\omega}l_0+1)]\omega}\frac{r}{\|A\|}\right\}, \tag{14.6.14}$$

then for $j-1=m$ the assumption (14.6.7) of Lemma 14.6.1 is satisfied. Using (14.6.2), we compute the matrices Q_{m+1} and R_{m+1}. We obtain the nonsingular matrices $Q_{m+1}^{(0)} + LQ_{m+1}$ and R_{m+1}. This allows us to obtain the same estimates for μ_{m+1}, ν_{m+1}, and l_{m+1} as the estimates for μ_m, ν_m, and l_m. In other words, if ε satisfies (14.6.14), then (14.6.13) holds for all $m = 1, 2, 3, \dots$.

The equality (see (14.6.10))

$$Y_q = C^q \mathcal{L} Z_0 + \sum_{i=0}^{q-1} C^i [\Delta_{21}^{(q-i)} Z_{q-i-1} + \Delta_{22}^{(q-i)} Y_{q-i-1}]$$

and the estimates (see (14.6.4), (14.6.6), (14.6.9), and (14.6.10))

$$\begin{aligned} \|C^k\| &\le \sqrt{\omega} e^{-k/(1+\omega)}, \\ \|\mathcal{L}\| &= l_0, \\ \|Z_0\| &\le 4\sqrt{\omega}(\sqrt{\omega} l_0 + 1), \\ \|\Delta_{21}^{(q-i)} Z_{q-i-1} + \Delta_{22}^{(q-i)} Y_{q-i-1}\| &\le \frac{\varepsilon\sqrt{2\omega}\|A\|}{r}(\mu_{q-i-1} + \nu_{q-i-1}) \\ &\le \varepsilon \frac{\sqrt{2\omega}\cdot 2\sqrt{\omega}(4\sqrt{\omega} l_0 + 3)}{r}\|A\|, \end{aligned}$$

$$\sum_{i=0}^{\infty} \|C^i\| < \sqrt{\omega}\frac{1}{1 - e^{-1/(1+\omega)}} = \sqrt{\omega}\frac{e^{1/(1+\omega)}}{e^{1/(1+\omega)} - 1} \le \sqrt{\omega}\frac{\sqrt{e}}{\sqrt{e} - 1}$$

imply the estimate

$$\begin{aligned} \|Y_q\| \le & \sqrt{\omega} l_0 \cdot 4\sqrt{\omega}(\sqrt{\omega} l_0 + 1) e^{-q/(1+\omega)} \\ & + \sqrt{\omega}\frac{\sqrt{e}}{\sqrt{e}-1}\varepsilon\frac{\sqrt{2\omega}\cdot 2\sqrt{\omega}(4\sqrt{\omega} l_0 + 3)}{r}\|A\|. \end{aligned}$$

Thus, we have shown that for a sufficiently small ε and all $q = 1, 2, 3, \dots$ we have

$$l_q = \|Y_q\| \le a e^{-q/(1+\omega)} + \frac{b}{r}\varepsilon,$$

where the constants a and b can be estimated in terms of ω, $\|A\|$, and l_0. In other words, applying the orthogonal-power method $(A + \Delta_j) Q_{j-1} = Q_j R_j$ with errors $\|\Delta_j\| < \varepsilon\|A\|$, we obtain the basis Q_q, which can be regarded as an approximation of the basis for the invariant subspace if the corresponding eigenvalues are separated from the rest of eigenvalues (i.e., ω is not too large) and the initial approximation Q_0 is not too bad (i.e., l_0 is finite). This assertion means that the orthogonal-power method is stable under perturbations caused by unavoidable round-off errors.

14.7. Bases for almost invariant subspaces

Equivalent formulations and the analysis of the above results. A matrix written in the bases obtained by the orthogonal-power method. QR algorithm. The case of Hessenberg matrices.

As in §§14.5 and 14.6, we assume that an $N \times N$-matrix A admits the representation

$$A = U^* \begin{bmatrix} I_{N_0} & -L \\ 0 & I_{N-N_0} \end{bmatrix} \begin{bmatrix} rB & 0 \\ 0 & rC \end{bmatrix} \begin{bmatrix} I_{N_0} & L \\ 0 & I_{N-N_0} \end{bmatrix} U,$$

where the matrix U is unitary ($U^*U = I_N$) and the spectra of the blocks B and C lie outside and inside the disk $|\lambda| < 1$, respectively. In addition, the projections P

and $I-P$ onto the invariant subspaces corresponding to the eigenvalues $|\lambda_j(A)| > r$ and $|\lambda_j(A)| < r$ commute with A and are given by the formulas

$$P = U^* \begin{bmatrix} I_{N_0} & L \\ 0 & 0 \end{bmatrix} U, \quad I - P = U^* \begin{bmatrix} 0 & -L \\ 0 & I_{N-N_0} \end{bmatrix} U.$$

The following estimates hold:

$$\|L\| \le \sqrt{1+\|L\|^2} \le \sqrt{\omega(r)},$$
$$\|C^j\| \le \sqrt{\omega(r)} e^{-j/[1+\omega(r)]},$$
$$\|B^{-j}\| \le \sqrt{\omega(r)} e^{-j/[1+\omega(r)]}.$$

In particular, $e^{-j/[1+\omega(r)]} < 1$ implies $\|B^{-1}\| \le \sqrt{\omega}$ and $\|C\| \le \sqrt{\omega}$. Recall that the parameter $\omega = \omega(r)$ characterizes the dichotomy quality for the spectrum of the matrix A by the circle $|\lambda| = r$ and can be found by the formula $\omega(r) = \|H\|$, where

$$H = \frac{1}{2\pi} \int_0^{2\pi} (A^* - re^{-i\varphi} I)^{-1} (r^2 I + A^*)(A - re^{i\varphi})^{-1} d\varphi$$

and $\|P\| = \|I-P\| \le \sqrt{\omega(r)}$.

Let a matrix Q be composed of N_0 columns and let each column contain N components of some basis vector. Thus, all possible linear combinations $x = Qy$ of these columns form an N_0-dimensional linear subspace of the N-dimensional space. The columns of the matrix Q are assumed to be orthonormal: $Q^*Q = I_{N_0}$. Suppose that the linear subspace defined by the basis Q is a sufficiently exact approximation of the N_0-dimensional invariant subspace of A corresponding to the eigenvalues $|\lambda_j(A)| > r$, i.e., we assume that for any $x = Qy$ we have

$$\gamma \|Px\| \ge \|(I-P)x\|,$$

where $\gamma > 0$ is not too large (it is required that $\gamma\sqrt{\omega} < 1$). In other words, we assume that

$$\gamma \|PQy\| \ge \|(I-P)Qy\|$$

or, equivalently, the Hermitian matrix

$$Q^*P^*PQ - \frac{1}{\gamma^2} Q^*(I-P^*)(I-P)Q \ge 0$$

is nonnegative. As follows from §§14.5 and 14.6, the matrix Q satisfying the above conditions can be obtained as the result of several iteration steps of the orthogonal-power algorithm.

If the columns of the matrix Q form a basis for an invariant subspace, then $AQ = QK$, where K is an $N_0 \times N_0$-matrix. It is obvious that $K = Q^*QK = Q^*AQ$, i.e., $AQ = QQ^*AQ$. If the subspace spanned by the columns of the matrix Q is not invariant, then $AQ - QQ^*AQ \ne 0$. We estimate $\|AQ - QQ^*AQ\|$ in terms of γ. More exactly, we establish the estimate

$$\|AQ - QQ^*AQ\| \le \frac{\gamma}{1-\gamma\sqrt{\omega}} \|A\| \tag{14.7.1}$$

under the condition that $\gamma\sqrt{\omega} < 1$. Let

$$Q = U^* \begin{bmatrix} Q_0 \\ Q_1 \end{bmatrix} \begin{matrix} N_0 \text{ rows} \\ N - N_0 \text{ rows.} \end{matrix}$$
$$N_0 \text{ columns}$$

The equality $Q^*Q = I_{N_0}$, which is equivalent to the equality $Q_0^*Q_0 + Q_1^*Q_1 = I_{N_0}$, implies the inequality $\|Q_0\| \leq 1$. The following relations hold:

$$Px = PQy = U^* \begin{bmatrix} I_{N_0} & L \\ 0 & 0 \end{bmatrix} \begin{pmatrix} Q_0 \\ Q_1 \end{pmatrix} y,$$

$$\|Px\| = \|Q_0y + LQ_1y\| \leq \|Q_0y\| + \|L\|\,\|Q_1y\| \leq \|Q_0y\| + \sqrt{\omega}\|Q_1y\|,$$

$$(I - P)x = (I - P)Qy = U^* \begin{bmatrix} 0 & -L \\ 0 & I_{N-N_0} \end{bmatrix} \begin{pmatrix} Q_0 \\ Q_1 \end{pmatrix} y,$$

$$\|(I - P)x\| = \sqrt{\|Q_1y\|^2 + \|LQ_1\|^2} \geq \|Q_1y\|.$$

The assumption $\gamma\|Px\| \geq \|(I - P)x\|$ implies that

$$\|Q_1y\| \leq \gamma\|Q_0y\| + \gamma\sqrt{\omega}\|Q_1y\|,$$
$$\|Q_1y\| \leq \frac{\gamma}{1 - \gamma\sqrt{\omega}}\|Q_0y\| \leq \frac{\gamma}{1 - \gamma\sqrt{\omega}}\|y\|,$$
$$\|Q_1\| \leq \frac{\gamma}{1 - \gamma\sqrt{\omega}}$$

for $\gamma\sqrt{\omega} < 1$. It is convenient to introduce the notation $\Pi = QQ^* = \Pi^* = \Pi^2$. We transform the expressions for AQ and ΠAQ as follows:

$$\begin{aligned} AQ &= U^* \begin{bmatrix} I_{N_0} & -L \\ 0 & I_{N-N_0} \end{bmatrix} \begin{bmatrix} rB & 0 \\ 0 & rC \end{bmatrix} \begin{bmatrix} I_{N_0} & L \\ 0 & I_{N-N_0} \end{bmatrix} UU^* \begin{pmatrix} Q_0 \\ Q_1 \end{pmatrix} \\ &= U^* \begin{pmatrix} rB \\ 0 \end{pmatrix} Q_0 + U^* \begin{pmatrix} r[BL - LC] \\ rC \end{pmatrix} Q_1, \\ \Pi AQ &= U^* \begin{pmatrix} rB \\ 0 \end{pmatrix} Q_0 + U^* \begin{pmatrix} r[BL - LC] \\ 0 \end{pmatrix} Q_1. \end{aligned}$$

Then

$$AQ - \Pi AQ = -U^* \begin{pmatrix} 0 \\ rC \end{pmatrix} Q_1,$$
$$\|AQ - \Pi AQ\| \leq \|A\|\,\|Q_1\| \leq \frac{\gamma}{1 - \gamma\sqrt{\omega}}\|A\|.$$

Here we used the estimate $\|rC\| \leq \|A\|$, which is obvious in view of the representation

$$A = U^* \begin{bmatrix} rB & r[BL - LC] \\ 0 & rC \end{bmatrix} U.$$

Now, it is not difficult to prove the inequality (14.7.1):

$$\|QQ^*AQ - AQ\| = \|\Pi AQ - AQ\| \leq \frac{\gamma}{1 - \gamma\sqrt{\omega}}\|A\|.$$

This inequality is useful in the study of the following construction which has numerous applications. Let an orthonormal basis consist of vectors whose coordinates form the matrix Q with N_0 columns. The basis can be completed to an orthonormal basis for the entire N-dimensional space. Denote by V a square unitary matrix ($V^*V = VV^* = I$) corresponding to this completed basis. We can regard V as a composite matrix, i.e.,

$$V = [\underset{N_0 \text{ columns}}{Q} \quad \vdots \quad \underset{N - N_0 \text{ columns}}{\widehat{Q}}].$$

The first N_0 columns of V form the matrix Q. The matrix $\widehat{A} = V^*AV$ can be divided into blocks A_{ij} so that

$$A_{11} = Q^*AQ, \quad A_{21} = \widehat{Q}^*AQ,$$
$$\begin{bmatrix} A_{11} \\ A_{21} \end{bmatrix} = V^*AQ, \quad \begin{bmatrix} A_{11} \\ 0 \end{bmatrix} = V^*QQ^*AQ,$$
$$\begin{bmatrix} 0 \\ A_{21} \end{bmatrix} = V^*(AQ - QQ^*AQ),$$
$$\|A_{21}\| = \|QQ^*AQ - AQ\|.$$

We see that if a matrix Q formed by the first N_0 columns of the unitary matrix V form a basis for a subspace close to the invariant subspace corresponding to N_0 largest eigenvalues ($|\lambda_j| > r$), i.e., γ is small, then the matrix V^*AV is *almost block triangular* since the norm $\|A_{21}\|$ is small.

Suppose that the spectrum of the matrix A admits the division by circles of radii $r^{(1)}, r^{(2)}, \dots$:

$$|\lambda_1| \geq \cdots \geq |\lambda_{j_1}| > r^{(1)} > |\lambda_{j_1+1}| \geq \cdots \geq |\lambda_{j_2}| > r^{(2)} > |\lambda_{j_2+1}| \geq 0;$$

moreover, $j_{k+1} < N_0$. Denote by $P^{(s)}$ ($P^{(s)}A = AP^{(s)}$) the projection onto the invariant subspace of A corresponding to the eigenvalues $\lambda_1, \lambda_2, \dots, \lambda_{j_s}$ outside the circle of radius $r^{(s)}$. We also assume that the first j_s, $s = 1, 2, \dots, k$, columns of the $N \times N_0$-matrix Q_0 ($Q_0^*Q_0 = I_{N_0}$) form a submatrix $Q_0^{(s)}$ such that

$$[Q_0^{(s)}]^*[P^{(s)}]^*[P^{(s)}]^*Q_0^{(s)} - \frac{1}{[\gamma_0^{(s)}]^2}[Q_0^{(s)}]^*(I - [P^{(s)}]^*)[P^{(s)}]^*[Q_0^{(s)}] > 0,$$

where $0 < \gamma_0^{(s)} \leq \gamma_0$. These submatrices can be used as initial approximations in the orthogonal-power method in order to construct good approximations of bases for invariant subspaces of dimension $j_1, j_2, \dots, j_k$, where each j_i corresponds to the set of j_s largest eigenvalues of A. Let $|\omega(r_j)| \leq \omega^*$ and let ω^* be not too large.

Realizing the orthogonal-power algorithm, i.e., computing in accordance with the formulas $AQ_l = Q_{l+1}R_{l+1}$, where $Q_l^*Q_l = Q_{l+1}^*Q_{l+1} = I_{N_0}$, R_{l+1} is an upper triangular $N_0 \times N_0$-matrix, we see that the first j_s columns of the matrices Q_l form submatrices $Q_l^{(s)}$ connected by similar relations

$$AQ_l^{(s)} = Q_{l+1}^{(s)}R_{l+1}^{(s)}, \quad [Q_l^{(s)}]^*Q_l^{(s)} = [Q_{l+1}^{(s)}]^*Q_{l+1}^{(s)} = I_{j_s},$$

where the upper triangular matrix $R_{l+1}^{(s)}$ is the principal minor of the triangular matrix R_{l+1}. In other words, applying the orthogonal-power algorithm for bases in subspaces of dimension N_0, we thereby construct bases for all invariant subspaces that have dimension less than N_0 and correspond to the eigenvalues outside some disks realizing the dichotomy of the spectra. For this to happen, it is necessary that the above assumptions concerning the structure of the spectra and the initial basis are valid. If these assumptions hold, the submatrices $Q_l^{(s)}$ depend only on the first j_s columns $Q_0^{(s)}$ of the initial matrix Q_0. Consequently, the products $[Q_l^{(s)}]^* A Q_l^{(s)}$ turn out to be independent of the choice of columns and the number of additional columns which we add to $Q_0^{(s)}$ to obtain the matrix Q_0. We can assume that the order N_0 of the matrix Q_0 is as large as possible, i.e., $N_0 = N$, $Q_0^* Q_0 = I_N$. Since the submatrices $Q_l^{(s)}$ become better approximations of the corresponding invariant subspaces of dimension j_s as $l \to \infty$, we can apply the estimates to the matrices $Q_l^* A Q_l$. We see that for large l almost all elements, except ones marked by cross in the following scheme, become almost zeros:

$$Q_l^* A Q_l = \left[\begin{array}{ccccccccc} \times & \dots & \times & & & & & & \\ \vdots & \ddots & \vdots & & & & & & \\ \times & \dots & \times & & & & & & \\ & & & \times & \dots & \times & & & \\ & & & \vdots & \ddots & \vdots & & & \\ & & & \times & \dots & \times & & & \\ & & & & & & \times & \dots & \times \\ & & & & & & \vdots & \ddots & \vdots \\ & & & & & & \times & \dots & \times \end{array}\right] \begin{array}{l} r^{(1)} \\ \text{rows} \\ r^{(2)} - r^{(1)} \\ \text{rows} \\ \\ \end{array}$$

$$\begin{array}{cc} r^{(1)} & r^{(2)} - r^{(1)} \\ \text{columns} & \text{columns} \end{array}$$

These matrices tend to some block triangular matrices with square blocks of orders $r^{(1)}$, $r^{(2)} - r^{(1)}$, ... at the main diagonal. Although in the limit the matrix $Q_l^* A Q_l$ becomes a matrix of such structure, the matrices themselves do not necessarily tend to some limit. If starting with some $l = l^*$, in the first r columns of $Q_l^* A Q_l$, the elements of the $(r + 1)$th row and all next rows are zero, then the matrix A has an invariant r-dimensional subspace. With a certain accuracy, for a basis for this subspace we can use columns of some of the matrices Q_l $(l > l^*)$. The eigenvalues of the upper principal $r \times r$-minor of the matrix $Q_l^* A Q_l$ are approximations of some eigenvalues of the matrix A. As a rule, they coincide with its largest eigenvalues.

The above arguments lead to the conclusion that it is useful to compute the matrices $Q_l^* A Q_l$ simultaneously. The scheme of the algorithm is as follows:

$$\begin{gathered} Q_0 \text{ is given,} \\ A Q_l = Q_{l+1} R_{l+1} \\ (Q_l^* Q_l = I_{N_0}, R_{l+1} \text{ is an upper triangular matrix}), \\ A_{l+1} = Q_{l+1}^* A Q_{l+1}, \end{gathered}$$

where A_{l+1} are square $N_0 \times N_0$-matrices.

Usually, N_0 is taken to be different from N (especially, if N is very large and we are interested in invariant subspaces of small dimension corresponding to maximal eigenvalues). The choice of the initial $N \times N_0$-matrices Q_0 is quite difficult because it is impossible to have definite ideas about bases for invariant subspaces as well as the location of eigenvalues and spots of the ε-spectrum in the complex plane if we have only information about the form of A.

If N is not too large, for Q_0 the unitary matrix $Q_0 = I_N$ is often taken. The scheme of the algorithm takes the form

$$
\begin{aligned}
A &= C_1 R_1, \\
AC_l &= C_{l+1} R_{l+1}, \quad l = 1, 2, \dots, \\
C_l^* C_l &= I_N, \quad R_l \text{ is an upper triangular matrix}, \\
A_l &= C_l^* A C_l,
\end{aligned}
\tag{14.7.2}
$$

where C_l is used instead of Q_l. We changed the notation because we want to give a modification of the algorithm (14.7.2) which allows us to obtain the same matrices A_l by a slightly different method. But the notation Q_l is used for some other matrices in this modification. Using the notation C for a block in the canonical form of a matrix cannot lead to misunderstanding because the canonical forms will not be used later.

We set

$$
\begin{aligned}
C_1 &= Q_1, \\
C_2 &= Q_1 Q_2, \\
C_3 &= Q_1 Q_2 Q_3, \\
&\dots\dots\dots\dots \\
C_l &= Q_1 Q_2 \dots Q_{l-1} Q_l,
\end{aligned}
$$

so that $Q_{l+1} = C_l^* C_{l+1}$ and, in view of the equalities

$$
\begin{aligned}
C_l A_l C_l^* &= C_{l+1} A_{l+1} C_{l+1}^*, \\
A_{l+1} &= C_{l+1}^* C_l A_l C_l^* C_{l+1} = (C_l^* C_{l+1})^* A_l (C_l^* C_{l+1}),
\end{aligned}
$$

we can assert that

$$
\begin{aligned}
A_{l+1} &= Q_{l+1}^* A_l Q_{l+1}, \\
A_l &= C_l^* A C_l = C_l^* C_{l+1} R_{l+1} = Q_{l+1} R_{l+1}, \\
A_{l+1} &= Q_{l+1}^* A_l Q_{l+1} = Q_{l+1}^* [Q_{l+1} R_{l+1}] Q_{l+1} = R_{l+1} Q_{l+1}.
\end{aligned}
$$

Thus, we conclude that

$$
\begin{aligned}
A \equiv A_0 &= Q_1 R_1, \quad A_1 = R_1 Q_1, \\
A_{l-1} &= Q_l R_l, \quad A_l = R_l Q_l, \\
Q_l Q_l^* &= I_N, \quad R_l \text{ is an upper triangular matrix}.
\end{aligned}
\tag{14.7.3}
$$

Each step described by formula (14.7.3) consists in the decomposition of the matrix A_{l-1} obtained at the previous step into the product of an orthogonal (unitary) Q_l matrix and a triangular R_l matrix (QR decomposition) and the computation of A_l in the form of the product of the same factors taken in the reverse order. This computational scheme was suggested by Kublanovskaya [**22**] in 1961 and Francis [**8**] in 1959. It is called the *QR algorithm.* This algorithm and its modifications are very popular. We do not discuss these modifications (see, for example, [**33**]) which, in particular, allow one to accelerate the procedure for separating eigenvalues. In many respects, the QR algorithm is close to the orthogonal sweep method published in the same 1961.

PROBLEM 14.7.1. Taking a chain of two-dimensional rotations of Q, we can realize the QR decomposition for a Hessenberg matrix A (see §3.3). Prove that the product RQ is a Hessenberg matrix.

In view of Problem 14.7.1, we may conclude that if we (before starting of the QR algorithm) reduce the initial matrix A_0 to a Hessenberg matrix by a similar orthogonal transformation, then $A_1, A_2, A_3, \dots$ are also Hessenberg matrices and all orthogonal Q_j's can be taken as chains of rotations. This decreases the computations (see §3.4) and facilitates the analysis of the results of the computation. Indeed, since A_l are Hessenberg matrices, we conclude that below the main diagonal only the elements of the Hessenberg diagonal can be nonzero. We use the term the *Hessenberg diagonal* for the elements located immediately below the main diagonal.

If with certain accuracy A_l turns out to be a block upper triangular matrix, then its neighboring diagonal blocks must be separated by the zero or almost zero elements of the Hessenberg diagonal. This significantly simplifies the control of whether the diagonal blocks are distinguished after some number of iterations. Indeed, it suffices to watch only the elements of the Hessenberg diagonal.

Bibliography

[1] A. A. Abramov, *On numerical solution of some algebraic problems appearing in the stability theory*, Zh. Vychisl. Mat. i Mat. Fiz. **24** (1984), no. 3, 339–347; English transl. in U.S.S.R. J. Comput. Math. and Math. Phys. **24** (1984).

[2] I. A. Adamskaya and S. K. Godunov, *The method of spherical harmonic in the problem about critical parameteres*, Zh. Vychisl. Mat. i Mat. Fiz. **4** (1964), no. 3, 555–570; English transl. in U.S.S.R. J. Comput. Math. and Math. Phys. **4** (1964).

[3] A. Ya. Bulgakov and S. K. Godunov, *Circle dichotomy of the spectrum of a matrix*, Sibirsk. Mat. Zh., **29** (1988), no. 5, 59–70; English transl., Siberian Math. J., **29** (1988), no. 5, 734–744.

[4] A. Ya. Bulgakov, *An effectively calculable parameter for the stability property of a system of linear differential equations with constant coefficients*, Sibirsk. Mat. Zh., **21** (1980), no. 3, 32–41; English transl., Siberian Math. J. **21** (1980), no. 3, 339–346

[5] ———, *The basis of guaranteed accuracy in the problem of separation of invariant subspaces for non-self-adjoint matrices*, Trudy Inst. Mat. Sibirsk. Otdel. Akad. Nauk SSSR, **15** (1989), 12–92; English transl., Siberian Advances in Math., **1** (1991), no. 1, 64–108; no. 2, 1–56.

[6] Yu. L. Daletskii and M. G. Krein, *Stability of solutions of differential equations in a Banach space*, "Nauka", Moscow, 1970; English transl., Amer. Math. Soc., Providence, RI, 1974.

[7] E. I. Davison and F. T. Man, *The numerical solution of* $A^*Q + QA = -C$, IEEE Trans. Automatic Control, **AC-13** (1968), 448–449.

[8] J. G. F. Francis, *The QR-transformation. The unitary analogue to the QR-transformation.* I, II, Comput. J., **4** (1961), 265–271; 332–345.

[9] F. R. Gantmacher, *The theory of matrices.* vol. I, II, Chelsea, New York, 1959.

[10] S. K. Godunov *On numeric solution of boundary-value problems for systems of linear ordinary differential equations*, Uspekhi Mat. Nauk **16** (1961), no. 3, 171–174. (Russian)

[11] ———, *Problems of guaranteed accuracy in numeric methods of linear algebra*, Proc. Intern. Congress Mat. 1986 (August 3–11, 1986, Berkeley, California), vol. 2, Amer. Math. Soc., Providence, RI, 1986, pp. 1353–1361.

[12] ———, *Problem of the dichotomy of the spectrum of a matrix*, Sibirsk. Mat. Zh., **27** (1986), no. 5, 24–37; English transl., Siberian Math. J., **27** (1986), no. 5, 649–659.

[13] ———, *Spectral portaits of matrices and criteria of spectral dichotomy*, Proc. Conf. Oldenburg, Germany (October, 1991), J. Herrberger and L. Atanasova (eds.) North-Holland and JMACS, Amsterdam, 1991.

[14] ———, *Equations of mathematical physics*, "Nauka", Moscow, 1979. (Russian)

[15] S. K. Godunov, A. G. Antonov, O. P. Kirilyuk, and V. I. Kostin, *Guaranteed accuracy for solutions of system of linear equations in Euclidean spaces*, "Nauka", Novosibirsk, 1988; English transl., Kluwer, Dordrecht, 1993.

[16] S. K. Godunov, O. P. Kirilyuk, and V. I. Kostin, *Spectral portraits of matrices*, Preprint Inst. Mat. Sibirsk. Otdel. Akad. Nauk SSSR, no. 3, Novosibirsk, 1990. (Russian)

[17] S. K. Godunov, V. I. Kostin, and A. D. Mitchenko, *Computation of an eigenvector of a symmetric tridiagonal matrix*, Sibirsk. Mat. Zh., **26** (1985), no. 5, 71–85; English transl., Siberian Math. J., **26** (1985), no. 5, 684-695.

[18] S. K. Godunov and S. V. Kuznetsov, *Estimates for convengences of the orthogonal-power method*, Trudy Inst. Mat. Sibirsk. Otdel. Ross. Akad. Nauk, **26** (1994), 20–41; English transl., Siberian Advances in Math., **5** (1995), no. 1, 16–41.

[19] I. C. Gohberg and M. G. Krein, *Introduction to the theory of linear nonselfadjoint operators*, "Nauka", Moscow, 1965; English transl., Amer. Math. Soc., Providence, RI, 1964.

[20] M. V. Keldysh, *On eigenvalues and eigenfunctions of some classes of non-self-adjoint equations*, Dokl. Akad. Nauk SSSR. **77** (1951), no. 1, 11–14. (Russian)

[21] M. V. Keldysh and V. B. Lidskii, *Questions of the spectral theory of non-self-adjoint operators*, Proc. Fourth All-Union Mathematical Congress, vol. 1, 1963, pp. 101–120. (Russian)

[22] V. N. Kublanovskaya, *On some algorithm of solving the full problem of eigenvalues*, Zh. Vychisl. Mat. i Mat. Fiz. **1** (1961), no. 4, 555–570; English transl. in U.S.S.R. J. Comput. Math. and Math. Phys. **1** (1961).

[23] A. G. Kurosh, *Higher algebra*, "Nauka", Moscow, 1971. (Russian)

[24] B. Ya. Levin, *Distribution of roots of entire functions*, Gostekhizdat, Moscow, 1956; English transl., Amer. Math. Soc., Providence, RI, 1964.

[25] V. B. Lidskii, *On the summability of series over principal vectors of non-self-adjoint operators*, Trudy Moskov. Mat. Obshch. **11** (1962), 3–35. (Russian)

[26] A. N. Malyshev, *Guaranteed accuracy in spectral problems of linear algebra*, Trudy Inst. Mat. Sibirsk. Otdel. Ross. Akad. Nauk, **17** (1991), 19–103; English transl., Siberian Advances in Math., **2** (1992), no. 1, 144–197.

[27] ———, *Introduction in the computational linear algebra*, "Nauka", Novosibirsk, 1991. (Russian)

[28] ———, *Parallel algorithm for solving some spectral problems*, Linear Algebra Appl., **188/189** (1993), 489–520.

[29] A. I. Markushevich, *The theory of functions of a complex variable*, "Nauka", Moscow, 1967; English transl., Chelsea, New York, 1977.

[30] J. D. Roberts, *Linear model reduction and solution of the algebraic Riccati equation by use of the sign-function*, Int. J. Control., **32** (1980), no. 4, 677–687.

[31] R. A. Sarybekov, *Extremal quadratic Lyapunov functions of systems of second-order equations*, Sibirsk. Mat. Zh., **18** (1977), no. 5, 1159–1167; English transl., Siberian Math. J., **18** (1977), no. 5, 823–828.

[32] L. N. Trefethen, *Pseudospectra of matrices*, Numerical Analysis 1991 (Dundee, 1991), D. F. Griffits and G. A. Watson (eds.), Longman Sci. Tech., Harlow, 1992, pp. 234–266.

[33] J. H. Wilkinson, *The algebraic eigenvalue problem*, Clarendon Press, Oxford, 1965.

Index